SURGERY - PROCEDURES, COMPLICATIONS, AND RESULTS

MINIMALLY INVASIVE SKULL BASE SURGERY

PRINCIPLES AND PRACTICE

SURGERY - PROCEDURES, COMPLICATIONS, AND RESULTS

Additional books in this series can be found on Nova's website under the Series tab.

Additional e-books in this series can be found on Nova's website under the e-book tab.

NEUROSCIENCE RESEARCH PROGRESS

Additional books in this series can be found on Nova's website under the Series tab.

Additional e-books in this series can be found on Nova's website under the e-book tab.

SURGERY - PROCEDURES, COMPLICATIONS, AND RESULTS

MINIMALLY INVASIVE SKULL BASE SURGERY

PRINCIPLES AND PRACTICE

MONCEF BERHOUMA, M.D.
EDITOR

New York

Copyright © 2013 by Nova Science Publishers, Inc.

All rights reserved. No part of this book may be reproduced, stored in a retrieval system or transmitted in any form or by any means: electronic, electrostatic, magnetic, tape, mechanical photocopying, recording or otherwise without the written permission of the Publisher.

For permission to use material from this book please contact us:
Telephone 631-231-7269; Fax 631-231-8175
Web Site: http://www.novapublishers.com

NOTICE TO THE READER

The Publisher has taken reasonable care in the preparation of this book, but makes no expressed or implied warranty of any kind and assumes no responsibility for any errors or omissions. No liability is assumed for incidental or consequential damages in connection with or arising out of information contained in this book. The Publisher shall not be liable for any special, consequential, or exemplary damages resulting, in whole or in part, from the readers' use of, or reliance upon, this material. Any parts of this book based on government reports are so indicated and copyright is claimed for those parts to the extent applicable to compilations of such works.

Independent verification should be sought for any data, advice or recommendations contained in this book. In addition, no responsibility is assumed by the publisher for any injury and/or damage to persons or property arising from any methods, products, instructions, ideas or otherwise contained in this publication.

This publication is designed to provide accurate and authoritative information with regard to the subject matter covered herein. It is sold with the clear understanding that the Publisher is not engaged in rendering legal or any other professional services. If legal or any other expert assistance is required, the services of a competent person should be sought. FROM A DECLARATION OF PARTICIPANTS JOINTLY ADOPTED BY A COMMITTEE OF THE AMERICAN BAR ASSOCIATION AND A COMMITTEE OF PUBLISHERS.

Additional color graphics may be available in the e-book version of this book.

Library of Congress Cataloging-in-Publication Data

ISBN: 978-1-62808-567-9

Library of Congress Control Number: 2013944995

Published by Nova Science Publishers, Inc. † New York

Contents

Foreword 1	*Edward R. Laws, Jr.*	ix
Foreword 2	*Paolo Cappabianca*	xi
Foreword 3	*James Tait Goodrich*	xv
Preface		xxv
Chapter I	Epistemology of Minimally Invasive Skull Base Surgery *Moncef Berhouma*	1
Chapter II	Endoscopic Endonasal Anatomy and Approaches to the Anterior Skull Base *Bashar Abuzayed*	19
Chapter III	Expanded Endoscopic Endonasal Approaches to the Orbit and Lateral Anterior Skull Base *Pornthep Kasemsiri, Ricardo L. Carrau, Daniel M. Prevedello, Jun Muto, Danielle de Lara, Leo F. S. Ditzel Filho, Bradley A. Otto and Amin B. Kassam*	35
Chapter IV	Endoscopic Repair of Anterior Cranial Base Defects *Sudheer Ambekar, Chiazo Amene, Justin Haydel, Osama Ahmed, Anil Nanda and Bharat Guthikonda*	49
Chapter V	Neurosurgical Overview of Minimally Invasive Resection of Anterior Skull Base Malignancy: A Comparison to Traditional Craniofacial Resection *Burak Sade, Pete S. Batra, Martin J. Citardi and Joung H. Lee*	71

Contents

Chapter VI	Expanded Endoscopic Endonasal Approaches to the Middle Cranial Fossa *Danielle de Lara, Daniel M. Prevedello, Ricardo L. Carrau, Leo F. S. Ditzel Filho, Jun Muto, Pornthep Kasemsiri, Bradley A. Otto, Ricardo L. Carrau and Amin B. Kassam*	83
Chapter VII	The Endoscopic Endonasal Transtuberculum-Transplanum Approach *Luigi Maria Cavallo, Domenico Solari, Alessandro Villa, Michelangelo de Angelis, Teresa Somma, Felice Esposito and Paolo Cappabianca*	93
Chapter VIII	Craniopharyngiomas: The Endonasal Endoscopic Approach *Daniel M. S. Raper, Robert M. Starke, John A. Jane and Ricardo J. Komotar*	111
Chapter IX	Endoscopic Optic Nerve Decompression *D. D. Sommer, M. S. Gill, K. Reddy and A. Vescan*	123
Chapter X	Endoscopic Approach and Management of Parasellar Lesions *Mahmoud Messerer, Moncef Berhouma, Marc Sindou and Emmanuel Jouanneau*	137
Chapter XI	Extended Endoscopic Endonasal Approaches to the Pterygopalatine and Infratemporal Fossae *Bashar Abuzayed*	153
Chapter XII	Management of Middle Fossa CSF Leak *Chiazo Amene, Sudheer Ambekar, Cedric Shorter, Osama Ahmed, Justin Haydel, Anil Nanda and Bharat Guthikonda*	165
Chapter XIII	The Epidural Anterior Petrosectomy: A Minimally Invasive Skull Base Approach to the Posterior Fossa *P. H. Roche, L. Troude, A. Melot and R. Noudel*	181
Chapter XIV	Pediatric Endoscopic Skull Base Surgery *Smriti Nayan, Kesava Reddy, Adam M. Zanation, and Doron D. Sommer*	193
Chapter XV	Endoscopic Transnasal Removal of Midline Skull Base Tumors under the Side-Viewing Scopes *Masaaki Taniguchi, Nobuyuki Akutsu, Kohkichi Hosoda and Eiji Kohmura*	213
Chapter XVI	Percutaneous Biopsy of Parasellar Lesions Through the Foramen Ovale *Mahmoud Messerer and Marc Sindou*	225

Chapter XVII	Purely Endoscopic Keyhole Supraorbital Approaches for Anterior and Middle Skull Base Tumors *E. Jouanneau, M. Berhouma, T. Jacquesson, M. Messerer,* *E. Bogdan, A. Gleizal, G. Raverot and F. Barral-Clavel*	237
Chapter XVIII	Mini-Invasive Microvascular Decompression for Posterior Fossa Neurovascular Conflicts *Francesco Acerbi, Morgan Broggi, Marco Schiariti,* *Melina Castiglione, Giovanni Broggi and Paolo Ferroli*	251
Chapter XIX	Radiosurgical Management of Trigeminal Neuralgia *Constantin Tuleasca, Marc Levivier, Romain Carron,* *Anne Donnet and Jean Regis*	273
Chapter XX	Minimally Invasive Approaches to Cranial Nerves for Microvascular Decompression: Hearing Preservation *Emile Simon and Marc Sindou*	293
Chapter XXI	Endoscopic Transchoroidal Fissure Approach to the Posterior Part of the Third Ventricle and Posterior Fossa *Kheireddine A. Bouyoucef, Mohamed Si Saber, A. Youssef Kada,* *Sofiane Imekraz, Rebiha Baba-Ahmed and Michael H. Cotton*	311
Chapter XXII	Optimally Invasive Skull Base Surgery for Large Benign Tumors *Roy Thomas Daniel, Constantin Tuleasca,* *Mahmoud Messerer, Laura Negretti, Mercy George,* *Philippe Pasche and Marc Levivier*	325
Chapter XXIII	Anterior Craniovertebral Junction Tumors: Successful Resection Through Simple Approaches *Mario Ammirati and Varun R. Kshettry*	357
Chapter XXIV	Endoscopic-Assisted Transoral Approach to the Clivus and the Craniovertebral Junction. Transnasal or Transoral? A Clinical and Experimental Issue *Massimiliano Visocchi*	369
Chapter XXV	Endoscopic Endonasal Approaches to the Paramedian Posterior Skull Base *Jun Muto, Danielle de Lara, Leo F. S. Ditzel Filho,* *Pornthep Kasemsiri, Bradley A. Otto, Ricardo L. Carrau* *and Daniel M. Prevedello*	389
Editor Contact Information		399
Index		401

Foreword 1

Edward R. Laws, Jr.
MD, FACS
Professor of Surgery, Harvard Medical School
Brigham& Women's Hospital
Boston, MA, US

This ambitious work reflects the excitement surrounding the evolving field of minimally invasive skull base surgery. Professor Berhouma has assembled a stellar international group of contributors. Each is an authoritative leader in a segment of the astounding variety of practical and effective neurosurgical procedures that have been refined and advanced using novel technology and new concepts.

The basic principles of Skull Base Surgery are reflected in each chapter, and form the conceptual framework within which minimally invasive concepts can evolve:

Respect for and knowledge of surgical anatomy
Removal of obstructing bone

Minimal or no brain retraction
Preservation of vessels, nerves and brain structures
Microsurgical technique
Optimal visualization and exposure of the pathology
Early devascularization of the lesion
Functional and cosmetic reconstruction

The wide range of disease entities that can now be treated with technologically advance minimally invasive techniques is extraordinary. Many of the cutting edge procedures are based on endoscopy and endoscopic anatomical approaches. Others utilize the advantages of radiosurgery. Excellent discussions of the latest advances are provided for: the removal of anterior skull base malignancies; treatment of parasellar lesions; repair and prevention of cerebrospinal fluid leaks; treatment of trigeminal neuralgia; decompression of the optic nerve; and the management of a variety of pediatric cranio-facial lesions.

The international scope of this volume is truly impressive. It attests to the significance and importance of the subject. Surely it will spark the development of more advances in our field, and will be a continuing resource and inspiration.

Edward R. Laws, Jr., MD, FACS
Professor of Surgery, Harvard Medical School
Brigham& Women's Hospital
Boston, MA, US

Foreword 2

Paolo Cappabianca
M.D.
Department of Neurological Sciences, Division of Neurosurgery,
Università degli Studi di Napoli Federico II, Naples, Italy

I have to admit that the book title itself sounds provocative, since the use of the endoscope has revolutionized, in this last decades, some possibilities and many strategies of accessing the intracranial neurovascular structures and pathologies, with intrusive operations and extensive surgeries.

Starting mostly in the old Continent, i.e. Europe, with endoscopic surgery of the paranasal sinuses and neuroendoscopy of the cerebral ventricles and then taking advantage from the sparkling and energetic vitality of the new world, USA first of all, the use of the endoscope has moved from transsphenoidal surgery to the pituitary toward regions of the skull base around it. This has determined the possibility to face pathologies and perform operations that are not minimal in terms of extension and target, but really maximally invasive, considering the structures at the site of surgery and those crossed to reach them. The boundaries have widened and the cooperation with other specialties, ENT and Head and Neck surgeons first, has enriched everyone, harboring a team approach attitude, this being easier and easier with increased travel exchanges around the world, and thanks to the progress and explosion of our digital times. Furthermore, after the active resistance of many surgeons of the old generations, very effective and skilled with established instrumentations and techniques, the impulse of few no-age innovators and the determinant contribute of young generations has fueled the progress and opened new horizons. And it is not surprising that many of the names of the contributors of the present volume have jumped into the neurosurgical scenario just few years ago, as it currently happens in the main literature, where the influence of young generations to promote innovation is crucial.

The endoscope has brought the eyes of the surgeon closer to the relevant anatomy than other visual instrumentation, such as the operating microscope, and has offered a wider look up around hidden corners. But the concepts and the methods of the microsurgical era, first of all the technique of bimanual dissection and the respect for the minimal feeders of every structure, being no one noble or poor, has been incorporated and not lost at all. The same has occurred in regards of the oncologic concept of cranio-facial malignancies removal, where surgery respects just the same boundaries certified by the pathologist during the operation, either performed with the microscope or the endoscope, with an open access or an endonasal corridor. What has gained renewed impulse is anatomy, not obviously being different when inspected by means of an endoscope, but because seen under a different perspective (i.e. endonasal vs subfrontal, as in suprasellar area; close to the target vs along its whole trajectory, as in the cerebello-pontine angle, etc.). This has promoted many new studies, further lesion understanding, definition of scientific districts in some new approaches, inspections of structures from opposite accesses, new classifications of areas and lesions in respect to the tailored approach.

And the need to overcome specific difficulties such as the reconstruction of the way gone through, including the skull base defect, preventing complications like CSF leaks has been an opportunity to design new techniques and flaps, to define and underline efficacy of tissues harvested from the patient him/herself, to study autologous and synthetic tissues and materials, to develop and produce new instruments.

What cannot be denied is the contribute of endoscopy, namely endonasal, in fixing new standards of treatment for midline skull base lesions, i.e. most of the anterior cranial base CSF leaks, clivus chordomas, and even upper CVJ conditions, once not considered amenable via such routes. And transsphenoidal endoscopy has offered new possibilities to complement the management of difficult tumors, such as craniopharyngiomas, both in adults and in children. Furthermore, as for the anterior and middle cranial base, and posterior fossa, in the coronal plane too regions not on the midline, i.e. paramedian, such as the pterygopalatine and infratemporal fossa have been accessed and the treatment paradigms are continuously evolving.

Besides the applications of the endoscope in skull base surgery, the book is enriched by chapters where percutaneous transnasal thermorhizotomy or lesion biopsy are nicely described, and chapters stimulating debate including modern effective techniques and technologies such as microvascular decompression (MVD) for posterior fossa neurovascular conflicts and/or radiosurgery. The recent contemporary development of endoscopy has favored productive controversies not expected to give conclusive remarks, but just to better define indications and landmarks for pterional vs subfrontal vs supraorbital vs transnasal approaches for conditions like anterior cranial fossa meningiomas and middle cranial fossa tumors.

All of this and new ideas and debates that this interesting book will stimulate are herein extensively seeded and may constitute the scaffold for further creative discussion and progress.

Paolo Cappabianca
Napoli, Italy

Foreword 3

Minimally Invasive Skull Base Surgery
A Foreword

James Tait Goodrich
M.D., Ph.D., D.Sci (Hon)
Director, Division of Pediatric Neurosurgery
Professor of Clinical Neurological Surgery,
Pediatrics, Plastic and Reconstructive Surgery
Leo Davidoff Department of Neurological Surgery
Albert Einstein College of Medicine
Children's Hospital at Montefiore
Montefiore Medical Center
111 East 210th Street; Bronx, New York 10467 US

In this remarkable work Dr. Berhouma and colleagues have put together an up to date monograph on minimally invasive approaches and techniques dealing with surgery of the skull base. Virtually every approach and the technical aspects have been nicely outlined by this select group of authors, all recognized as authorities in this arena of surgery. Subjects such as anatomy, narrow and expanded approaches, repairs using the endoscope, etc., all have been addressed in the various chapters. But having said that it remains most interesting to me how our field of surgery and especially skull based surgery has evolved over such a short period of time.

In an historical analysis of skull base surgery it is interesting to note that anatomical images of the skull base region only date back to the Renaissance and Leonardo da Vinci (1452-1519). Leonardo was the quintessential Renaissance man. Multi-talented, recognized as an artist, an anatomist, and a scientist, Leonardo went to the dissection table to better understand surface anatomy and in turn used those studies to better illustrated his artistic paintings. In Leonardo's remarkable anatomical codices appear the earliest illustrated example of the skull base and in particular his anatomical drawings of the cranial nerves. [1, 2] (Figure 1,2) While Leonardo's drawings were relatively crude by today's standards, they nevertheless provide us with early and important examples of skull base anatomy. Leonardo was not a surgeon, but he gave an important impetus to the study of human anatomy and the definition of correct anatomical relationships, a reversal of the earlier crude medieval views. Unfortunately Leonardo's great opus on anatomy, which was to be published in some 120 volumes, never appeared. [2] His anatomical manuscripts circulated in Italy within the art community throughout the sixteenth century, only to be lost and then rediscovered in the eighteenth century by the English surgeon and anatomist William Hunter (1718-1783).

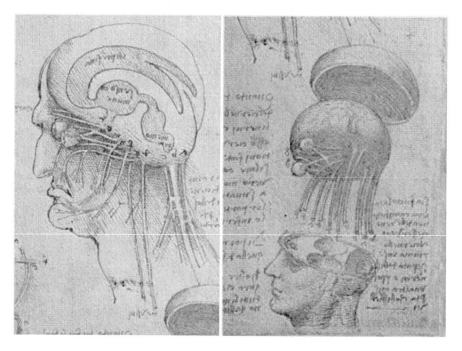

Figure 1a-b. From Leonardo da Vinci's anatomical codices – an illustration outlining the brain, cranial nerves, and the skull base anatomy as it relates to the cranial nerves.

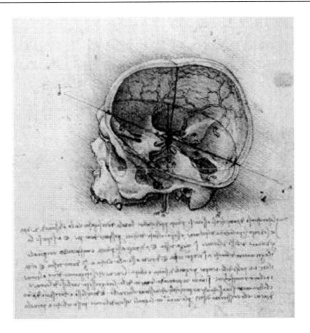

Figure 2. Anatomical drawing by Leonardo da Vinci detailing, for the first time an anatomically correct illustrated skull base. Drawings of this type of exacting illustrated detail only began to appear with the works of Leonardo.

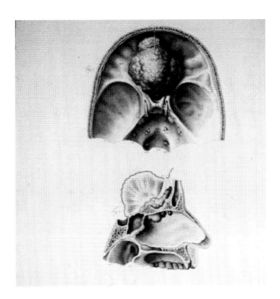

Figure 3a. A hand colored illustration from Cruveilhier's remarkable pathological atlas illustrating here a series of meningiomas. These two figures are some of the earliest known illustrated examples of skull base lesions involving the anterior fossa and the olfactory groove.

In looking back over the history of skull base surgery we find it a relatively recent development with some of the first surgeries being done in skull base region dating only to the 1860s. It really took the introduction of antisepsis and anesthesia to allow a surgical team to approach these regions of the skull base safely and with significantly reduced morbidity and mortality. The concept of a pathological basis of a disease dates from the late eighteenth century. Skull based tumors were first accurately illustrated in several nineteenth century anatomical atlases the most important which was published by Jean Cruveilhier (1791-1874) Cruveilhier was a surgeon, anatomist and pathologist and using these combined skills produce a most remarkable atlas of pathology on tumors and diseases of the human body. To Cruveilhier we owe the earliest illustrated examples of skull-based tumors. In Cruveilhier's atlas we find examples meningiomas of frontal fossa and olfactory groove along with epidermoid tumors of the skull base.

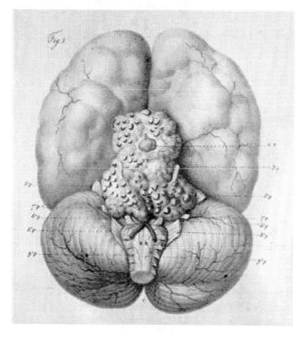

Figure 3b. Cruveilhier's elegant illustration of an epidermoid tumor involving the skull base and encasing the pontine and subtemporal region. No surgical options were offered just a post mortem finding.

With the arrival of the 20th century surgeons were becoming more aggressive and exploring territories normally not explored within the head, neck and skull base. Along with these surgeries came more sophisticated instrumentation to view these surgical regions and their unique pathologies. At the end of the nineteenth century the most common tools used to perform a craniectomy were nothing more than a hammer and chisel. A "typical" surgical technique for removal of an acoustic neuroma in the cerebello-pontine angle (CPA) was described by Fedor Krause (1857-1937) in 1912. (Figure 4) The surgical technique for removing the tumor was accomplished by simply placing the index finger into the angle and removing the lesion. This technique unfortunately often included the VIIth and VIIIth cranial nerves along with the tumor.

Surgeons first began to adopt the use of endoscopes at the very beginning of the 20th century, mostly adopting endoscopes from our urology and veterinary colleagues. In looking back at these devices they were quite primitive by today's standards. Lighting and surgical field illumination was by an incandescent bulb, the optics quite poor with minimal magnification. Surgeons were also limited in the range of available endoscopic instrumentation; the combination of poor optics, lighting and tools led most surgeons away their use.

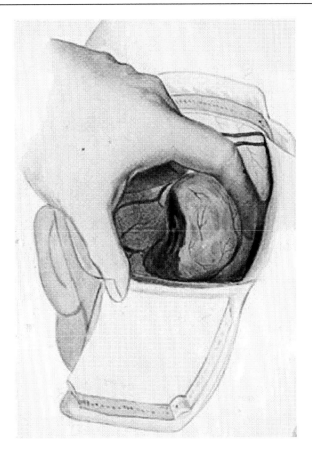

Figure 4. Fedor Krause's technique for removing an acoustic neuroma in the cerebello-pontine angle, note the ungloved hand – circa 1912.

Figure 5a. An endoscope manufactured in the USA about 1920. The design concept was basically a urological scope but also adapted for use by neurosurgeons for exploring the brain and ventricles and by ENT surgeons for transnasal work. From the author's collection.

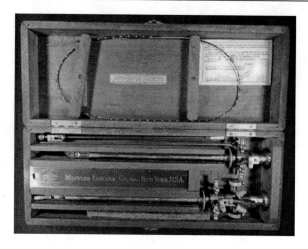

Figure 5b. Another example of an American made endoscope used primarily for urological investigations but similar to what was adopted by other surgeons in this early era of endoscopic surgery and treatment. . From the author's collection.

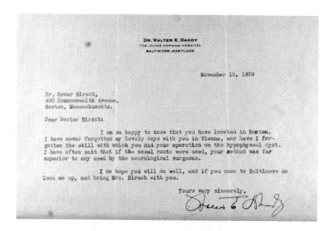

Figure 6. An interesting letter from Walter Dandy to Oscar Hirsch written in 1939 discussing the value of Hirsch's techniques using the endoscope. It has taken another sixty plus years to bring these surgical techniques into the mainstream of neurosurgery and ENT surgery. Letter from the personal collection of the author.

A pioneer in the endoscopic approach to the sphenoid sinus and pituitary gland was a Vienese rhinologist by the name of Oskar Hirsch (1877-1965) The endonasal transsphenoidal operation was first developed by Dr. Hirsch. In the period 1907-1909 Hirsch worked on a number of approaches to the sphenoid region using various techniques that required transposing the entire nose; hinging it laterally, downward or upward depending on skin incisions. These procedures however proved to be quite mutilating though the morbidity and mortalities results were better than the then standard transcranial approaches to the pituitary. By 1910 Hirsch had perfected his unique endonasal approach to the sphenoid region using a midline dissection and a speculum exposure. His surgical positioning was quite unique by today's standard. The patient was seated upright, secured to the chair and with the head firmly fixed. The patient was kept awake, the nasopharyngeal surface was cocainized and the nasal mucosa was infiltrated with local anesthesia. Wearing an ENT reflective mirror Hirsch used this reflective light to see up into the surgical site. Through a speculum placed the sphenoid sinuses and sella turcica could be adequately exposed with a much reduced morbidity and mortality. Hirsch's assistant was a faithful dwarf and ex-patient by the name of Shostel. In addition to assisting his job was to operate a foot-driven suction device. (Figure 6).

The earliest known endoscopic exploration of the human brain was by V.L. Lespinasse of Chicago, IL. Lespinasse used a small cystoscope to fulgurate the choroids plexus in two infants with hydrocephalus. One child died postoperatively and the other survived for five years. It is of interest to note that Lespinesse was a urologist by training! In the first half of the 20th century in the field of neurosurgery there appeared several talented individuals including Walter Dandy, M.D. (1886-1946) at Johns Hopkins University and J. Lawrence Pool, M.D. (1906-2004) at New York Neurological Institute who popularized the use of endoscopes. Endoscopes were used as early as 1922 by Dandy to explore the cerebral ventricles for tumors and treatment of the hydrocephalus by cauterizing the choroid plexus. Dr. Dandy had been encouraged by a prominent Johns Hopkins gynecological surgeon – Dr. Howard A. Kelly (1858-1943) - to investigate its use. Dr. Pool developed an interesting "myeloscope" for looking into the spine and spinal cord. Pool was the first to have a direct visualization of the spinal roots and cauda equina. The myeloscope was passed through a lumbar puncture needle into the subarachnoid space.

Mostly due to technical disadvantages surgeons gave up the use of the endoscopes by the late 1930s and reverted back to more open techniques. As a resident in the 1980s I had only one attending who used an endoscope and remembering back it was rather crude instrument with a poor light source and awkward to handle. We used the system on pituitary tumors through a sublabial approach and on occasion in the cerebral ventricles. In each case a lot of muttering went on about the visualization and the lack of adequate tools to do the dissections. At the same time there was also much controversy in the neurosurgical community on the use of the operating microscope. The neurosurgical operating microscope came into practice in the early 1960s with an early pioneer also being J. Lawrence Pool, M.D.. At first the operating microscope was not warmly accepted as some surgeons considered the scope too big and too bulky and prefer to clip aneurysms and do intracranial surgeries with just eyeglasses on. To the older crowd the scope just slowed you down! As the microscopes developed better optics and more flexibility more surgeons adopted its use to where now an operating microscope is a routine fixture in the operating room.

In the time frame of surgery the last 170 years of surgical history have been a remarkable period to review. So much technology has been added that now aids various surgical teams. Imagine operating on a wide-awake patient with no anesthesia other than maybe some opium or alcohol – general anesthesia (ether, nitrous oxide and chloroform) only came about in the 1840s. In the 1850s bewhiskered and top-hatted surgeons routinely operated with unwashed hands and in blood encrusted operating room clothing. For the surgeon the sign of a blood encrusted and stained surgical coat was exemplary in that it clearly reflected a skilled surgeon. Even more remarkable was the concept that pus was considered a normal finding in the postoperative wound. It was Joseph (later Lord) Lister (1827-1912), a surgeon at the Glasgow Royal Infirmary, who introduced the concept of "antiseptic surgery". In the 1860s he introduced the use of a carbolic sprayer to sterilize the patient operation site and also the operating room environs. Surgical instruments were for the first time cleaned and sterilized after each case. Instruments manufacturing went from wood and non-sterilizing materials to metals and steel. It is interesting to reflect back on Lister and realize that his system of sterilization was initially fought off with a vengeance by surgeons and not really full accepted until the 1890s.

In 1895 the Roentgen Ray (X-ray) was introduced by Wilhelm Roentgen ( ) after a serendipitous discovery in his laboratory while working with a Crookes tube. The "ray" he was observing he could not identify so he decided to call it the "X-ray". With this discovery the physician could see "inside" the human body. From this primitive ray we have now rapidly advanced to MR and CT scanning. Over the last twenty-five years frameless stereotactic technology has become routine in the operating room. Extensive endoscopic approaches to the skull base were really not possible with any degree of safety until this technology became available. I was involved in one of the first frameless systems in the late 1980s. At the time this was truly exciting technology but looking back the instrumentation and technology was quite primitive with wide error range and a very slow ability to refresh images in real time.

It is amazing to reflect back historically and see how far technology has advanced, particularly with computer-assisted devices. The generation of surgeons coming up under me can no longer enter an operating room for a case unless a image guided frameless system is available and this situation exists in what I would call "simple" cases such as the placement of a ventriculo-peritioneal shunt. In the manufacturing industry the robotic designers have done a remarkable of job of replacing humans at the production line and instead use robots and robotic arms to do the assembly. Surgeons are now routinely using devices such as the RoboDoc, the DaVinci robotic arm, etc. for surgical assistance in the operating room. The ability to be able to turn an arm 360 degrees and with no tremor is a great asset. With frameless guided approaches the next logical surgical scenario is the combination of a stereotactic frameless system conjoined with a robotic arm to perform the "ultimate" precise surgical intervention. With CT/MR data and computers this ability to work and be guided in 3-D space is already a reality. The airline industry has already done this in that computers are now flying commercial airplanes with pilots being just data entering personnel. Airline executives have long thought that plane crashes are mostly due to pilot error hence the change in philosophy to go from pilot to computer.

To provide a perspective of how far we have come I have added a paragraph of a description of a "modern" operating room in 1930s. Paul Bucy (1902-1992), one of our pioneer's in neurosurgery was a young visiting surgeon to Queens Square in London and described the following scene:

> ". . . at the National Hospital were Sir Percy Sargent and Mr. Donald Armour. They were both poor surgeons, unbelievably crude in their surgical procedures. On one occasion (Gordon) Holmes told me to go with a patient to the operating theatre and tell Sargent that because the lesion was probably an arteriovenous malformation, he should use great care in exposing it. I did tell Sargent, but he paid no attention and proceeded to open the dura mater with a pair of sharp pointed scissors. In doing so he ripped the malformation wide open, resulting in a severe hemorrhage and the patient's death. On another occasion Armour performed an occipital craniectomy with hammer and chisel. This patient also did not survive the operation. There was a story current at the National Hospital that Denny-Brown, then a house officer, when assisting Armour in an operation would often remark that the blood had reached the drain in the floor on the far side of the room and that perhaps it would be wise to terminate the operation." [1]

This quote is not meant to reflect on the modern Queen's Square, a bastion of outstanding surgery at this point, but only to reflex on how recently the mere surgical exposure of the brain could be highly dangerous!

In reviewing the history of minimally invasive surgery of the skull base it is not a long period of time. The nineteenth century brought the introduction of anesthesia, antisepsis and cerebral localization. The later half of the 19th century produced strong surgical personalities, adventurous enough to do surgery on the formidable skull base region. By the 1980s skull base surgery had become a subspecialty field of neurosurgery in collaboration with our ENT colleagues. It took a number of pioneering surgeons with strong personalities to make these advancements. The rapid advance in diagnostic techniques has made it much easier for skull base surgeons to localize a lesion and the details of its surrounding environs. In 2013 the skull base surgeon can now operate painlessly with a much-reduced risk of infection and rarely in the wrong location. This is a far cry from our 19th and 20th century forefathers who only 90 years ago were reporting surgical mortality rates of greater than 50%! The next question to ask is how long we will be able to remain at the side of the operating table? Following the philosophies of the airlines will we surgeons be just GPS data entering individuals in the near future?

In reflecting back over the history of endoscopic surgery is useful to keep in mind a quote by the great medical writer and proponent of antisepsis - Oliver Wendell Holmes (1809-1894).

"There is a dead medical literature, and there is a live one.
The dead is not all ancient, and the live is not all modern." [7]

References

[1] Leonardo da Vinci. (1911–1916) Quaderni d'Anatomia. Christiania: Jacob Dybwad.

[2] Leonardo da Vinci (1979) Corpus of the anatomical studies in the collection of Her Majesty the Queen at Windsor Castle. Edited by Kenneth D. Keele and Carlo Pedretti. Harcourt Brace Jovanovich, New York.

[3] Cruveilhier J. (1829–42.) Anatomie Pathologique du Corps Humain. Paris: J.-B. Baillière.

[4] Krause, Fedor. (1909-1912) Surgery of the Brain and Spinal Cord Based on Personal Experiences. Translated by H. Haubold and M. thorek. Rebman Co. New York.

[5] Bucy P: (ed.). Neurosurgical Giants: Feet of Clay and Iron New York: Elsevier, 1985 pp. 5.

[6] Holmes OW. (1911) Address delivered at the dedication of the hall of the Boston Medical Library Association, on December III, MDCCCCLXXVII. Minneapolis, MN, H.W. Wilson. See pp 3.

Preface

Classically defined as the art of curing by the hand, hand intended as the organ of the possible and positive certitude according to Paul Valery, surgery is shifting toward a scientific discipline with a very high technological valence. Neurosurgery in general, and skull base surgery in particular do not stave off this natural evolution. Obviously, technological advances have driven the tremendous progresses in both diagnosis (CT scan, MRI, angiography…) and therapeutic fields (ultrasonic aspiration, radiosurgery…). This technological aspect should not hide the humanistic remnant of the modern neurosurgeon, who should propose the less invasive technique in his possession to treat the most efficiently his patient, keeping in mind the quality of life above all. The compromise between the invasiveness of the surgical approach to the skull base and the main goal of the surgery has shed the light on the recent concept of minimally invasive skull base surgery. This concept has been conspicuously initiated by Axel Perneczky in the late 1980's under the descriptive "keyhole neurosurgery", especially through the renowned eyebrow supra-orbital mini-craniotomy and the implementation of endoscope-assisted microneurosurgery. A decade after, Jho and others introduced the endoscopic endonasal approaches to the skull base, with a perpetual development and an exponential rhythm of scientific publications. This recent paradigm shift toward a minimal approach-related iatrogeny coupled with a maximally efficient surgical target is not so clear cut, as pioneering neurosurgeons such as Cushing, Dandy or Dott among others already adopted this philosophy of work, limited by the technology available at that time that did not permit their minimally invasive expectations. This has been possible only with the progresses made in the fields of imaging, surgical instrumentation, illumination technologies (microscope and endoscope), radiosurgery, and neuroanesthesia. In this book, the authors are honored to gather many of the world's experts in the field of minimally invasive skull base surgery, to review the impact of recent technological progresses especially endoscopy on the management of skull base pathologies. Introduced by world-renowned experts (namely Edward Laws, Paolo Cappabianca and James Tait Goodrich), this volume has made possible the meeting of leading skull base surgeons worldwide to expose didactically the quintessence of minimally invasive skull base surgery.

Moncef Berhouma MD; Consultant Neurosurgeon
Associate Professor of Neurosurgery;
Minimally-invasive Neurosurgery Program – Department of Neurosurgery
Pierre Wertheimer Hospital; 59 Boulevard Pinel 69500 Bron; Lyon – France
Correspondence: berhouma.moncef@yahoo.fr

In: Minimally Invasive Skull Base Surgery
Editor: Moncef Berhouma

ISBN: 978-1-62808-567-9
© 2013 Nova Science Publishers, Inc.

Chapter I

Epistemology of Minimally Invasive Skull Base Surgery

Moncef Berhouma[*]
Department of Neurosurgery – Pierre Wertheimer Hospital, Lyon, France

"The fact that an opinion has been widely held is no evidence whatever that it is not utterly absurd; indeed in view of the silliness of the majority of mankind, a widely spread belief is more likely to be foolish than sensible."

Bertrand Russell
Marriage and Morals (Liveright Publishing Corporation, 1970)

Abstract

How have the neurosurgical knowledge and current skull base surgery corpus been constituted? How was it possible to extract such a multidisciplinary transversal supra-specialty such as skull base surgery from the field of general neurosurgery? How can neurosurgical residents receive training on skull base surgery while work hours are becoming limited? These are some of the questions raised in this chapter reporting a brief epistemology of skull base surgery, as it applies through the discipline of the philosophy of the sciences. After an overview of the historical heritage of anatomists and pioneering neurosurgeons, evidence-based skull base surgery, education problematic and the emerging concept of minimally invasive neurosurgery are discussed.

Introduction

Skull base surgery (SBS) represents the typical discipline in surgery requiring at least the close collaboration of two specialists, a neurosurgeon and an ENT or a maxillofacial surgeon. A multidisciplinary approach is of paramount importance even during the diagnostic phase

---

[*] E-mail address: berhouma.moncef@yahoo.fr

(morphologic and functional neuroradiology, diagnostic and interventional angiography, nuclear imaging...). To understand the epistemological evolution of SBS, it is obviously necessary to give an overview of the historical milestones in the development of SBS, beginning as challenging case reports during the first half of the 20th century before shifting to an individualized supra-specialty at the end of the last century. This shift toward a specific discipline was not possible without the core technological progresses made during the same period, such as: the surgical microscope, bipolar coagulation, imaging, neuronavigation, drills, neurophysiologic monitoring and neuro-anesthesia as well as asepsis. The recent development of dedicated training and research laboratories completed the emergence of SBS as a discipline by itself. In this rich accumulation of technological advances and shift toward minimally invasive techniques, can we really base our management on an evidence-based skull base surgery? This is one of the interrogations raised up in this chapter.

A Tremendous Heritage

Many anatomical misconceptions from the Greco-Roman era were still widespread until the 15th century, before the emergence of pioneering anatomists among whom Leonardo da Vinci (1452-1519) [1–4], Berengario da Carpi (1470-1550) [5–7] and Andreas Vesalius (1514-1564) [8] (Figures 1-4). Da Vinci depicted very detailed diagrams of the cranial nerves and particularly optic tracts. These Renaissance pioneers revolutionized the human anatomical corpus and settled down the premises of cranial surgery. During the 17th century, parallel to individual scientific discoveries, the concept of scientific societies and corporatist associations emulated scientific endeavor and facilitated scientific communication. It is in this same century that Thomas Willis (1621-1675) described the eponym arterial circle of the skull base [9] (Figure 5). Throughout the 18th century, several surgeons attempted neurosurgical procedures basing their surgical decisions upon detailed semiological examinations made either by themselves or by skillful neurologists: Antoine Louis (1723-1792) resected a large pterional meningioma [10], François-Sauveur Morand (1697-1773) removed a temporal abscess with a mastoidectomy. Jean Cruveilhier (1791-1874) should be remembered as one of the fathers of neuropathology and the originator of what will be later called the anatomo-clinical method [11] (Figures 6a and 6b). The development of asepsis parallel to anesthesia will lead to new surgical attempts in dedicated professional operative rooms [12, 13]. Skull base surgery history is directly tributary of pituitary surgery's one. Despite the first trans-cranial pituitary surgery performed by Sir Victor Horsley in 1889 via a trans-frontal approach, Schloffer is considered as the father of modern pituitary surgery (figures 7 and 8). He described the trans-sphenoidal approach to sella turcica in 1906, which he performed the following year via a nasal translocation [14–19]. In parallel, Oskar Hirsh, an otolaryngologist, developed a trans-septal trans-sphenoidal approach to the pituitary fossa [20, 21]. The father of modern neurosurgery, Harvey Cushing used Schloffer's approach for his first pituitary surgery in 1909, but shifted rapidly to Hirsh's technique using a sublabial incision and a headlamp (Figures 9-11). This way, Cushing did 231 procedures with a relatively low mortality for the epoch mostly due to meningitis (5.6%) [20, 22–28]. The sublabial route was thereafter abandoned during three to four decades because mainly of the lack of control of hemorrhage and the CSF leak issue. Norman Dott, a British neurosurgeon, learned the technique from Cushing and in his turn showed it to Gerard Guiot, a French neurosurgeon who resuscitated the sublabial trans-sphenoidal approach by adding the use of fluoroscopy and endoscopy [21, 29, 30]. In 1967, Jules Hardy, a fellow of Guiot, dramatically improved the technique by introducing the microscope and micro-instrumentation. Walter Dandy remains certainly the name to be kept in mind when searching for the initiation of modern craniofacial surgery (Figure 12) [31–36]. In 1941, Dandy described the trans-frontal approach to orbital tumors, an approach still used nowadays [37]. Klopp, Smith and Williams described in 1954 probably the first craniofacial resection for malignancy, followed by Malecki in 1959 [38]. Along with Sir Charles Ballance (1856-1936), Fedor Krause (1857-1937) developed cerebello-pontine surgery and described the supra-cerebellar infra-tentorial approach to pineal region [39, 40]. Thierry de Martel (1875-1940) designed the first electric trephine and introduced the sitting position for posterior fossa surgery and operative photography [41]. These are only a few names of pioneers in neurosurgery and skull base surgery, which contributed to the construction of the corpus of knowledge in the field at the beginning of the second half of the 20th century [42].

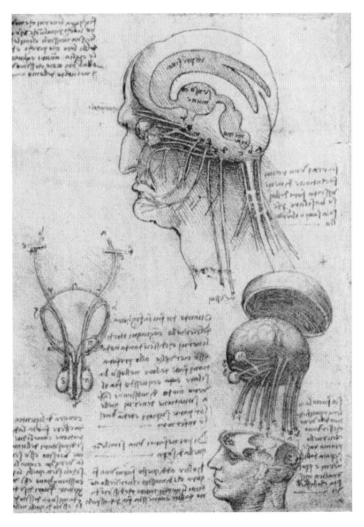

Figure 1. Leonardo Da Vinci drawings detailing the cranial nerves and their foramina, as well as a beautiful schema of the ventricular system.

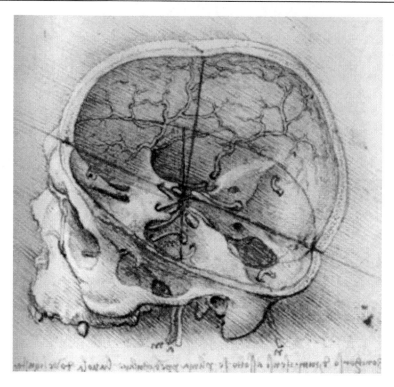

Figure 2. Leonardo Da Vinci represented beautifully the skull base and the cranial nerves foramina.

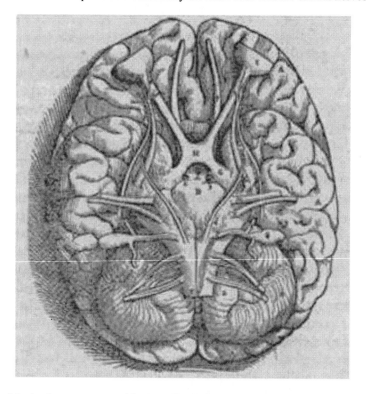

Figure 3. Base of the brain as represented by Vesalius in his Fabrica (*De humani corporis fabrica libri septem*, 1543).

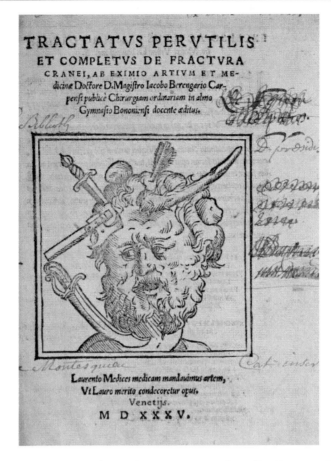

Figure 4. Contribution of Berengario da Carpi to neurotraumatology (*De Fractura Calvae sive Cranei*, Bologna, 1518).

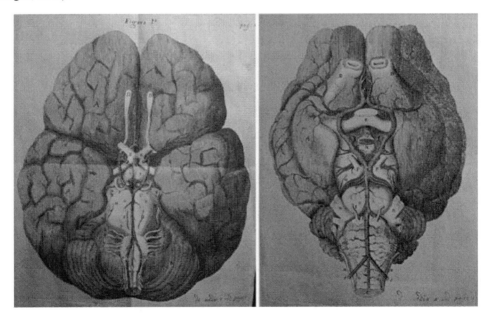

Figure 5. Thomas Willis arterial circle of the base of the skull (17th century AD).

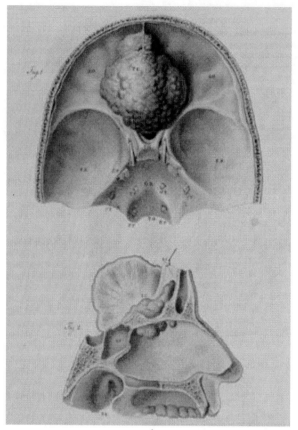

Figures 6a and 6b. Jean Cruveilhier was the pioneer of evidence-based neuropathology, with exceptional drawings of skull base tumors. Example of an olfactory groove meningioma (6a) and a cerebellopontine tumor (6b). (From. *L'anatomie pathologique du corps humain*. Paris J.B. Bailliere, 1829-1842).

Figure 7. Sir Victor Horsley (1857-1916).

Figure 8. Hermann Schloffer (1868-1937).

Figure 9. Harvey Cushing (1869-1939).

Figure 10. Cushing's masterwork on pituitary disorders.

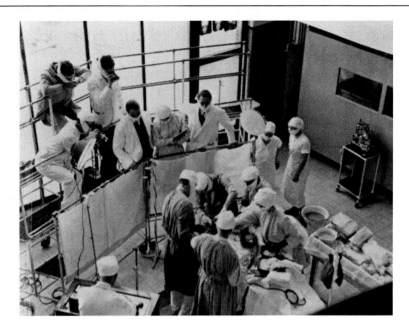

Figure 11. Cushing's operative room.

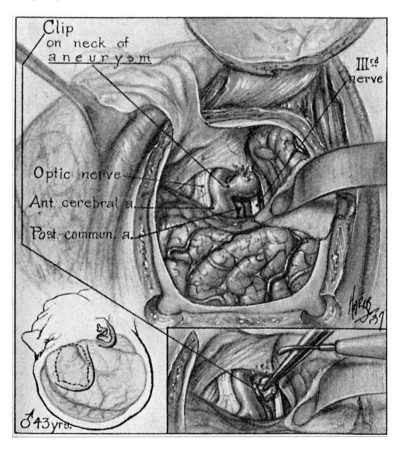

Figure 12. Frontal craniotomy for skull base approach as described by Walter Dandy, the ancestor of the frontal supra-orbital mini-craniotomy (1937).

Toward an Evidence-based Skull Base Surgery

Evidence-based medicine (EBM) is defined as the integration of the best research evidence with clinical expertise and patient values [43]. According to Haines, the term was coined in 1992 by Gordon Guyatt (McMaster University) [44, 45]. There has been a shift toward EBM in all medical specialties during the last 2 decades. The difficulty of randomized controlled trials in surgical research and particularly in neurosurgery raised the question of the applicability of EBM in neurosurgery [46]. The question of the access to extensive neurosurgical databases at any moment of the decision making process is also called when dealing with EBM, as well as the available tools to critically abstract data from systematic reviews and meta-analysis. During the last decade, an increasing number of high-quality scientific evidence in the field of neurosurgery is available especially in the Cochrane Database of Clinical Trials. Already in 2003, Haines identified in this database more than 300 randomized clinical trials in the field of neurosurgery [44]. The daily practice of evidence-based skull base surgery requires, in addition to having a critical mind, a real-time access to Internet updated sources such as MEDLINE, Sciencedirect and SCOPUS. The skull base surgeon in particular and the neurosurgeon in general must produce a personal effort to assemble dedicated evidence repositories where the information should be systematically and rigorously classified by thematic and levels of evidence. A regular and close follow-up of evidence repositories is mandatory, such as the Cochrane Library, the Scientific Statements of the American Heart Association, and the National Guideline Clearing House of the US department of Health.

An effort should also be made at an institutional or academic level in order to implement courses and workshops dedicated to educate young surgeons to evidence-based medicine principles.

Elsewhere, a local policy of feedback audits should be regularly set within each individual hospital as to compare surgical results with national standards and therefore correct and optimize continuously the practice locally and regionally. These audits may also compare inter-departments and intra-departments results and guidelines.

Training in Skull Base Surgery

The familiarization with skull base techniques is a legitimate part of the neurosurgical training of residents and young neurosurgeons. During the last decade, neurosurgical education has undergone many changes especially concerning duty-hours restrictions, limiting therefore their direct contact with patients and senior neurosurgeons in their daily practice [47].

This time restriction and accordingly shortened training opportunities come along with the increasing patients' requirements and procedural tendency. The latter shed light on the necessity of developing new educational models in neurosurgery and skull base surgery in particular due to the complex pathologies it deals with. To overcome the problem of training on real patients in daily practice and go beyond lab training on normal anatomical samples, skull base tumor models have been described [48–50]. These allow training in the context of distorted skull base anatomy by the injection of polymerized tumor models. The continuing medical education pre and postgraduate should also naturally include regular participations to specific workshops encompassing live surgeries from master educators. The setting of specifically designed skull base laboratories is certainly a high value mean of training out of the operating room and out of non-dedicated general anatomy laboratories [51].

The role of neurosurgical simulation will certainly gain an increasing place within the education armamentarium in the coming years, to optimize patient safety and fulfill medicolegal requirements [52]. This training must include transversal rotations in the other specialties required in this multidisciplinary specialty, such as ENT and maxillofacial rotations.

In the coming years, the spread of augmented reality training including haptics and robotics will certainly reduce the need of human or animal cadaver use, even if such installations require expensive investments [53–59].

The Concept of Minimally Invasive Skull Base Surgery: Toward a Minimum Approach-related Iatrogeny Coupled with a Maximally Efficient Surgical Target

The concept of minimally invasive neurosurgery can be reasonably attributed to Axel Perneczky in the late nineties [60–66]. He pioneered the clinical applications of endoscope-assisted microneurosurgery, particularly for aneurysm management [63]. The term "minimally invasive" may lead to miscomprehension of the concept itself as it may lead to understand that the final treatment will be minimal. This misconception still exists nowadays. It should be categorically rejected, as this qualification of "minimally" reports exclusively to the trauma that may be occasioned during the surgical approach, while the final surgical target is exactly the same and sometimes superior to what is obtained during so-called classic neurosurgical approaches. One can resume this concept as achieving the most efficient surgical target while minimizing the trauma due exclusively to the approach. This concept has only been possible with the development of preoperative imaging to plan the optimal less traumatic approach, the progresses in visualization and enlightenment of the surgical field particularly with the advent of high definition endoscopes, and the setting of dedicated microsurgical instruments (Figures 13 and 14). The obvious risk in the comprehension of this concept is to evolve toward very small craniotomies that do not allow achieving completely the surgical goal. This risk can be avoided by rigorously planning preoperatively the surgical approach without forgetting the primary surgical target (aneurysm clipping, tumor removal…) [66]. The evolution of neurosurgery from a branch of general surgery at the end of the 19th century toward a specific technically demanding surgical specialty has been accompanied by the permanent aim to minimize invasiveness. This has been possible only by improving the technological support of neurosurgery, apart of personal surgeons' skills.

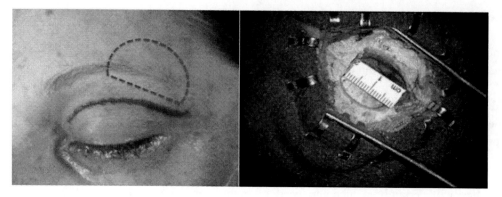

Figure 13. Mini-invasive supra-orbital craniotomy via hidden trans-palpebral incision to access anterior and middle cranial base.

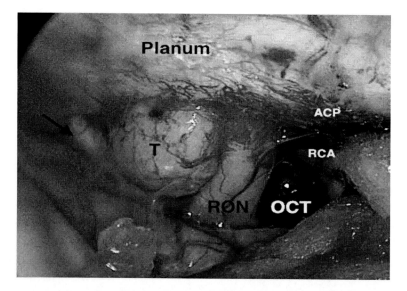

Figure 14. Same surgical keyhole approach as Fig 13. Purely endoscopic approach to tuberculum sellae meningioma (T) elevating the ispsilateral right optic nerve (RON). Arrow. left optic nerve, OCT. optico-carotid triangle, RCA. right internal carotid artery, ACP. anterior clinoid process.

At the beginning of neurosurgery, at least five factors have justified the resort to large craniotomies (and spinal approaches as well):

- The lack of diagnostic tools (especially imaging) and the topographic localization of the lesion based only on neurological assessment.
- The large size of tumors at the moment of diagnosis because of the lack of medical education of the populations.
- The limited possibility of bringing light in the surgical field frequently limited to an external light source or a forehead source.
- The absence of dedicated and purposely shaped instruments. Most of instruments used in neurosurgery at that time derived directly from general surgery and orthopedics
- The need to involve two or more surgical assistants to sponge, to aspirate…

The progressive resolution of these drawbacks, particularly in the fields of diagnosis tools and advances in illumination devices, allowed the reduction of invasiveness of neurosurgical approaches. One must keep in mind the first ventriculographies brought by Walter Dandy in 1918 and angiography by Moniz in 1927, later improved by Seldinger in 1953. In the field of instrumentation, the introduction of the Gigli saw in 1898 revolutionized the osteoplastic craniotomy possibilities instead of craniectomies, as well as hemostasis means developed by Harvey Cushing at the beginning of the 19th century changed the course of intra-operative bleeding (electrosurgery and silver clips) [67]. Nevertheless, one of the most critical revolutions in microneurosurgery was the introduction of the surgical microscope in the operative rooms in the early 1960's (Yasargil, Parkinson, Jannetta) [68]. These are only few examples of the technological revolutions in the field of neurosurgery. Finally, one should also not forget the importance of the evolution of neuroanesthesia especially the development of the concept of chemical brain retraction and neuroprotection, which considerably improved both operative morbidity and mortality [69–71].

Table 1. Technological and human factors enabling skull base surgery as a discipline by itself

Conclusion

The epistemological evolution of skull base surgery has followed that of neurosurgery in general [72]. This evolution has not been linear, but marked by acute (r)evolutions especially in the technological field: CT scan, MRI, angiography, operative microscope, endoscope, neuronavigation, drill, ultrasonic aspirator…This epistemological evolution has in the meantime shed the light on the value of neurosurgical education and skull base training based mainly on mentorship and lab training as well as augmented reality. Elsewhere, the notion of skull base surgery itself obviously requires transversal scientific interactions between different specialties involved in the management of specific complex patients. The importance of multidisciplinary team work in skull base surgery (ENT, Plastic surgeon, Maxillofacial surgeon, Neurosurgeon, Neuroradiologist, Radiosurgeon, Endovascular surgeon…) has been clearly proven during the last two decades, and allowed a significant improvement of scientific and surgical skills exchanges as well as the creation of reference multidisciplinary centers dedicated to skull base pathologies [73].

References

[1] Cavalcanti DD, Feindel W, Goodrich JT, Dagi TF, Prestigiacomo CJ, Preul MC. Anatomy, technology, art, and culture: toward a realistic perspective of the brain. *Neurosurg Focus.* 2009 Sep;27(3):E2.

[2] Del Maestro RF. Leonardo da Vinci: the search for the soul. *J. Neurosurg.* 1998 Nov;89(5):874–87.

[3] Goodrich JT. A millennium review of skull base surgery. *Childs Nerv Syst.* 2000 Nov;16(10-11):669–85.

[4] Pevsner J. Leonardo da Vinci's contributions to neuroscience. *Trends Neurosci.* 2002 Apr;25(4):217–20.

[5] Di Ieva A, Gaetani P, Matula C, Sherif C, Skopec M, Tschabitscher M. Berengario da Carpi: a pioneer in neurotraumatology. *J. Neurosurg.* 2011 May;114(5):1461–70.

[6] Mazzola RF, Mazzola IC. Treatise on skull fractures by Berengario da Carpi (1460-1530). *J Craniofac Surg.* 2009 Nov;20(6):1981–4.

[7] Merlini L, Tomba P, Viganò A. Berengario da Carpi, a pioneer in anatomy, rediscovered by Vittorio Putti. *Neuromuscul. Disord.* 2003 Jun;13(5):421–5.

[8] Saban R. [The beginnings of physiology of the human brain, from antiquity to the Renaissance]. *Vesalius.* 1999 Jun;5(1):41–7.

[9] Rengachary SS, Xavier A, Manjila S, Smerdon U, Parker B, Hadwan S, et al. The legendary contributions of Thomas Willis (1621-1675): the arterial circle and beyond. *J. Neurosurg.* 2008 Oct;109(4):765–75.

[10] Lee JH. *Meningiomas: Diagnosis, Treatment, and Outcome.* Springer; 2008.

[11] Berhouma M, Dubourg J, Messerer M. Cruveilhier's legacy to skull base surgery: Premise of an evidence-based neuropathology in the 19th century. *Clin Neurol Neurosurg.* 2012 Aug 18;

[12] Steimle RH. [Operating rooms during the second half of the 20th century and its change with surgical advances]. *Hist Sci Med.* 2011 Jun;45(2):187–95.

[13] Carter KC. SemmelWeis and his predecessors. *Med Hist.* 1981 Jan;25(1):57–72.
[14] Lindholm J. A century of pituitary surgery: Schloffer's legacy. *Neurosurgery.* 2007 Oct;61(4):865–867; discussion 867–868.
[15] Liu JK, Das K, Weiss MH, Laws ER Jr, Couldwell WT. The history and evolution of transsphenoidal surgery. *J. Neurosurg.* 2001 Dec;95(6):1083–96.
[16] Schmidt RF, Choudhry OJ, Takkellapati R, Eloy JA, Couldwell WT, Liu JK. Hermann Schloffer and the origin of transsphenoidal pituitary surgery. *Neurosurg Focus.* 2012 Aug;33(2):E5.
[17] Tan T-C, Black PM. Sir Victor Horsley (1857-1916): pioneer of neurological surgery. *Neurosurgery.* 2002 Mar;50(3):607–611; discussion 611–612.
[18] Uff C, Frith D, Harrison C, Powell M, Kitchen N. Sir Victor Horsley's 19th century operations at the National Hospital for Neurology and Neurosurgery, Queen Square. *J. Neurosurg.* 2011 Feb;114(2):534–42.
[19] Case of Acusticus Tumour (Right); Operation by Sir Victor Horsley in 1912; Removal of Tumour; Recovery. *Proc. R. Soc. Med.* 1923;16(Otol Sect):31–2.
[20] Liu JK, Cohen-Gadol AA, Laws ER Jr, Cole CD, Kan P, Couldwell WT. Harvey Cushing and Oskar Hirsch: early forefathers of modern transsphenoidal surgery. *J. Neurosurg.* 2005 Dec;103(6):1096–104.
[21] Lanzino G, Laws ER Jr. Pioneers in the development of transsphenoidal surgery: Theodor Kocher, Oskar Hirsch, and Norman Dott. *J. Neurosurg.* 2001 Dec;95(6):1097–103.
[22] Aron DC. The path to the soul: Harvey Cushing and surgery on the pituitary and its environs in 1916. *Perspect. Biol. Med.* 1994;37(4):551–65.
[23] Cohen-Gadol AA, Liu JK, Laws ER Jr. Cushing's first case of transsphenoidal surgery: the launch of the pituitary surgery era. *J. Neurosurg.* 2005 Sep;103(3):570–4.
[24] Cohen-Gadol AA, Laws ER, Spencer DD, De Salles AAF. The evolution of Harvey Cushing's surgical approach to pituitary tumors from transsphenoidal to transfrontal. *J. Neurosurg.* 2005 Aug;103(2):372–7.
[25] Harvey AM. Harvey Williams Cushing: the Baltimore period, 1896-1912. *Johns Hopkins Med J.* 1976 May;138(5):196–216.
[26] Jay V. The extraordinary legacy of Dr. Harvey Cushing. *J Insur Med.* 2006;38(4):303–6.
[27] Pendleton C, Adams H, Salvatori R, Wand G, Quiñones-Hinojosa A. On the shoulders of giants: Harvey Cushing's experience with acromegaly and gigantism at the Johns Hopkins Hospital, 1896-1912. *Pituitary.* 2011 Mar;14(1):53–60.
[28] Savitz SI. Cushing's contributions to neuroscience, part 2: Cushing and several dwarfs. *Neuroscientist.* 2001 Oct;7(5):469–73.
[29] Guiot J, Rougerie J, Fourestier M, Fournier A, Comoy C, Vulmiere J, et al. [Intracranial endoscopic explorations.]. *Presse Med.* 1963 May 18;71:1225–8.
[30] Guiot G, Bouche J, Hertzog E, Vourc'h G, Hardy J. [Hypophysectomy by trans-sphenoidal route.]. *Ann Radiol* (Paris). 1963;6:187–92.
[31] Dandy W. Practice of surgery. *The brain.* Hagerstown; 1932. p. 247–52.
[32] Goodrich JT. Reprint of "The Operative Treatment of Communicating Hydrocephalus" by Walter E Dandy, MD. 1938. *Childs Nerv Syst.* 2000 Sep;16(9):545–50.
[33] Hsu W, Li KW, Bookland M, Jallo GI. Keyhole to the brain: Walter Dandy and neuroendoscopy. *J Neurosurg Pediatr.* 2009 May;3(5):439–42.

[34] Kilgore EJ, Elster AD. Walter Dandy and the history of ventriculography. *Radiology.* 1995 Mar;194(3):657–60.

[35] Kretzer RM, Coon AL, Tamargo RJ. Walter E. Dandy's contributions to vascular neurosurgery. *J. Neurosurg.* 2010 Jun;112(6):1182–91.

[36] Pearce JMS. Walter Edward Dandy (1886-1946). J Med Biogr. 2006 Aug;14(3):127–8.

[37] Horwitz NH. Library: historical perspective. Walter E. Dandy (1886-1946). *Neurosurgery.* 1997 Mar;40(3):642–6.

[38] Donald PJ. History of Skull Base Surgery. *Skull base surgery.* 1991 Jan;1(1):1–3.

[39] Machinis TG, Fountas KN, Dimopoulos V, Robinson JS. History of acoustic neurinoma surgery. *Neurosurg Focus.* 2005 Apr 15;18(4):e9.

[40] Ramsden RT. The bloody angle: 100 years of acoustic neuroma surgery. *J R Soc Med.* 1995 Aug;88(8):464P–468P.

[41] Pecker J. Thierry de Martel. 1875-1940. *Surg Neurol.* 1980 Jun;13(6):401–3.

[42] Greenblatt G. A History of Neurosurgery. *American Association of Neurological Surgeons* (AANS); 1996.

[43] Sackett DL, Rosenberg WM, Gray JA, Haynes RB, Richardson WS. Evidence based medicine: what it is and what it isn't. *BMJ.* 1996 Jan 13;312(7023):71–2.

[44] Haines SJ. Evidence-based neurosurgery. *Neurosurgery.* 2003 Jan;52(1):36–47; discussion 47.

[45] Evidence-based medicine. A new approach to teaching the practice of medicine. *JAMA.* 1992 Nov 4;268(17):2420–5.

[46] Bandopadhayay P, Goldschlager T, Rosenfeld JV. The role of evidence-based medicine in neurosurgery. *J Clin Neurosci.* 2008 Apr;15(4):373–8.

[47] Fargen KM, Chakraborty A, Friedman WA. Results of a national neurosurgery resident survey on duty hour regulations. *Neurosurgery.* 2011 Dec;69(6):1162–70.

[48] Aboud E, Al-Mefty O, Yaşargil MG. New laboratory model for neurosurgical training that simulates live surgery. *J. Neurosurg.* 2002 Dec;97(6):1367–72.

[49] Berhouma M, Baidya NB, Ismaïl AA, Zhang J, Ammirati M. Shortening the learning curve in endoscopic endonasal skull base surgery: A reproducible polymer tumor model for the trans-sphenoidal trans-tubercular approach to retro-infundibular tumors. *Clin Neurol Neurosurg.* 2013 Mar 4;

[50] Gragnaniello C, Nader R, van Doormaal T, Kamel M, Voormolen EHJ, Lasio G, et al. Skull base tumor model. *J. Neurosurg.* 2010 Nov;113(5):1106–11.

[51] Salma A, Chow A, Ammirati M. Setting up a microneurosurgical skull base lab: technical and operational considerations. *Neurosurg Rev.* 2011 Jul;34(3):317–326; discussion 326.

[52] Ganju A, Aoun SG, Daou MR, El Ahmadieh TY, Chang A, Wang L, et al. The Role of Simulation in Neurosurgical Education: A Survey of 99 United States Neurosurgery Program Directors. *World Neurosurg.* 2012 Nov 24;

[53] L'Orsa R, Macnab CJB, Tavakoli M. Introduction to haptics for neurosurgeons. *Neurosurgery.* 2013 Jan;72 Suppl 1:139–53.

[54] Chan S, Conti F, Salisbury K, Blevins NH. Virtual reality simulation in neurosurgery: technologies and evolution. *Neurosurgery.* 2013 Jan;72 Suppl 1:154–64.

[55] Choudhury N, Gélinas-Phaneuf N, Delorme S, Del Maestro R. Fundamentals of Neurosurgery: Virtual Reality Tasks for Training and Evaluation of Technical Skills. *World Neurosurg.* 2012 Nov 23;

[56] Agarwal N, Schmitt PJ, Sukul V, Prestigiacomo CJ. Surgical approaches to complex vascular lesions: the use of virtual reality and stereoscopic analysis as a tool for resident and student education. *BMJ Case Rep.* 2012;2012.

[57] Alaraj A, Charbel FT, Birk D, Tobin M, Luciano C, Banerjee PP, et al. Role of cranial and spinal virtual and augmented reality simulation using immersive touch modules in neurosurgical training. *Neurosurgery.* 2013 Jan;72 Suppl 1:115–23.

[58] Robison RA, Liu CY, Apuzzo MLJ. Man, mind, and machine: the past and future of virtual reality simulation in neurologic surgery. *World Neurosurg.* 2011 Nov;76(5):419–30.

[59] Yc Goha K. Virtual reality applications in neurosurgery. *Conf Proc IEEE Eng Med Biol Soc.* 2005;4:4171–3.

[60] Conrad J, Welschehold S, Charalampaki P, van Lindert E, Grunert P, Perneczky A. Mesencephalic ependymal cysts: treatment under pure endoscopic or endoscope-assisted keyhole conditions. *J. Neurosurg.* 2008 Oct;109(4):723–8.

[61] Fischer G, Stadie A, Reisch R, Hopf NJ, Fries G, Böcher-Schwarz H, et al. The keyhole concept in aneurysm surgery: results of the past 20 years. *Neurosurgery.* 2011 Mar;68(1 Suppl Operative):45–51; discussion 51.

[62] Hopf NJ, Reisch R. Axel Perneczky, 1.11.1945-24.1.2009. *Minim Invasive Neurosurg.* 2009 Feb;52(1):1–4.

[63] Perneczky A, etc, Mueller-Forell W, Lindert EV. *The Keyhole Concept in Neurosurgery: With Endoscope-assisted Microsurgery and Case Studies.* Thieme Publishing Group; 1998.

[64] Perneczky A, Reisch R. *Keyhole Approaches in Neurosurgery: Concept and Surgical Technique.* 1st ed. Springer Verlag GmbH; 2008.

[65] Reisch R, Perneczky A. Ten-year experience with the supraorbital subfrontal approach through an eyebrow skin incision. *Neurosurgery.* 2005 Oct;57(4 Suppl):242–255; discussion 242–255.

[66] Stadie AT, Kockro RA, Reisch R, Tropine A, Boor S, Stoeter P, et al. Virtual reality system for planning minimally invasive neurosurgery. Technical note. *J. Neurosurg.* 2008 Feb;108(2):382–94.

[67] Bulsara KR, Sukhla S, Nimjee SM. History of bipolar coagulation. *Neurosurg Rev.* 2006 Apr;29(2):93–96; discussion 96.

[68] Uluç K, Kujoth GC, Başkaya MK. Operating microscopes: past, present, and future. *Neurosurg Focus.* 2009 Sep;27(3):E4.

[69] Samuels SI. History of neuroanesthesia: a contemporary review. *Int Anesthesiol Clin.* 1996;34(4):1–20.

[70] Ravussin P, Mustaki JP, Boulard G, Moeschler O. [Neuro-anesthetic contribution to the prevention of complications caused by mechanical cerebral retraction: concept of a chemical brain retractor]. *Ann Fr Anesth Reanim.* 1995;14(1):49–55.

[71] Ravussin P. Anaesthesia for neurosurgical procedures. *Minerva Anestesiol.* 1992 Oct;58(10):861–4.

[72] Greenblatt SH, Dagi TF, Epstein MH. *A history of neurosurgery: in its scientific and professional contexts.* Thieme; 1997.

[73] McLaughlin N, Carrau RL, Kelly DF, Prevedello DM, Kassam AB. Teamwork in skull base surgery: An avenue for improvement in patient care. *Surg Neurol Int.* 2013;4:36.

In: Minimally Invasive Skull Base Surgery
Editor: Moncef Berhouma

ISBN: 978-1-62808-567-9
© 2013 Nova Science Publishers, Inc.

Chapter II

Endoscopic Endonasal Anatomy and Approaches to the Anterior Skull Base

Bashar Abuzayed
Department of Neurosurgery, Al-Bashir Governmental Hosp*ital, Amman, Jord*an

Abstract

Recently, endoscopic endonasal approaches to the anterior skull base are gaining popularity among neurosurgeons, and the details of the endoscopic anatomy and approaches are highlighted from the neurosurgeons' point of view, correlated with demonstrative cases.

After resection of the superior portion of the nasal septum, bilateral middle and superior turbinates, bilateral anterior and posterior ethmoidal cells resection can be performed either by anterior to posterior approach or posterior to anterior approach. Then we obtain full exposure of the anterior skull base. After resecting the anterior skull base bony structure and the dura between the 2 medial orbital walls, we visualize the olfactory nerves, interhemispheric sulcus, and gyri recti. With dissecting the interhemispheric sulcus, we expose the first (A1) and second (A2) segments of the anterior cerebral artery, anterior communicating artery, and Heubner arteries.

Extended endoscopic endonasal approaches are sufficient in providing wide exposure of the bony structures, and the extradural and intradural components of the anterior skull base and the neighboring structures providing more controlled manipulation of lesions.

Abbreviations

A1:   pre-communicating segment of the anterior cerebral artery.
A2:   post-communicating segment of anterior cerebral artery.

AEA: anterior ethmoidal artery.
ACoA: anterior communicating artery.
CSF: cerebrospinal fluid.
CT: computed tomography.
ESS: endoscopic sinus surgery.
PEA: posterior ethmoidal artery.

Introduction

The midline skull base is an anatomic area that covers the upper nasal cavity and the sphenoidal sinus and is composed of the crista galli and the cribriform plate of the ethmoid bone and the sphenoid body [1]. This region is a site for a variety of pathologies, which are considered surgically challenging because of its critical location and neighboring neurovascular structures. Many transfacial and transcranial approaches were proposed [2-6], which are considered aggressive because of large incisions and flaps, facial skeleton disturbance, cosmetic problems, wide neurovascular manipulation and retraction, and patients' difficulty in tolerating with bad general condition, especially for those with oncologic pathologies. Surgeons started to seek more minimally invasive techniques to avoid these difficulties. With the introduction of functional endoscopic sinus surgery in the 1980s, endoscopic techniques and approaches were developed starting from diagnosing and treating chronic inflammatory diseases. With the evolution of the endoscopic equipments, vast endoscopic anatomic studies, collaboration between neurosurgeons and ENT surgeons, and the wide experience obtained with time, endoscopic techniques to the anterior skull base lesions became more identified and even standardized for some pathologies, such as CSF leaks [7,8], meningoencephaloceles [9, 10], paranasal sinus and anterior skull base neoplasms [11-15]. Now, indeed, these techniques can compete with or complete the microscope in surgical treatment in many pathologies. Endoscopic visualization from below has offered the surgeons the possibility to reach the midline anterior skull base without skin incision and brain retraction and with minimal neurovascular manipulation [7, 12].

In this chapter, we demonstrate the endoscopic anatomy and approaches to the anterior skull base with documentation of many important structures and landmarks, which will help us to understand the neurovascular relations and surgical planning to avoid possible complications.

Anatomic Considerations

The ethmoid sinus is the most compartmentalized paranasal sinus, located lateral to the olfactory cleft and fossa, between the lateral nasal wall and the medial orbital wall, the ethmoid sinus. At birth, only a few cells are pneumatized, but in adulthood their number can go beyond 15 cells [16]. The frontal bone in its posterior extension covers the roof of the ethmoid sinus, forming the so-called ethmoidal fovea [16]. The width of the ethmoid increases from anterior to posterior because of the cone-like structure of the orbit. The ethmoid sinus is referred to as the ethmoid labyrinth because of the complexity of its anatomy, due to the honeycomb-like appearance of its air cells with intricate passageways and blind alleys. Rhinologists have tried to simplify its difficult anatomy by considering the sinus as a series of five obliquely oriented parallel lamellae (Figure 1). These derive from the ridges in the lateral nasal wall of the fetus called ethmoturbinals [16]. The lamellae are relatively constant and easy to recognize intraoperatively.

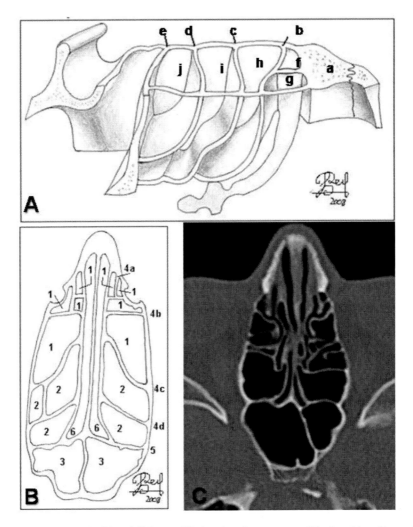

Figure 1. Schematic drawing (A,B) and CT-scan (C) showing the anatomy of the basal lamellae and the ethmoidal cells.
a: Basal lamella of the uncinate process.
b: Basal lamella of the ethmoid bulla.
c: Basal lamella of the middle turbinate.
d: Basal lamella of the superior turbinate.
e: Supreme turbinate.
1: Anterior ethmoid cells.
2: Posterior ethmoid cells.
3: Sphenoid sinus.
4a: First basal lamella.
4b: Second basal lamella.

f: space with ethmoid infundibulum.
g: Anterior middle meatus.
h: Ethmoid bulla.
i and j: Posterior ethmoid.

4c: Basal lamella of middle turbinate (third).
4d: Basal lamella of superior turbinate (fourth).
5: Posterior ethmoid wall.
6: Sphenoethmoid recess.

The first and most anterior lamella corresponds to the uncinate process, which embryologically represents the basal lamella of the first ethmoturbinal. The second lamella is the ethmoid bulla, or bulla ethmoidalis as first referred to by Zuckerkandl, the largest and most constant anterior ethmoid cell [17]. It has a round shape with thin walls, extending from the lamina papyracea laterally and bulging into the middle meatus medially. Rarely, when non-pneumatized, a bony projection from the lamina papyracea results and is referred to as the torus lateralis. The most important lamella is the third one, the ground or basal lamella of the middle turbinate, not only defining the anatomical separation between the anterior and the posterior ethmoid cells, but also creating a bony septation that dictates the drainage pattern of the ethmoid cells into the middle meatus (for the anterior ethmoid cells) and the superior and supreme meati (for the posterior ethmoid cells). It thus represents the surgical posterior limit for an anterior ethmoidectomy. Its S-shaped insertion with a sagittal anterior third and a vertical middle third gives the middle turbinate its mechanical stability, and its posterior horizontal part at the level of the tail of the middle turbinate represents the most straightforward way for entry into the posterior ethmoids. The fourth lamella is the basal lamella of the superior turbinate and the fifth is the basal lamella of the supreme turbinate [16, 17]. Besides the bony lamellae, particular groups of ethmoid cells have been identified, and recognizing them helps understanding the pathophysiology and spread of sinus disease, as well as performing the most complete ethmoid surgery with the least surgical complications. The agger nasi cells are the most anterior ethmoid cells, and are endoscopically visualized as a prominence anterior to the insertion of the middle turbinate. From Latin, *agger* means mound or eminence, and agger nasi refers to the pneumatized superior remnant of the first ethmoturbinal which persists as a mound anterior and superior to the insertion of the middle turbinate. Rarely, the pneumatization can extend inferiorly to involve the anterosuperior part of the uncinate process, which as described previously derives from the descending portion of the first ethmoturbinal. The agger nasi pneumatization can also have a significant impact on the uncinate process insertion, as well as on the patency of the frontal recess. [16, 17]. Accurate identification of the agger nasi is the key to the surgical access to the frontal recess. Conchal cells are ethmoid air cells that invade the middle turbinate in its anterior aspect, whereas interlamellar cells, originally described by Grunwald, arise from pneumatization of the vertical lamella of the middle turbinate from the superior meatus [16]. The supraorbital ethmoid cells (also referred to as suprabullar cells) are anterior ethmoid cells that arise immediately behind the frontal recess and extend over the orbit through pneumatization of the orbital plate of the frontal bone. They can compromise posteriorly the frontal sinus drainage, in a similar way as the agger nasi anteriorly. During ESS, supraorbital cells can be mistaken for the frontal sinus by inexperienced surgeons. Transillumination of these cells with a telescope reveals the light in the inner canthal area, rather than the supraorbital area when the frontal sinus is transilluminated [17]. The sphenoethmoid cells, or Onodi cells, are another important group of ethmoid cells. In this case, the posterior ethmoid cells extend superiorly or laterally to the sphenoid sinus, and the pneumatization can reach the posterior clinoid process. The sphenoethmoid cell is intimately related to the optic nerve, whether the latter is prominent or not in its lateral wall. Also, if large enough, the carotid artery can bulge through its posterior wall. Thus, attempts to open the sphenoid through a sphenoethmoid cell can result in serious damage to the optic nerve or the carotid artery. These important structures are usually related to the lateral wall of the sphenoid sinus; however, accurate identification of these structures and possibly Onodi cells on a preoperative CT scan is the best way to avoid such severe complications.

The anterior and posterior ethmoid arteries, terminal branches of the internal carotid artery via the ophthalmic artery, run along the roof of the ethmoid from lateral to medial (Figure 2). The anterior ethmoidal artery (AEA) runs from lateral to medial and obliquely forward along the skull base. In approximately 40% of cases the artery is up to 2 mm from the skull base, lying free or attached to a "mesentery", and in 60% of cases it courses directly on the skull base [16]. Laterally, the bone at the level of the lamina papyracea forms a small funnel. The superior ethmoid in front of the artery almost always extends over the orbit to some small degree; the portion posterior to the artery does so only in rare cases. The vessel can be clearly visualized in 80% of cases. AEA is located by slow dissection of the anterior wall of the bulla in the direction of the skull base. Most commonly the artery is located 1-2 mm behind the bulla lamella. Its average distance from the posterior portion of the middle turbinate is 20 [17-25] mm [18]. In front of the artery the, the skull base slopes gently upward (15 degree) and then steepens more anteriorly as it joins with the posterior wall of the frontal sinus. This point is located an average of 9 mm in front of the artery, and a final ethmoid cell is frequently encountered in this area [19]. The medial part of the AEA is located behind the globe of the eye in the coronal plane. The posterior ethmoidal artery (PEA) has a larger caliber than the anterior artery but is usually closer to the skull base. The distance between optic canal and PEA ranges from 8 to 16 mm (mean, 11.08 mm), and the distance between PEA and anterior ethmoid artery ranges from 10 to 17 mm (mean, 13 mm) [7]. A third ethmoidal artery is present in 30% of cases [16, 17].

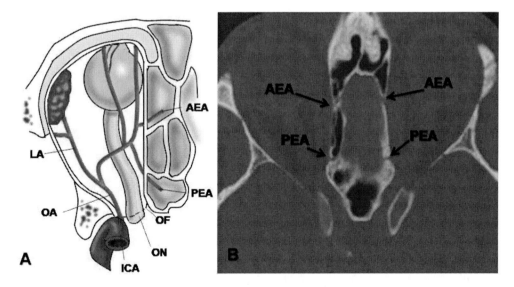

Figure 2. Schematic drawing (A) and CT-scan (B) showing the anatomy of the ethmoidal arteries. AEA: anterior ethmoidal artery, PEA: posterior ethmoidal artery, OF: optic foramen, ON: optic nerve, ICA: internal carotid artery, OA: ophthalmic artery, LA: lacrymal artery.

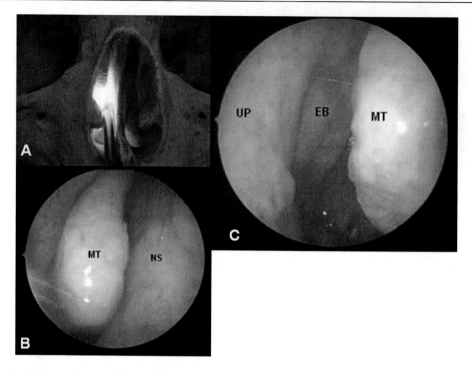

Figure 3. A) The endoscope is inserted into the right nasal cavity of a cadaver showing the roof of the right nasal cavity. B) Endoscopic view of the right nasal cavity after inserting the telescope. C) The middle turbinate is retracted medially to expose the ethmoid bulla. NS: nasal septum, MT: middle turbinate, UP: uncinate process, EB: ethmoid bulla.

Surgical Technique

Surgical Position and Preparations

The patient is placed in supine position, with the head in neutral position and 10 to 15 degrees adducted toward the left shoulder. This position is more suitable for the surgeon in that it does not disturb surgical orientation and avoids the lateral bending of the surgeon's trunk toward the patient, causing less fatigue during the surgical procedures [7]. We also recommend not covering the whole face of the patient. Instead, leaving the nose and the eyes exposed with transparent drape will help the surgeon to preserve the orientation during the surgery. Intraoperative C-arm fluoroscopy can be used to check the position of the telescope during the approach. Moreover, neuronavigation and intraoperative imaging techniques are now available for more controlled surgery, especially with computed tomography (CT) for better visualization of the bony parasinuses and skull base structures.

Surgical Corridor Preparation (Nasal Step)

The endoscope is introduced from the right nostril (Figure 3). The inferior and middle turbinates and the nasal septum are identified. The middle turbinate is retracted medially to expose the uncinate process and the ethmoid bulla. By using the concha scissors, the middle turbinate is resected with caution not to injure the ethmoidal roof. The superior portion of the nasal septum is resected, making the contralateral anterior skull base visible. The same steps are done in the left side from the left nostril.

Anterior and Posterior Ethmoidectomy

The main procedure to expose the anterior skull base and the medial orbital wall is the anterior and posterior ethmoidectomy (bilateral). Two approaches can be defined to perform this procedure, the anterior-posterior and the posterior-anterior approaches.

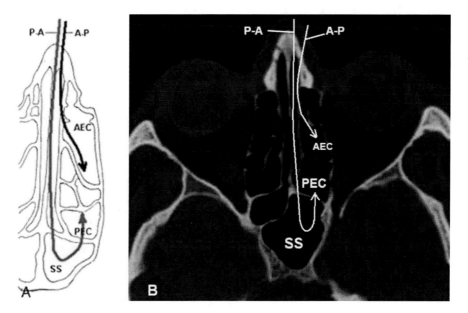

Figure 4. A drawing (A) and brain CT scan (B) demonstrating the anterior-posterior and posterior-anterior approaches for ethmoidectomy. AEC: anterior ethmoid cells, A-P: anterior-ethmoidal approach, PEC: posterior ethmoidal cells, P-A: posterior-anterior approach, SS: sphenoid sinus.

The anterior-posterior approach: In this approach, the anterior ethmoidal cells are resected first, and dissection extends posteriorly performing posterior ethmoidectomy (Figure 4). After the surgical corridor preparation in the nasal step, the mucosa covering the uncinate process is dissected and the bony structure of the uncinate process is resected or drilled, therefore exposing the maxillary ostium and entering the ethmoidal bulla. Maxillary antrostomy is performed as next step (Figure 5). This step is important in that it permits ventilation of the maxillary sinus and defines the medial wall of the orbit, which lies at the same vertical plane as the maxillary ostium [7, 20, 21]. The inferoanterior wall of the ethmoidal bulla is opened, and the anterior ethmoidal cells are entered and resected. The lateral wall of the ethmoid bulla is the lamina papyracea, which is the lateral limit of anterior skull base exposure. This step of the procedure was completed through the opening of the agger nasi and the suprabullar recess (Figure 6). The next step is directing the dissection posteriorly by resecting the basal lamella of the middle turbinate to enter the posterior ethmoidal cells. The posterior ethmoidal cells may be entered safely through the most horizontal portion of the middle turbinate lamella. Endoscopically, there is an imaginary line perpendicular to the nasal septum from the posterior medial orbital roof, another line along the vertical antrostomy ridge, and a third line along the free edge of middle turbinate basal lamella, forming a triangle (Figure 7). This triangle demarcates the zone of safe entry into the inferior aspect of posterior ethmoidal cells [7, 20]. Dissection is continued posteriorly until facing the sphenoid sinus. The direction of the dissection is posterior-superior, leading to the optic and the carotid protuberances located in the sphenoid sinus roof. This approach is performed bilaterally to expose the whole anterior skull base from the frontal sinus anteriorly to sphenoid sinus roof posteriorly and between the lamina papyracea bilaterally.

The posterior-anterior approach: In this approach, first the sphenoid sinus is opened, and retrograde dissection from posterior ethmoidal cells to the anterior ethmoidal cells is performed (Figure 4). After preparation of the surgical corridor in the nasal step, the ostia of the sphenoid sinus are identified in the sphenoethmoidal recess. The sphenoid ostium can be identified at an average distance of 15 mm superior to choanae, in the midway between the nasal septum and superior turbinate, and with 15-degree angle of the endoscope to the hard palate [22]. Posterior and superior parts of the nasal septum are resected to expose the contralateral sphenoid ostium and the contralateral anterior wall of the ethmoid sinuses, respectively. Anterior sphenoidectomy is performed, and the sphenoid sinus roof is exposed. Using the anatomic landmarks in the sphenoid sinus roof, the posterior ethmoidal cells located anterosuperior to sphenoid sinus (Figure 8) are resected on both sides until the lamina papyracea laterally, the medial orbital wall medially, and the middle turbinate's basal lamella anteriorly (Figure 9A). The next step is bilateral resection of the ethmoid bulla and anterior ethmoidal cells until the frontal sinus anteriorly (Figure 9B).

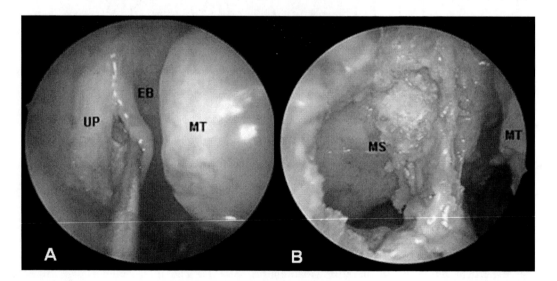

Figure 5. Endoscopic view showing right infundibulotomy. A) Uncinate process identified lateral to middle turbinate and mucosal incision made and mucosa was dissected to expose bone. B) After exposing the maxillary sinus entered by diamond burr drill. Antrostomy completed and maxillary sinus exposed. UP: uncinate process, EB: ethmoid bulla, MT: middle turbinate, MS: maxillary sinus.

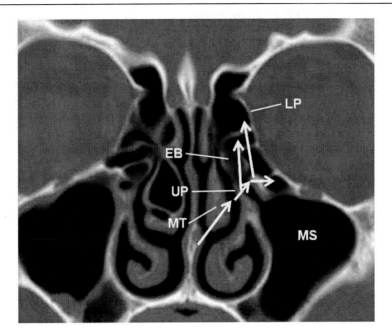

Figure 6. Paranasal CT scan demonstrating the pathway from left nasal cavity to anterior ethmoidal cells in anterior-posterior approach. EB: ethmoid bulla, LP: lamina papyracea, MS: maxillary sinus, MT: middle turbinate, UP: uncinate process.

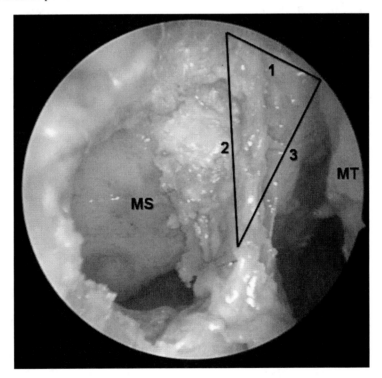

Figure 7. Endoscopic view of the right nasal cavity demonstrating the triangle of safe entry zone to posterior ethmoidal cells. "1" is the line perpendicular to nasal septum from the posterior medial orbital roof, "2" is the line along the vertical antrostomy ridge, and "3" is the line along the free edge of middle turbinate basal lamella. MS: maxillary sinus, MT: middle turbinate.

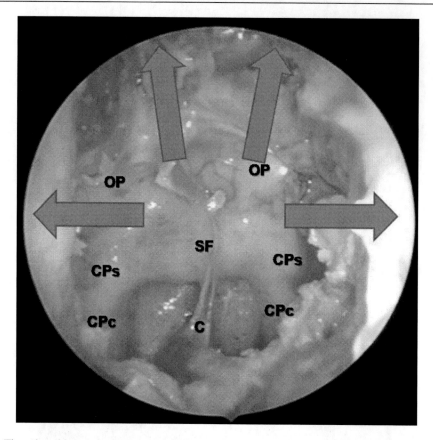

Figure 8. The sphenoid sinus roof is exposed after anterior sphenoidectomy. Arrows indicate the direction of dissection toward the posterior ethmoidal cells. C: clivus, CPc: clival part of carotid protuberance, CPs: sellar part of carotid protuberance, OP: optic protuberance, SF: sellar floor.

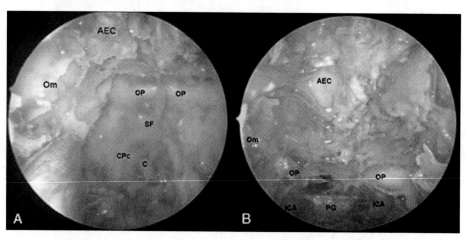

Figure 9. A) The right posterior ethmoidal cells are removed, and the medial orbital wall is exposed. B) The right anterior ethmoidal cells are opened, and its mucosa is dissected. AEC: anterior ethmoidal cells, C: clivus; CPc: clival part of carotid protuberance, Om: medial orbital wall, OP: optic protuberance, PG: pituitary gland, SF: sellar floor. *(Source: 7. Abuzayed B et al. Endoscopic endonasal anatomy and approaches to the anterior skull base: a neurosurgeon's viewpoint. J Craniofac Surg 21(2):529–537, 2010. Editor: Mutaz B. Habal. Used with permission).*

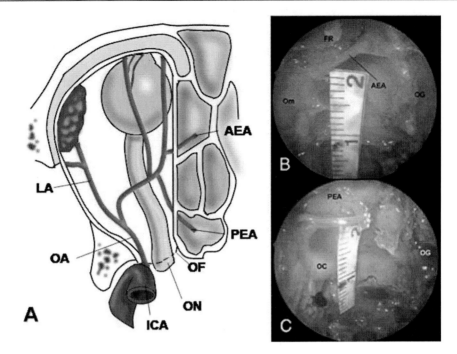

Figure 10. A) A drawing showing the anatomy of the AEA and PEA. B) Endoscopic view demonstrating the right AEA and its course from the medial orbital wall toward the olfactory groove. C) Endoscopic view demonstrating the right PEA and its course from the medial orbital wall toward the olfactory groove. FR: frontal recess, OC: optic canal, OG: olfactory groove, Om: medial orbital wall. *(Source: 7. Abuzayed B et al. Endoscopic endonasal anatomy and approaches to the anterior skull base: a neurosurgeon's viewpoint. J Craniofac Surg 21(2):529–537, 2010. Editor: Mutaz B. Habal. Used with permission).*

Identification of the Anterior and Posterior Ethmoidal Artery

The anterior ethmoidal artery (AEA) runs from lateral to medial direction, and obliquely along the skull base (Figure 10A). The bone at the level of the lamina papyracea forms a small funnel laterally. Intraoperative injury of the AEA may lead to retraction of this artery into the orbit at that site, resulting in a dangerous retrobulbar hematoma. Opening the agger nasi cell exposes the posterior wall of the frontal recess. This recess is bordered medially by the anterior part of the middle turbinate, laterally by the lamina papyracea, and posteriorly by the agger nasi cell. The AEA is identified by careful dissection of the anterior wall of the bulla in the direction of the skull base (Figure 10B). The skull base slopes gently upward (15 degree) in front of this artery and then steepens more anteriorly as it joins with the posterior wall of the frontal sinus. This point is located approximately 9 mm in front of the artery, and final ethmoid cells are frequently encountered in this area [19].

The superior portion of the lamina papyracea is gently removed to expose the PEA. In this point, the artery is exiting the orbit to enter its osseous canal and could be easily ligated (Figure 10C). Identification of posterior and anterior ethmoidal arteries is completed bilaterally, and the anterior skull base is fully exposed (Figure 11).

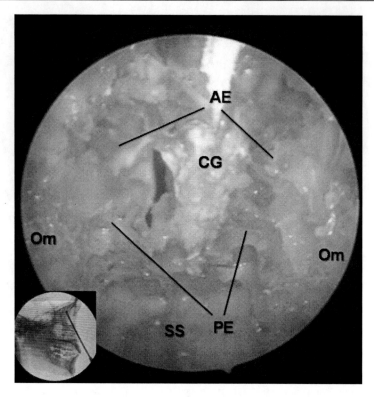

Figure 11. Endoscopic view of anterior skull base, demonstrating the relations of the bony and vascular structures. The position and orientation of the endoscope are checked by fluoroscopy (inset). CG: crista galli, Om: medial wall of orbit, SS: sphenoid sinus, AE: anterior ethmoid cells, PE: posterior ethmoid cells.

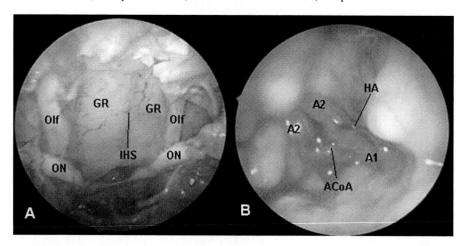

Figure 12. A) Endoscopic view of anterior fossa after removing the bony anterior skull base and opening the dura. B) Endoscopic view of opened interhemispheric sulcus and related vasculature. A1: the first segment of the anterior cerebral artery, A2: the second segment of the anterior cerebral artery, HA: Heubner artery, GR: gyrus rectus, HIS: interhemispheric sulcus, ON: optic nerve, Olf: olfactory nerve. (Source: 7. Abuzayed B et al. Endoscopic endonasal anatomy and approaches to the anterior skull base: a neurosurgeon's viewpoint. J Craniofac Surg 21(2):529–537, 2010. Editor: Mutaz B. Habal. Used with permission).

Intradural Exposure of the Anterior Skull Base

The anterior skull base bony structure between the 2 medial orbital walls and optic nerves bilaterally, the sella anteriorly, and frontal recesses anteriorly, is resected using a 2-mm Kerrison rongeur and microdrill, exposing the anterior fossa's dura. To mobilize the dura in this region, a portion of the crista galli is removed such that its attachment to the falx cerebri can be divided. It is through this maneuver that the dura is no longer "tented" away from the surgeon, thus permitting safe and adequate dural incisions [23]. This can be done with the use of an angled irrigating, diamond burr under endoscopic visualization.

With an endoscopic knife, dural incision is made, and dura is retracted laterally to expose the intradural structures of anterior fossa (olfactory nerves, interhemispheric sulcus, and gyri recti) (Figure 12A). With gentle dissection, the interhemispheric sulcus is opened, and the gyri recti are lateralized. The first (A1) and second (A2) segments of anterior cerebral artery, anterior communicating artery (ACoA), and Heubner arteries are exposed (Figure 12B).

Skull Base Reconstruction

Many techniques are defined for the reconstruction of the skull base after endoscopic approaches. These include nasoseptal flap [24], pedicled middle and inferior turbinate flaps [7], fat and/or fascia grafts [7] (Figure 13). The reconstruction can be architected in one layer [7], bi-layer [7, 25] or tri-layer fashions [26]. The choice of the technique depends upon the surgeon's experience, the size of the bone and/or dural defect, the surgical approach and the viability of the tissues. The reconstruct can be enforced with biologic glues and inflation of intranasal balloon to hold the construct and prevent it from collapsing. Also, Postoperative lumbar CSF drainage (48–72 h) or repeated daily spinal taps can be applied when contra-indications are absent (such as pneumocephalus).

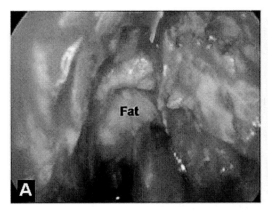 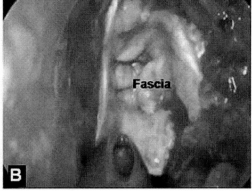

Figure 13. Intraoperative endoscopic view demonstrating defect reconstruction, first by in-layer fat graft (A), covered with out-layer fascia latta graft (B). *(Source: 7. Abuzayed B et al. Endoscopic endonasal anatomy and approaches to the anterior skull base: a neurosurgeon's viewpoint. J Craniofac Surg 21(2):529–537, 2010. Editor: Mutaz B. Habal. Used with permission).*

Complication

Despite the advantages of the endoscopic technique, there are still some important drawbacks and difficulties encountered during this approach:

1. The operative field is relatively small with the need to use and manipulate many instruments in a narrow surgical tunnel. This difficulty can be overcome by using binostril route and more practice to use the endoscope, which is a less familiar tool to neurosurgeons than ENT surgeons.
2. Disturbed surgical field with blood due to bleeding arteries and venous sinuses. This difficulty can be overcome by knowing the vascular anatomy related to the surgical corridor and pathology, with preoperative planning to determine vessels that must be coagulated before performing any step in the surgical procedure. Moreover, intraoperatively ruptured AEA and/or PEA can retract into the orbit causing retrobulbar hematoma. This can be avoided by early identification and ligation (or coagulation) of these arteries.
3. The limited working space embarrasses bone grafting and dural suturing, with consequent high risk of postoperative CSF leakage. Although the postoperative CSF leakage is the most common complication after endoscopic endonasal approaches to the skull base ($\approx$9%), one must not forget that endoscopic endonasal surgery is an effective treatment of CSF fistulas of the anterior skull base [7, 27]. Because of the difficulty of suturing the defects through the nose, many methods for reconstruction of the skull base defects have been described. [7, 24, 25, 26]

References

[1] Cavallo LM, Messina A, Cappabianca P, Esposito F, de Divitiis E, Gardner P, Tschabitscher M : Endoscopic endonasal surgery of the midline skull base: anatomical study and clinical considerations. *Neurosurg Focus* 19:E2, 2005.
[2] Al-Mefty O: Supraorbital-pterional approach to skull base lesions. *Neurosurgery* 1987;21:474-477.
[3] Blacklock JB, Weber RS, Lee YY, Goepfert H: Transcranial resection of tumors of the paranasal sinuses and nasal cavity. *J Neurosurg* 71:10-15, 1989.
[4] Cheesman AD, Lund VJ, Howard DJ: Craniofacial resection for tumors of the nasal cavity and paranasal sinuses. *Head Neck Surg* 8:429-435, 1986.
[5] McCutcheon IE, Blacklock JB, Weber RS, DeMonte F, Moser RP, Byers M, Goepfert H: Anterior transcranial (craniofacial) resection of tumors of the paranasal sinuses: surgical technique and results. *Neurosurgery* 38:471-479, 1996.
[6] Abuzayed B, Canbaz B, Sanus GZ, Aydin S, Cansiz H: Combined craniofacial resection of anterior skull base tumors: long-term results and experience of single institution. *Neurosurg Rev* 34:101–113, 2011.
[7] Abuzayed B, Tanriover N, Gazioglu N, Sanus GZ, Ozlen F, Biceroglu H, Kafadar AM, Akar Z: Endoscopic endonasal anatomy and approaches to the anterior skull base: a neurosurgeon's viewpoint. *J Craniofac Surg* 21(2):529–537, 2010.

[8] Locatelli D, Rampa F, Acchiardi I, Bignami M, De Bernardi F, Castelnuovo P: Endoscopic endonasal approaches for repair of cerebrospinal fluid leaks: nine-year experience. *Neurosurgery* 58:ONS246-256, 2006.

[9] Hao SP, Wang HS, Lui TN: Transnasal endoscopic management of basal encephalocele-craniotomy is no longer mandatory. *Am J Otolaryngol* 16:196-199, 1995.

[10] Komotar RJ, Starke RM, Raper DM, Anand VK, Schwartz TH: Endoscopic Endonasal versus Open Repair of Anterior Skull Base CSF Leak, Meningocele, and Encephalocele: A Systematic Review of Outcomes. *J Neurol Surg A Cent Eur Neurosurg* 2012 [Epub ahead of print]

[11] Carrau RL, Snyderman CH, Kassam AB, Jungreis CA: Endoscopic and endoscopic-assisted surgery for juvenile angiofibroma. *Laryngoscope* 111:483-487, 2001.

[12] Thaler ER, Kotapka M, Lanza DC, Kennedy DW: Endoscopically assisted anterior cranial skull base resection of sinonasal tumors. *Am J Rhinol* 13:303-310, 1999.

[13] Mccoul E, Anand VK, Schwartz T: Improvements in site-specific quality of life 6 months after endoscopic anterior skull base surgery: a prospective study. *J Neurosurg* 117:498–506, 2012.

[14] Ransom ER, Doghramji L, Palmer JN, Chiu AG: Global and disease-specific health-related quality of life after complete endoscopic resection of anterior skull base neoplasms.*Am J Rhinol Allergy* 26(1):76-79, 2012.

[15] Padhye V, Naidoo Y, Alexander H, Floreani S, Robinson S, Santoreneos S, Wickremesekera A, Brophy B, Harding M, Vrodos N, Wormald PJ: Endoscopic endonasal resection of anterior skull base meningiomas. *Otolaryngol Head Neck Surg*147(3):575-582, 2012.

[16] Basak S, Karaman CZ, Akdilli A, Mutlu C, Odavasi O, Erpek G: Evaluation of some important anatomical variations and dangerous areas of the paranasal sinuses by CT for safer endonasal surgery. *Rhinology* 8:162-167, 1998.

[17] Terrier F, Terrier G, Rufenacht D, Friedrich JP, Weber W: Die Anatomie der Siebbeinregion: topographische, radiologische und endoskopische Leistrukturen. *Therapeutische Umschau* 44:75-85, 1987.

[18] Lee WC, Ming KU PK, Van Hasselt CA: New guidelines for endoscopic localization of the anterior ethmoidal artery: a cadaveric study. *Laryngoscope* 110:417-421, 2000.

[19] .Hosemann W, Brob R, Gode U, Kuhnel TH: Clinical anatomy of the nasal processof the frontal bone "spina nasalis interna". Otolarungol. *Head Neck Surg.* 125:60-65, 2001.

[20] Abuzayed B, Tanriover N, Gazioglu N, Eraslan BS, Akar Z: Endoscopic endonasal approach to the orbital apex and medial orbital wall: anatomic study and clinical applications. *J Craniofac Surg* 20:1594-1600, 2009.

[21] Schaefer SD, Li JCL, Chan EK, Wu ZB, Branovan DI: Combined anterior-to-posterior and posterior-to-anterior approach to paranasal sinus surgery: an update. *Laryngoscope* 116:509-513, 2006.

[22] Abuzayed B, Tanriover N, Ozlen F, Gazioğlu N, Ulu MO, Kafadar AM, Eraslan B, Akar Z: Endoscopic endonasal transsphenoidal approach to the sellar region: results of endoscopic dissection on 30 cadavers. *Turk Neurosurg* 19: 237-244, 2009.

[23] Lee JM, Ransom E, Lee JYK, Palmer JN, Chiu AG: Endoscopic Anterior Skull Base Surgery: Intraoperative Considerations of the Crista Galli. *Skull Base* 21(2): 83-86, 2011.

[24] Hadad G, Bassagasteguy L, Carrau RL, Mataza JC, Kassam A, Snyderman CH, Mintz A: A novel reconstructive technique after endoscopic expanded endonasal approaches: vascular pedicle nasoseptal flap. *Laryngoscope* 116:1882–1886, 2006.

[25] Anderson Eloy J, Choudhry OJ, Christiano LD, Ajibade DV, Liu JK: Double flap technique for reconstruction of anterior skull base defects after craniofacial tumor resection: technical note. *Int Forum Allergy Rhinol.* 2012, doi: 10.1002/alr.21092. [Epub ahead of print].

[26] Eloy JA, Patel SK, Shukla PA, Smith ML, Choudhry OJ, Liu JK: Triple-layer reconstruction technique for large cribriform defects after endoscopic endonasal resection of anterior skull base tumors. *Int Forum Allergy Rhinol.* 2012, doi: 10.1002/alr.21089. [Epub ahead of print].

[27] Greenfield JP, Anand VK, Kacker A, Seibert MJ, Singh A, Brown SM, Schwartz TH: Endoscopic endonasal transethmoidal transcribriform transfovea ethmoidalis approach to the anterior cranial fossa and skull base. *Neurosurgery* 66(5): 883-892, 2010.

Chapter III

Expanded Endoscopic Endonasal Approaches to the Orbit and Lateral Anterior Skull Base

Pornthep Kasemsiri[1], Ricardo L. Carrau[1], Daniel M. Prevedello[2]*, Jun Muto[2], Danielle de Lara[1,2], Leo F. S. Ditzel Filho[2], Bradley A. Otto[1] and Amin B. Kassam[3]

[1]Department of Otolaryngology – Head and Neck Surgery, The Ohio State University, Columbus, OH, US
[2] Department of Neurological Surgery, The Ohio State University, Columbus, OH, US
[3]Department of Neurological Surgery, University of Ottawa, Ottawa, ON, Canada

Abstract

Endoscopic endonasal approaches (EEAs) were developed to access the skull base with minimal brain disruption. EEAs can be oriented in the sagittal (median) or coronal (paramedian) planes. The main feature of endoscopic endonasal approaches is that they provide the most direct access to median and paramedian skull base lesions with the least manipulation of neurovascular structures.

Surgical modules in the coronal or paramedian plane (extend laterally from the midline), must be considered in three different depths. From anterior to posterior, these include: the orbits and their roofs (anterior), the floor of the middle cranial fossa (middle), and the petrous bone and jugular foramen (posterior).

Endoscopic endonasal approaches in the anterior coronal plane, provide access to orbital and periorbital pathologies via a transorbital approach with or without a supraorbital extension (e.g. extensive inverted papilloma, fibro-osseous lesion,

---

* Corresponding Author: Daniel M. Prevedello, M.D., Department of Neurological Surgery, The Ohio State University; N-1011 Doan Hall, 410 W. 10th Avenue, Columbus, H, 3210. Phone: 614-293-7190; E-mail: dprevedello@gmail.com.

meningioma, hemangioma, and schwannoma). Transorbital approaches may be further categorized in extraconal and intraconal. However, these approaches are restricted by critical anatomical structures including the optic nerve and ophthalmic artery; therefore, disease extension laterally beyond the mid-orbital or optic nerve axes constitutes a contraindication for transorbital approaches.

Orbital and optic canal decompressions are the most common indications of an extraconal transorbital approach. Conversely, an intraconal approach provides access to lesions that are inferior and/or medial to the optic nerve. A natural gap between the inferior and medial rectus muscles provides a corridor to this space, while preserving extraocular muscle function.

A supraorbital extension requires the removal of the bony medial wall of orbit and lateral-inferior displacement of the orbital soft tissues to visualize the orbital roof. Its removal provides access to the lateral aspect of the anterior cranial fossa. This modification extends the lateral reach of a transcribiform or transplanum approach optimizing the resection of lesions that involve the medial half of the orbital roof.

Optimal surgical outcomes and avoidance of complications require meticulous surgical technique and the identification and preservation of critical anatomical structures. Ultimately, the expanded endonasal approaches provide a safe and effective corridor for the surgical treatment of pathology that extends laterally over the anterior skull base but remains medial to the meridian of the orbit and optic nerve.

Introduction

The skull base is a complex anatomical area with abundant critical neurovascular structures. Endoscopic endonasal surgical approaches can be oriented in the sagittal (median) and coronal (paramedian) planes. Surgical modules in the coronal plane must be considered in three different depths. From anterior to posterior, these include: the medial orbits (anterior), the floor of the middle cranial fossa (middle), and the petrous bone and jugular foramen (posterior) [1].

Well-established external approaches to the anterior coronal plane include the transconjunctival approach, subtarsal approach, subciliary approach, LeFort I orbitotomy [2], and orbitozygomatic craniotomy. In addition, the endoscopic- assisted techniques were introduced to provide better visualization and preservation of cosmesis. Since the 1990s, endoscopic endonasal approaches (EEAs) became part of the skull base surgeon's armamentarium [3]. Expanding experience, better understanding of the endoscopic anatomy and advances in technology are helping to refine the techniques and their indications. In general, EEAs provide the most direct access with the least manipulation of neurovascular structures, avoiding external incisions and obviating the need for the translocation of the maxillofacial skeleton. However, EEAs are restricted by the relationship of the pathology (target lesion) with critical anatomical structures. In the anterior skull base, the optic nerve and ophthalmic artery form the lateral limit of the corridor; thus, any pathology extending beyond the mid-orbital roof or optic nerve are better accessed by lateral orbitotomies and/or craniotomies [4].

A thorough knowledge of orbital anatomy and its adjacent structures is necessary to perform endoscopic endonasal surgery in the parasagittal anterior skull base. This is critical to avoid complications and achieve optimal surgical outcomes. This chapter briefly discusses the salient orbital and periorbital anatomy, and principles of endoscopic endonasal techniques.

Anatomical Considerations

Components of the anterior skull base include the orbital plate of the frontal bone, the cribiform plate and crista galli of the ethmoid bone, and the lesser wing of the sphenoid bone. Anatomically, the parasagittal (paramedian) anterior skull base corresponds to the orbit and its contents.

The orbit is a four-sided pyramidal shaped compartment that becomes three-sided near the apex and has a total volume of approximately 30 ml [5]. The orbital plate of the frontal bone and the lesser wing of the sphenoid bone form the orbital roof from anterior to posterior. The lateral wall consists of the greater wing of the sphenoid bone and the frontal process of the zygomatic bone. The posterolateral wall is separated from the roof by the superior orbital fissure, which transmits the cranial nerves (CN) III, IV, VI, and branches of V_1 between the orbit and cavernous sinus. The orbital plate of the maxilla, the orbital surface of zygomatic bone, and the orbital process of the palatine bone form the orbital floor. The posterior part of the floor is separated from the lateral wall by the inferior orbital fissure, which allows the passage of tributaries from the maxillary nerve (V_2), and associated branches, the zygomatic branches and infraorbital nerve. The frontal process of the maxilla, the lacrimal bone, the orbital plate of the ethmoid bone, and body of the sphenoid bone form the medial wall of the orbit. This wall is very thin, especially in the area of the orbital plate of the ethmoid bone, hence, its name *lamina papyracea*. At the orbital apex, the medial wall bifurcates around the optic strut and its superior portion forms the medial wall of the optic canal, whereas the inferior forms the medial wall of the superior orbital fissure. The optic canal allows the optic nerve and opthalmic artery to reach the orbit from the subarachnoid space. The superior orbital fissure allows cranial nerves III, IV and VI to enter the orbit from the cavernous sinus and the ophthalmic branch of the trigeminal nerve to leave the orbit to enter the cavernous sinus.

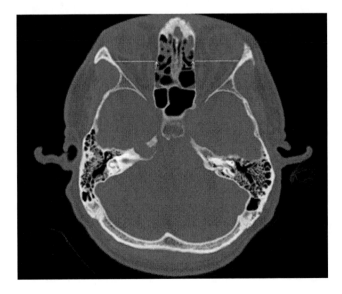

Figure 1. Computed tomographic scan showing the level of the third lamella (White arrow) corresponding to an imaginary line just posterior to the globe; therefore, it is a useful landmark for the localization of the retrobulbar space.

Soft-tissue contents of the orbit include the globe, extraocular muscles, cranial nerves II, III, IV, VI and ophthalmic vessels and nerves (V_1) and, along with its branches, the lacrimal apparatus and the periorbital fat. These soft tissues are contained within the orbital periosteal lining, which is continuous with the periosteal layer of the dura mater surrounding the brain at the optic canal and superior orbital fissure.

Each orbit contains six extraocular muscles including the superior rectus muscle, inferior rectus muscle, lateral rectus muscle, medial rectus muscle, superior oblique muscles, and inferior oblique muscles. These muscles coordinate physiologic eye movement. The recti muscles originate at the annulus of Zinn, a fibrous tendon that separates orbital apex structures into an intra-conal and extra-conal compartment.

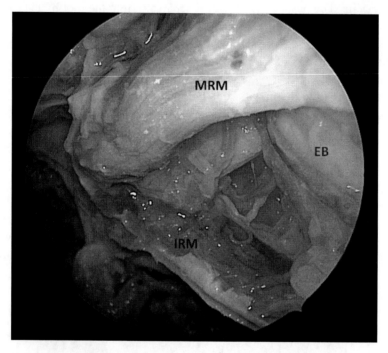

Figure 2. Corridor between medial rectus muscle and inferior rectus muscle provides safe access to the intraconal space (blue area). (MRM = medial rectus muscle; IRM = inferior rectus muscle; EB = eye ball).

The sinonasal corridor utilized by EEAs comprises important structures that serve not only as sinonasal landmarks but also as general intraorbital landmarks. The sinonasal cavity is roughly compartmentalized by bony lamellae. The first through fourth are relatively constant lamellae that include the uncinate process, ethmoid bulla, basal lamella of the middle turbinate, and the superior turbinate. Although detailed review of sinonasal anatomy is beyond the scope of this chapter, these landmarks are worth mention. The third lamella, the basal lamella of the middle turbinate, divides the ethmoid cavity into anterior and posterior compartments. It also serves as a useful landmark to identify the retrobulbar space, as its attachment is generally located at the posterior limit of the globe. Hence, the basal lamella of the middle turbinate is a useful landmark for the localization of the retrobulbar space (Figure 1). The gap between medial rectus muscle and inferior rectus muscle is wide and is optimally positioned for endonasal access to the intra-conal space (Figure 2). This area correlates to the ethmoid–maxillary plate (Figure 3); therefore, the intra-conal space can be accessed by an endonasal transorbital approach through the ethmoid–maxillary plate in order to avoid an injury to the medial rectus muscle [6]. The anterior and posterior ethmoid arteries also serve as important landmarks. Ligation of these arteries not only improves the ability to distract orbital tissue, but also reduces the risk of avulsion injury and hemorrhage during intraorbital dissection. The arteries pass through the respective foramina at the frontoethmoidal suture, the junction of orbital roof and medial wall. Approximate distances of the anterior lacrimal crest to the anterior ethmoidal foramen, the anterior ethmoidal foramen to the posterior ethmoidal foramen and the posterior ethmoidal foramen to the optic canal are '24-12-6 mm'[7]. From the endonasal viewpoint, the mean distance from the axilla of the middle turbinate to the anterior ethmoidal artery is 17.5 mm and from the anterior ethmoidal artery to the posterior ethmoid artery is 14.9 mm [8].

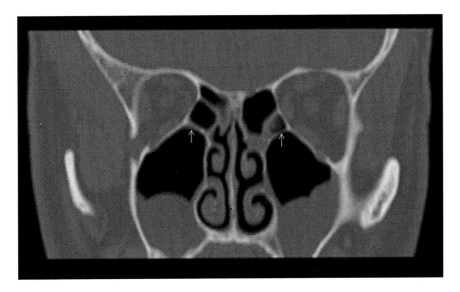

Figure 3. Computed tomographic scan, coronal view, bony algorithm showing the relationship of the gap between the medial rectus muscle and inferior rectus muscle with the ethmoid–maxillary plate (white arrow). Intra-conal transorbital approach can access via ethmoid–maxillary plate for avoiding medial rectus muscle injury.

Preoperative Planning

All patients undergo a preoperative workup that includes a neurologic assessment, with emphasis on the ophthalmologic status, and an endoscopic assessment of the sinonasal cavity to evaluate for lesions and anatomical variations. Endoscopic evaluation allows the surgeon to identify and treat preoperative inflammation or infection and to develop an understanding of any anatomical variance, such as septal deviation or concha bullosa that could affect the procedure. Computed tomographic scan allows for an approximation of the extension of the lesion, although are most suited in determining the salient bony anatomy and the presence of bone involvement. Magnetic resonance imaging (MRI) provides excellent soft tissue discrimination and is often useful in discriminating intraorbital lesions from important adjacent structures. MRI is especially critical when lesions extend intracranially. CT and MRI scans can be fused for intraoperative guidance, providing simultaneous bony and soft tissue discrimination relative to the lesion of interest and relevant nearby anatomy.

We administer broad-spectrum prophylactic perioperative antibiotics, generally a fourth generation cephalosporin. In penicillin allergic patients, however, we administer a combination of vancomycin and aztreonam. We recommend that surgeons consult their institutional epidemiologist and antibiograms to better choose preoperative prophylactic antibiotics.

Somatosensory evoked potentials are routinely monitored. However, cranial nerves III, IV and VI are monitored if superior orbital fissure or cavernous sinus dissection is anticipated. Transorbital approaches with intraconal dissection cause artifacts due to muscle manipulation and EMG monitoring is not routinely used in these cases.

Operative Setup

Surgery proceeds under general anesthesia with the patient in supine position and with the head resting on a horseshoe or fixed by Mayfield headpins. Three main scenarios require stabilization of the head with a three-pin fixation system to avoid complications: an expected long surgery; need for extensive drilling; and the need to avoid muscle paralysis to allow EMG monitoring. Neck position is neutral (aligned with the horizon) but tilted to the left and turned to the right.

Cottonoids soaked with oxymethazoline 0.05% or adrenaline 1/10,000-1/20,000 are inserted in both nasal cavities. A solution of lidocaine 1% and epinephrine 1/100,000 is infiltrated into both middle meati/turbinates and nasal septum.

Endoscopic endonasal approaches to the orbit in the anterior parasagittal plane may be didactically divided in extra-conal and intra-conal approaches. The endonasal supraorbital approach involves a superolateral extension of an extra-conal approach and will be explained in detail independently.

Transorbital Approach

A transorbital approach is defined by the removal of the lamina papyracea and/or medial optic nerve canals. As described above, this approach can be further divided in extraconal and intraconal approaches [1].

1. Extra-conal Transorbital Approach

The extraconal transorbital approach is used mainly for orbital and optic canal decompression. However, it can be used as a corridor to reach the lateral anterior skull base above the orbit, which is then denominated as a supraorbital extension.

1.1. Orbital Decompression

Orbital decompression is most commonly performed for Graves' disease. Approximately 65% of patients with Graves' opthalmopathy, however, can have spontaneous resolution[9]. Therefore, an orbital decompression is only indicated to prevent or treat for patients with optic neuropathy, proptosis with exposure keratopathy,or unacceptable cosmesis that does not respond to medications. Additionally, surgery may play a role in situation where contraindications preclude medical therapy [10-11].

Orbital decompression should also be considered to release pressure while awaiting the effect of medical therapy in inflammatory orbital diseases with optic neuropathy (i.e. granulomatous disease, and orbital pseudotumor) [12]. Furthermore, traumatic and/or iatrogenic conditions may also indicate an orbital decompression, (e.g. anterior ethmoid artery injury causing a retro-orbital hemorrhage).

Operative Technique

Following the operative set-up, the uncinate process is removed with a back-biting rongeur and/or a microdebrider, exposing the natural ostium of the maxillary sinus. The ostium should be widened posteriorly and inferiorly to prevent blockage by periorbital fat, which often protrudes significantly after decompression. Furthermore, a large antrostomy facilitates removal of the floor of the orbit when necessary. A complete ethmoidectomy and sphenoidotomy exposes the lamina papyracea from the orbital apex to the nasolacrimal duct. Bone removal during endoscopic orbital decompression often includes the entire lamina papyracea and the medial portion of the orbital floor (infraorbital nerve is the lateral limit of the decompression). Its superior limit is the frontoethmoidal suture line (i.e. anterior and posterior ethmoidal neurovascular bundles). Anterior dissection is limited to the junction between lamina papyracea and lacrimal bone. Removal of the bone beyond the frontal recess may lead to protrusion of periorbital fat potentially obstructing the frontal sinus; therefore, we avoid it. The lamina papyracea is elevated posteriorly up to the orbital apex, while preserving the periorbita.

Angled-lens endoscopes (30^0, 45^0) and curved instruments facilitate the dissection. After removal of entire lamina papyracea and medial orbital floor, the periorbita is fully exposed and a sickle knife is used to incise it, while avoiding injury to the medial rectus muscle. The periorbital incision should be initiated at its most superior aspect and from posterior to anterior to prevent orbital fat protrusion, which would obscure the surgical the field posterior to the herniation. Techniques for the periorbital incision are numerous; however, we prefer the orbital sling technique, which has been developed to decrease the incidence of lateral displacement of the globe and/or extraocular muscles with subsequent postoperative diplopia [13]. Two incisions are performed at the superior and inferior margins of the medial rectus muscle, parallel to the ethmoid roof and orbital floor, respectively (Figure 4). A ball-tipped probe and sickle knife may be used to identify and incise remaining fibrous bands. The sling of periorbita serves to support the medial rectus and prevent its medial prolapse.

Postoperative Considerations

Nasal packing should be avoided to ensure maximal decompression of exposed orbital contents. Oral antibiotics and nasal irrigations are initiated early after surgery, and nasal crusting is cleaned one week postoperatively.

Complications

Diplopia is the most common complication of orbital decompression with 18% to 62% of patients reporting postoperative new-onset diplopia or worsening of pre-existing symptoms [14, 15]. However, several techniques have been developed to decrease postoperative diplopia including the preservation of a inferomedial bone strut [15], and orbital sling technique [13]. Other complications include epistaxis, sinusitis, mucocele due to orbital fat blocking the sinonasal drainage, orbital emphysema associated with coughing, sneezing or nose blowing, and epiphora related to nasolacrimal duct injury.

1.2. Optic Nerve Decompression

Optic nerve decompression is indicated for optic neuropathy secondary to compression of the optic nerve caused by trauma or various other pathologies such as neoplasm, inflammation, infection, and fibrous osseous lesion [16]. Optic nerve decompression as an initial management for visual loss associated with trauma is still controversial particularly with the option of medical management with high dose corticosteroids. However, the presence of bony fragments impinging on the optic nerve favors optic nerve decompression [17-19]. Furthermore, other indications for optic nerve decompression are prophylactic preservation of optic nerve function and the restoration of visual function if affected by any of the conditions listed above.

Operative Technique

Patients are prepared for surgery as previously described. Complete anterior and posterior ethmoidectomies are performed. The sphenoid sinus rostrum is opened widely to allow identification of the optic canal at the lateral wall of the sphenoid sinus. However, the optic canal may be identified initially in the posterior ethmoid cell (Onodi cell). When present, this cell should be completely opened to access the optic canal. The lamina papyracea over the orbital apex anterior to the optic canal is removed carefully. The medial wall of the optic canal is then drilled from the level of the annulus of Zinn, anteriorly, to the tuberculum sellae medial to the optico-carotid recess, posteriorly.

The region of the optic foramen includes the thick bone of the lesser wing of the sphenoid; therefore, this bone must be drilled with a diamond or hybrid burr while irrigating copiously to obviate a thermal injury. After the bone is appropriately thinned, a dissector is used to carefully fracture and remove the thinned bone; thus, exposing the optic nerve sheath (i.e. optic canal periosteum) (Figure 5). Incision of the optic sheath is rarely performed due to risk complication including nerve fibers or ophthalmic artery injury, CSF leak, and meningitis. However, some conditions, such as invasion by tuberculum sellae meningioma, intra-sheath hematoma, or significant papilledema, may require opening the optic nerve sheath [13]. The sheath is incised initially just anterior to the anulus of Zinn and the incision can be extended posteriorly using microscissors; or, in cases of a tuberculum sellae meningioma, the optic sheath can be opened with "back-cutting" 120^0 micro-scissors. The sheath should be entered in its superomedial quadrant to minimize risk to the ophthalmic artery.

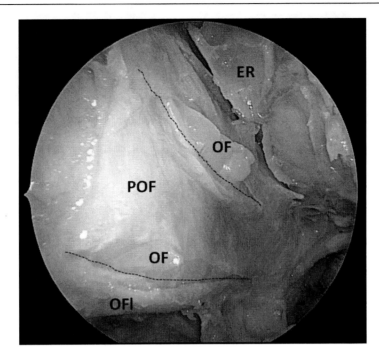

Figure 4. A sickle knife is used to incise the periorbita using an orbital sling technique for orbital decompression. The superior and inferior incisions (blue dashed lines) are parallel to the ethmoid roof and the floor of orbit, respectively. (ER = Ethmoid roof; OF = Orbital fat; POF = Periorbital fascia; OFl = Orbital floor).

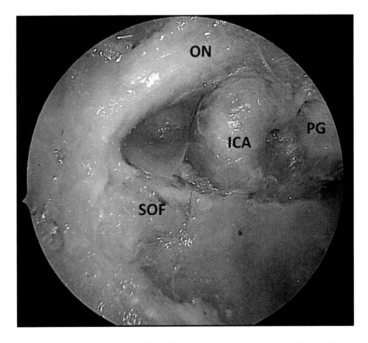

Figure 5. Cadaveric dissection demonstrating the right optic nerve after removal of its bony canal from the annulus of Zinn to the lateral optico-carotid recess.
(ON = Optic nerve; ICA = Internal carotid artery; PG = Pituitary gland; SOF = Superior orbital fissure).

Postoperative Considerations

Postoperative care is similar to that of endoscopic endonasal orbital compression.

Complications

Complications such as CSF leak, meningitis, and visual loss are concerns in endoscopic optic nerve decompression. These complications seem to occur significantly more frequently than in endoscopic orbital decompression [19-20].

1.3. Supraorbital Extension

The working corridor for the supraorbital extension requires the removal of the lamina papyracea, and lateral-inferior displacement of the orbital soft tissues. This approach can access pathologies at the orbital roof. Frequently, the lesion of interest extends from the sinonasal region (i.e. inverted papilloma, esthesioneuroblastoma, fibro-osseous lesion) or from intracranial origins (i.e. meningioma, meningoencephalocele, olfactory schwannoma); therefore, this approach may be combined with a transcribiform or transplanum approach to obtain margins that could extend laterally to include the medial half of the orbital roof. However, this approach is limited by critical structures including the globe and the optic nerve, which preclude further lateral access. This effect magnifies at the orbital apex, where the periorbita cannot be easily lateralized as it occurs more anteriorly in the orbit. At the orbital apex, the optic nerve and the densely packed superior orbital fissure and annulus of Zinn do not permit or tolerate lateral dislocation. Thus, lesions that extend on the top of the anterior clinoid cannot be directly accessed via endonasal approaches.

Operative Technique

Patients are prepared for endoscopic surgery as previously described. A middle turbinectomy and complete anterior and posterior ethmoidectomy allow visualization of anterior and posterior ethmoid arteries. These arteries are identified and coagulated by bipolar cautery. The lamina papyracea is removed with a Cottle elevator and forceps. During removal of the lamina papyracea, the periorbita should remain intact to prevent herniation of the orbital fat that may obstruct the visualization of the surgical field. The periorbita is elevated from the orbital roof and is displaced infero-laterally to expose the medial orbital roof, which can be removed with drills, straight or Kerrison roungeurs or even with angled frontal sinus instruments (Figure 6). The supraorbital dura is removed as lateral as possible in order to encompass the entire disease. In a malignancy, such as an esthesioneuroblastoma the resection continues until the margins are negative by histological analysis. In benign cases, as such as meningiomas, the dura is opened lateral to the tumor in to expose normal brain.

Postoperative Complications

Routine postoperative care includes antibiotics and nasal irrigation to prevent infection and nasal crusting. Endoscopic nasal examination and debridement start one week after surgery.

Complications

Complications following endoscopic endonasal supraorbital approaches are rare include CSF leak, retrobulbar hemorrhage, and visual disturbances. Double vision associated with a CN IV palsy can result from excessive lateral retraction at the orbital apex.

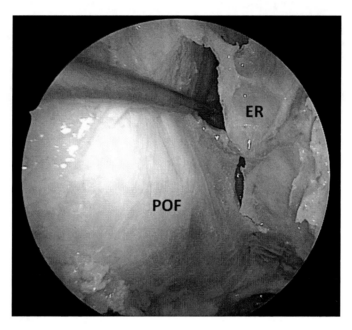

Figure 6. The periorbital fascia (POF) is elevated from the orbital roof using a Freer elevator. Infero-lateral displacement of the orbital orbital soft tissues (intact periorbital fascia) allows access to the medial orbital roof. (POF = Periorbital fascia; ER= Ethmoid roof)).

2. Intra-Conal Transorbital Approach

This approach is used for intra-conal lesions that are inferior and/or medial to the optic nerve. A corridor between the inferior and medial recti muscles (or occasionally the medial rectus and superior oblique muscles provides adequate access to this space and preserves extraocular muscle function.

The intra-conal transorbital approach provides access for a diagnostic biopsy or for tumor removal including retro-orbital lesions (i.e. lymphoma, vascular malformation, hemangioblastoma). Important anatomical structures related to this module include the optic nerves, the anterior and posterior ethmoidal arteries, and the ophthalmic artery with its branch (Figure 7). The optic nerve and ophthalmic artery with the central retinal artery constitute the lateral limits, as neither tolerates mobilization. Lesions that lateral to the optic nerve need to be accessed by a lateral corridor (lateral orbitotomy with or without craniotomy).

Operative Technique

The endoscopic endonasal approach preparation proceeds as previously described. To gain maximum exposure of the orbital floor, widening of the maxillary ostium extends inferiorly to the insertion of the inferior turbinate and posteriorly to the pterygoid process.

The medial orbital wall is exposed by a complete anterior and posterior ethmoidectomy; then, it is carefully resected. The entry through the lamina papyracea, should occur below the level of the ethmoidal foramina to avoid damage to the ethmoid arteries and reduce the risk of retrobulbar hemorrhage and vision loss. The periorbita is incised creating as U-shaped flap pedicled posteriorly whenever possible.

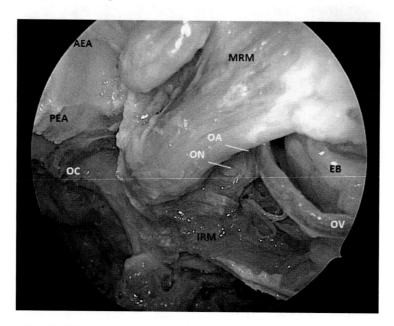

Figure 7. Intraconal transorbital approach is indicated for diagnostic biopsy or removal of retrobulbar lesions. Important anatomical structures related to this module include the optic nerve (ON), anterior ethmoidal artery (AEA), posterior ethmoidal artery (PEA), and ophthalmic artery (OA) with its branches. Its limitation is access to the lateral to optic nerve and ophthalmic artery.
(AEA = Anterior ethmoidal artery; PEA = Posterior ethmoidal artery; OC = Optic nerve canal; MRM = Medial rectus muscle; IRM = Inferior rectus muscle; OA = Ophthalmic artery; ON = Optic nerve; OV = Ophthalmic vein; EB = Eye globe).

The orbital soft tissues are freely mobile and loose, which complicates the access to the intraconal space. The medial and inferior recti muscles can be isolated with vessel loops as they insert on the globe through a transconjunctival approach. This allows to apply traction; thus, stretching the belly of the muscle and expanding the gap in between, in turn expanding the intra-conal corridor [21]. An alternative technique involves detaching the medial rectus muscle from the globe and tagging it with a long silk suture, which is then passed from the orbit into the nose. This step opens the medial orbit like a book, leaving the medial rectus muscle pedicled on the annulus of Zinn. A vessel loop is placed endoscopically around the contralateral posterior septum and the posterior medial rectus muscle, allowing medial retraction of the muscle to open the corridor to the intraconal space [21].

The tumor is identified and removed with bipolar electrocautery and sharp dissection. The dissection should occur between muscle groups rather than through individual muscles for preservation of function. Once the tumor is removed, meticulous hemostasis is obtained, and the rectus muscle vessel loops are released so that extraocular movements can be assessed. The rectus muscles are placed back into their native positions, and the medial rectus muscle is sutured back to the globe. The orbital fat is then pulled over the muscles to prevent scarring. When a nasoseptal flap is used, it is placed directly over the defect and bolstered in place with Surgicel (Ethicon, Inc., Somerville, NJ) and fibrin glue.

Postoperative Consideration

Typically, packing is avoided to eliminate the potential for raising the intraocular pressure in the case of a large orbital defect.

Conclusion

The endoscopic endonasal approach is an alternative safe and effective technique for the treatment of pathology that extends laterally at anterior skull base and orbit. However, its limitations include lateral extension of the disease beyond the midorbital or optic nerve axes. When applied appropriately, these approaches are efficient, effective and associated with minimal morbidity. As with any other approach, patients should be carefully selected, considering the lesion of interest, as well as the preservation of important surrounding neurovascular structures.

References

[1] Kassam AB, Prevedello DM, Carrau RL, Snyderman CH, Thomas A, Gardner P, et al. Endoscopic endonasal skull base surgery: analysis of complications in the authors' initial 800 patients. *J. Neurosurg.* 2011 Jun;114(6):1544–68.

[2] Dailey RA, Dierks E, Wilkins J, Wobig JL. LeFort I orbitotomy: a new approach to the inferonasal orbital apex. *Ophthal Plast Reconstr Surg.* 1998 Jan;14(1):27–31.

[3] Prevedello DM, Doglietto F, Jane JA Jr, Jagannathan J, Han J, Laws ER Jr. History of endoscopic skull base surgery: its evolution and current reality. *J. Neurosurg.* 2007 Jul;107(1):206–13.

[4] Snyderman CH, Pant H, Carrau RL, Prevedello D, Gardner P, Kassam AB. What are the limits of endoscopic sinus surgery?: the expanded endonasal approach to the skull base. *Keio J Med.* 2009 Sep;58(3):152–60.

[5] Chen C-T, Chen Y-R. Endoscopic orbital surgery. *Atlas Oral Maxillofac Surg Clin North Am.* 2003 Sep;11(2):179–208.

[6] Karaki M, Kobayashi R, Kobayashi E, Ishii G, Kagawa M, Tamiya T, et al. Computed tomographic evaluation of anatomic relationship between the paranasal structures and orbital contents for endoscopic endonasal transethmoidal approach to the orbit. *Neurosurgery.* 2008 Jul;63(1 Suppl 1):ONS15–19; discussion ONS19–20.

[7] Rootman J, Dr BS, Goldberg RA. *Orbital Surgery: A Conceptual Approach.* Lippincott Williams & Wilkins; 1995.

[8] Han JK, Becker SS, Bomeli SR, Gross CW. Endoscopic localization of the anterior and posterior ethmoid arteries. *Ann. Otol. Rhinol. Laryngol.* 2008 Dec;117(12):931–5.

[9] Perros P, Crombie AL, Kendall-Taylor P. Natural history of thyroid associated ophthalmopathy. *Clin. Endocrinol.* (Oxf). 1995 Jan;42(1):45–50.

[10] Bartalena L, Pinchera A, Marcocci C. Management of Graves' ophthalmopathy: reality and perspectives. *Endocr. Rev.* 2000 Apr;21(2):168–99.

[11] Lee HBH, Rodgers IR, Woog JJ. Evaluation and management of Graves' orbitopathy. *Otolaryngol. Clin. North Am.* 2006 Oct;39(5):923–942, vi.

[12] Robinson D, Wilcsek G, Sacks R. Orbit and orbital apex. Otolaryngol. *Clin. North Am.* 2011 Aug;44(4):903–922, viii.

[13] Metson R, Samaha M. Reduction of diplopia following endoscopic orbital decompression: the orbital sling technique. *Laryngoscope.* 2002 Oct;112(10):1753–7.

[14] Eloy P, Trussart C, Jouzdani E, Collet S, Rombaux P, Bertrand B. Transnasal endoscopic orbital decompression and Graves' ophtalmopathy. *Acta Otorhinolaryngol Belg.* 2000;54(2):165–74.

[15] Wright ED, Davidson J, Codere F, Desrosiers M. Endoscopic orbital decompression with preservation of an inferomedial bony strut: minimization of postoperative diplopia. *J Otolaryngol.* 1999 Oct;28(5):252–6.

[16] Pletcher SD, Metson R. Endoscopic optic nerve decompression for nontraumatic optic neuropathy. Arch. Otolaryngol. *Head Neck Surg.* 2007 Aug;133(8):780–3.

[17] Metson R, Pletcher SD. Endoscopic orbital and optic nerve decompression. *Otolaryngol. Clin. North Am.* 2006 Jun;39(3):551–561, ix.

[18] Pletcher SD, Sindwani R, Metson R. Endoscopic orbital and optic nerve decompression. *Otolaryngol. Clin. North Am.* 2006 Oct;39(5):943–958, vi.

[19] Levin LA, Beck RW, Joseph MP, Seiff S, Kraker R. The treatment of traumatic optic neuropathy: the International Optic Nerve Trauma Study. *Ophthalmology.* 1999 Jul;106(7):1268–77.

[20] Thakar A, Mahapatra AK, Tandon DA. Delayed optic nerve decompression for indirect optic nerve injury. *Laryngoscope.* 2003 Jan;113(1):112–9.

[21] McKinney KA, Snyderman CH, Carrau RL, Germanwala AV, Prevedello DM, Stefko ST, et al. Seeing the light: endoscopic endonasal intraconal orbital tumor surgery. *Otolaryngol Head Neck Surg.* 2010 Nov;143(5):699–701.

Chapter IV

Endoscopic Repair of Anterior Cranial Base Defects

Sudheer Ambekar, Chiazo Amene, Justin Haydel, Osama Ahmed, Anil Nanda and Bharat Guthikonda[*]

Department of Neurosurgery, LSU HSC Shreveport, Shreveport, LA, US

Abstract

Cerebrospinal fluid (CSF) rhinorrhea secondary to anterior cranial base defect occurs in various conditions, including spontaneous, post-traumatic, and post-operative. Ethmoidal roof, sphenoid sinus, cribriform plate and frontal sinus are the major sites of CSF leak in the anterior cranial fossa. CSF may also drain through the Eustachian tube and present as otorrhea. Left alone, CSF leak can predispose the patient to life threatening complications such as meningitis and may also lead to chronic low-pressure headache. The various radiological methods used to visualize the site of leak include thin cut spiral CT scan, MRI, CT cisternography, MR cisternography, and intrathecal fluoroscein insufflation during endoscopic surgery.

Traditionally, intradural repair of anterior cranial fossa has been the standard of treatment for many years. Since the first use of endoscope by Wigand to treat CSF rhinorrhea in 1981, endoscopic repair of cranial base defects has evolved as a safe, effective and minimally invasive method of visualizing the exact site of leak and treating this potentially life threatening condition. When compared to the open approach, endoscopic approach has the advantages of being minimally invasive, with shorter hospital stay, excellent visualization of the defect and possibility of a second attempt in case of recurrent leak. Anosmia and frontal lobe retraction, which are potential complications in the open approach, are largely avoided with endoscopic approach.

The broad categories of endoscopic repair techniques are overlay repair, inlay repair and repair using pedicled nasoseptal flap. There are certain limitations with endoscopic repair such as a steeper learning curve and difficulty in repairing frontal sinus leaks. We review the literature on current techniques of endoscopic repair of anterior cranial base

---

[*] Corresponding author: Bharat Guthikonda; bguthi@lsuhsc.edu.

defects and their outcome and present our experience with managing such cases. We also propose a management algorithm for a patient with suspected CSF rhinorrhea.

Introduction

Cerebrospinal fluid (CSF) rhinorrhea is the leakage of CSF to the exterior through a cranial base defect. It is a serious and potentially life threatening condition. If not treated appropriately and on time, it can lead to meningitis and possibly death of the patient. CSF leak occurs commonly secondary to trauma, surgery, neoplastic or inflammatory erosion of cranial base and congenital malformations. However, in a vast majority of patients, the cause is not known. Over the years, the treatment of CSF rhinorrhea has evolved from a conventional craniotomy based approach to a minimally invasive endoscopic approach. The present chapter describes the evaluation and management of patients with CSF rhinorrhea secondary to defects in the anterior cranial fossa using endoscopic techniques.

History

Galen first described CSF rhinorrhea in second century AD. He postulated that the water from brain was released into the nostrils via defects in the pituitary and ethmoidal regions [1]. In the middle ages, CSF rhinorrhea was considered to be associated with head trauma [2]. The condition was revisited in 1826 by Charles Miller who published the first case of CSF rhinorrhea in a hydrocephalic child with an intermittent discharge of nasal fluid. Autopsy revealed communications between the nasal and cranial cavities.[2]. In 1899, St. Clair Thompson reported the first series of patients with CSF leaks and coined the term rhinorrhea.

The first description of repair of a fistula was by Walter E Dandy in 1926. He described three cases; one patient with a frontal depressed fracture where the dural defect was repaired with fascia lata graft, another with a communication of mastoid air cells with intracranial space and the third with post traumatic pneumocephalus which resolved with conservative management [3]. This was followed by description of extracranial repair of anterior cranial fossa defect via a naso-orbital incision by Gusta Dohlman in 1948. This was then followed by the introduction of transnasal approaches by Oskar Hirsch in 1952 and later by Vrabec and Hallberg in 1964. The intra and extracranial approaches remained in vogue and were the standard of treatment until the early 1990's.

Endoscopic repair of cranial fossa defects was first described by Wigand in 1981 [4] and later by Mattox and Kennedy in 1990 [5]. Since then, there have been great strides in the technology and technique associated with endoscopic repair of CSF leaks to the extent that endoscopy has come to be the standard for treatment of CSF rhinorrhea.

Epidemiology

CSF leak most commonly occurs within three months following the head trauma. Other causes include postoperative defects, bone erosion secondary to tumor, inflammation and hydrocephalus and congenital defects of the skull base. In a few individuals, no specific cause can be found and these patients are categorized as having idiopathic spontaneous CSF leak. In a systematic review of endoscopic repair of CSF leaks, spontaneous fistulae were the most commonly reported type, followed by iatrogenic trauma and accidental trauma. Table 1 lists the common causes of CSF rhinorrhea with their incidence.

Table 1. Common causes of CSF leaks

Etiology	Percentage
Spontaneous	41.1
Iatrogenic trauma	30.1
Accidental trauma	23.2
Tumor related	5.0
Congenital	3.0

Etiology and Classification

CSF leaks can be classified based on etiology, anatomic site, temporal sequence with underlying etiology and intracranial pressure. Based on etiology, the CSF leak may be due to accidental trauma, surgical trauma, tumoral or inflammatory erosion of anterior cranial base, congenital defects of the skull base and hydrocephalus. CSF may leak from defect in the anterior cranial fossa or the middle cranial fossa. With respect to time duration after which CSF rhinorrhea occurs, it is classified as early or delayed. CSF leak may also occur in the setting of a tumor or hydrocephalus causing increased intracranial pressure and erosion of skull base. Taking all these factors into consideration, Ommaya et al. [6], in 1968, described eighteen patients with CSF leak and proposed a classification system. Figure 1 depicts the classification of patients with CSF rhinorrhea.

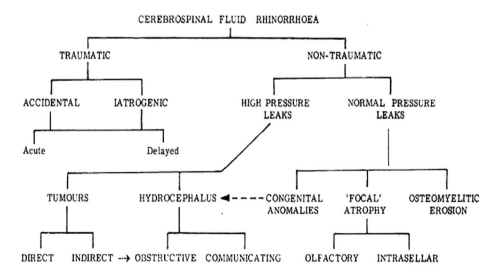

Figure 1. Etiological classification of CSF rhinorrhea [6].

A number of hypotheses have been proposed for pathogenesis of spontaneous leaks. As majority of these leaks occur in the region of cribriform plate, ethmoidal and sphenoid sinus, studies have proposed natural dehiscence in these regions as a cause of these leaks. Excessive pneumatization may also be a factor associated with bony dehiscence. Continuous brain pulsations have also been thought to cause thinning of bone in these areas. Ommaya proposed that ischemic focal atrophy of bone in the region of ethmoid, cribriform plate and sphenoid sinus may also contribute to the development of CSF leak [7].

Location of Leak

The most common site of leak is ethmoid/cribriform plate, followed by sphenoid, frontal and sella in that order. In a small proportion of patients, more than one site of leak may be present. Leaks through defects in the middle fossa also form a small proportion of cases. Table 2 lists the common sites of leak through the anterior fossa.

Table 2. Common sites of CSF leaks

Site	Percentage
Cribriform plate / ethmoid	57.2
Sphenoid	30.2
Frontal	67
Sella / clival region	21.2
Multiple sites	1.3

Clinical Features

Accurate identification of the presence of rhinorrhea and site of leak is paramount to optimal management of the condition. A good history and physical exam is pre-requisite. It is important to distinguish CSF rhinorrhea from nasal discharge due to local causes. Usually, a clear, watery unilateral or sometimes bilateral nasal discharge points towards CSF rhinorrhea. History of prior trauma, surgery and meningitis should be sought. There may be a history of increase in nasal discharge or postnasal drip while in supine position. The patient may also complain of salty taste in his or her mouth. Headache in may be due to either raised intracranial tension secondary to a tumor or hydrocephalus or intracranial hypotension. Sometimes, CSF may track through the Eustachian tube into middle ear and lead to otorrhea if the tympanic membrane is perforated (paradoxical rhinorrhea).

Physical examination in most patients is unremarkable. Anosmia may be present in patients presenting after trauma. Maneuvers that raise intracranial tension such as coughing, sneezing or compression of both jugular veins may precipitate leak. Another way is to have the patient to lean forward and strain. In posttraumatic leaks, CSF may be mixed with blood. Halo sign (central clearing when a drop of CSF mixed with blood is placed on a piece of filter paper) has been described to detect the presence of CSF in the nasal discharge, although the test is not sensitive or specific.

Evaluation

Evaluation of a patient with suspected CSF rhinorrhea consists of three aspects: confirming the presence of CSF leak, identifying the etiology and localizing the site of leak.

Although low protein (<2g/L), high glucose (>0.4g/L) and a specific gravity of 1.006 favor the presence of CSF, detection of β-2 transferrin in the fluid is the gold standard in determining whether CSF is present in the nasal discharge. β-2 transferrin was reported to have a sensitivity of near 100% and a specificity of about 95% in a large retrospective study [8]. Another biomarker, the β trace protein has also been reported as equally sensitive and specific in the diagnosis of CSF rhinorrhea.

The two methods used to localize the site of leak are high resolution computed tomography (CT) and magnetic resonance imaging (MRI). Presence of tumor and hydrocephalus can be detected on CT. Indirect signs of the site of leak include fracture involving anterior cranial base or frontal sinus, pooling of fluid in frontal/maxillary sinus and presence of pneumocephalus. High-resolution multiplanar spiral CT is the standard of evaluation in all suspected cases. On MRI, sometimes, a gliotic tissue herniating through the defect can be made out. On CISS 3D imaging, CSF can sometimes be seen tracking through the defect. The sensitivity of CT or MRI in detecting the site of leak is about 90% [9, 10]. Combining CT and MRI increases the sensitivity to 97% [10]. Intrathecal injection of contrast during CT or MRI is another technique with high sensitivity for detection of exact site of leak. However, due to their invasive nature, should be used only if there is discordance between other studies or the exact site of leak is not established. Radionuclide cisternography (using Technetium-99, I-131 and Indium-111) is another option, however due to its lower sensitivity and specificity than intrathecal CT or MRI, it is seldom used.

Endoscopic visualization allows not only identification of the exact site of leak, but treatment of the defect in the same sitting. Intrathecal injection of fluoroscein dye during surgery enhances the accuracy of identifying the site of leak. Typically, an intrathecal injection of a solution of 0.5%-1% Fluoroscein dye is performed. The patient is then examined about 30 minutes later by an endoscope. Fluoroscein mixed with CSF can be seen leaking through the defect. In most cases, filters are not required. When the amount leaking is small, a yellow filter is placed over the endoscope and a blue filter is placed over the light source to enhance visualization of the dye. Higher concentrations of fluoroscein, however, may lead to serious adverse effects such as seizures, pulmonary edema and even death.

Treatment Options

Management of a patient with CSF rhinorrhea should take into consideration numerous factors such as etiology of the leak, disturbances in intracranial pressure, presence of encephaloceles, the site and size of defect. The various treatment options in the management of CSF rhinorrhea include conservative management, use of lumbar drain, open surgical approach for repair of defect and endoscopic repair. Another area of controversy is the use of antibiotics in a patient with CSF rhinorrhea. Discussion about various management options is beyond the scope of this chapter and we will focus on endoscopic repair of anterior cranial fossa defects.

Endoscopic Repair

Preoperative Preparation

Topical decongesting with oxymetazoline (0.05%) followed by injection of 1% lidocaine with 1:100,000 adrenaline helps in hemostasis during surgery. The nose is also irrigated with antibiotic solution to reduce the number of bacteria within the operating field and decrease the potential for intracranial seeding. In some patients when the exact site of leak is not demonstrated with pre operative imaging, lumbar drain may be inserted for intraoperative insufflation of fluoroscein during the surgery. Surgical scrub and graft site preparation is done as per standard of care.

Grafts and Tissue Adhesives

A variety of autografts may be used during repair of the defect. The graft material chosen depends upon the size and location of defect, the anatomic character of the defect and the recipient bed, and presence of elevated intracranial tension. By far the most common graft used is autologous fat graft harvested from anterior abdominal wall. First used by Mayfield [11], autologous fat grafts serve as an excellent water sealant, prevent scar formation, do not adhere to neural tissues and are easy to harvest. The fat survives for a long time and gets revascularized. Other grafts that can be used are fascia lata and muscle. A free graft helps in wound healing by acting as a scaffold. They are adherent to the bone at one week and are replaced by fibrous tissue at three weeks. Besides fat, other grafting materials include cartilage, bone (septum, mastoid tip, middle turbinate), mucoperichondrium, septal mucosa, turbinate mucosa and/or bone, fascia (temporalis, fascia lata), abdominal fat, and pedicled septal or turbinate flaps [12]. The mucosal graft can be harvested from middle turbinate mucosa or inferior turbinate mucosa or the septal mucosa. Septal mucosal graft has the advantage of being thicker than the middle turbinate mucosal graft and larger than the inferior turbinate mucosal graft. Composite grafts consisting of turbinate bone and mucosa may also be used to provide multilayer closure.

A variety of tissue adhesives have been reported for repair of CSF leaks. Some of them are fibrin glue, collagen matrix, polymethylmethacrylate cement and multilayer acellular dermis. Fibrin sealants are two-component biological sealants consisting of a fibrinogen solution and thrombin, with a biological cofactor. The resulting mixture converts fibrin, factor XIII and calcium of fibrinogen monomer to fibrin polymer, thus creating a hemostatic adhesive clot. The fibrin sealant provides a temporary watertight closure and creates an additional barrier to CSF leakage during dural healing and development of local fibrosis [13].

In patients with CSF rhinorrhea following surgery, the defects can be large and hence, require a more elaborate and robust repair to prevent CSF leak. A vascular pedicled flap of the nasal septum mucoperiosteum and mucoperichondrium based on posterior nasoseptal artery (Hadad-Bassagasteguy flap) may be harvested during endonasal surgery. This flap is highly vascularized and sturdy, yet pliable. It provides a large surface area and so is helpful in repair of large skull base defects typically seen with expanded endoscopic approaches [14].

Techniques of Endoscopic Repair

The first step in endoscopic repair of CSF rhinorrhea is diagnostic evaluation. 0° and 30° rigid endoscopes are used in diagnosis and treatment of CSF leak repair. Diagnostic endoscopy is performed to visualize the probable area of defect. Egress of CSF through the defect is the only definitive method of demonstrating CSF leak. When in doubt, maneuvers that increase the intracranial pressure may be performed to demonstrate CSF leak. These include Valsalva maneuver and jugular vein compression. In cases where no definite site of leak is noted, intrathecal insufflation of 0.1 mL of 10% fluoroscein (Akorn Inc., Buffalo Grove, IL) is done slowly over 30 minutes. It takes about 60-90 minutes for the CSF to stain yellow-green. In most cases, a filter is not required; however, a blue light filter may be used to localize fluoroscein colored CSF in equivocal cases. In a series of 103 patients who underwent endoscopic repair of CSF rhinorrhea, the authors reported a sensitivity of 73.8% with the use of fluoroscein during surgery and a false-negative rate of 26.2% when a skull base defect existed [15]. The bottom-line is that lack of fluoroscein visualization should not rule out the presence of a CSF leak [15]. Another caveat to CSF leak repair is that besides identifying the site of leak, one should also examine for other sites of leak. Multiple sites of CSF leak are known and it is important to detect them to prevent recurrence of the leak.

Wormald and McDonough described the "bath-plug" technique which consists of introducing a fat plug with a secured vicryl suture into the intradural space [16]. The two basic techniques of repair of a defect are overlay and underlay repair. These techniques can be used separately, but more often, are used in combination. Multilayer closure is preferred over unilayered closure of the defect.

In the overlay repair, the graft is placed over the bone (extracranial). The soft tissue around the defect is dissected off so that the graft fixes to the bone. It is important not to plug the defect with graft. This could enlarge the bony borders of the defect and may interfere with interpretation of signals on postoperative imaging. Additionally, plugging the defect could potentially prevent osteoneogenesis and closure across the bony defect (Figure 2)

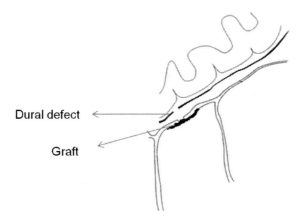

Figure 2. Technique of overlay grafting.

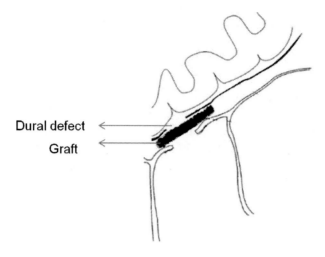

Figure 3. Technique of underlay grafting.

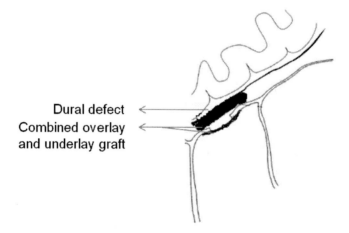

Figure 4. Combined overlay and underlay grafting.

In the underlay technique, the graft is placed in the epidural space between the bone and the dural defect (Figure 3). Care should be taken not to stuff the epidural space with excess graft. In all the cases of underlay repair, another portion of graft is fixed to the bone in an overlay fashion (two-graft technique). In the three-graft technique, the first graft is inserted under the dura, the second graft is inserted between the dura and the bone and the third graft is inserted over the bone. Multi-layer repair is preferred to single layer repair unless the defect is very small (Figure 4).

In defects involving sphenoid and frontal sinus, the obliteration technique is recommended. It involves removal of the sinus mucosa and packing the sinus with autologous abdominal fat. If an encephalocele is associated, it should be cauterized at its base prior to grafting. In CSF leaks following accidental trauma, presence of comminuted bone fragments can act as an impediment for the graft to seal the defect. These bone fragments have to be removed before the graft is placed. Another factor to be considered in leaks following accidental trauma is that the dural tear may be irregular and in many cases larger than the bony defect.

Hence, a multilayered repair is advocated in these cases. The gliotic tissue herniating through the defect also needs to be resected before doing a repair. External nasal packing with ribbon gauze or merocel is employed by many surgeons and it depends on the individual preference.

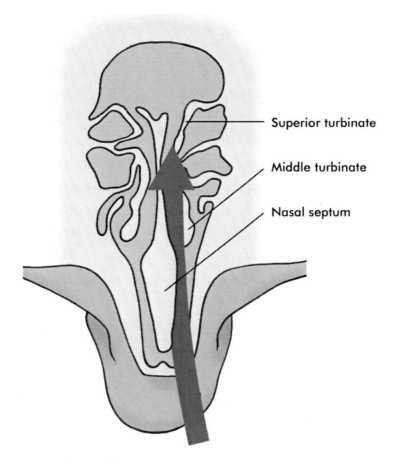

Figure 5. Direct paraseptal approach.

Endoscopic Approaches to Anterior Cranial Base

The different types of endoscopic approaches to anterior cranial base are

4. Direct paraseptal approach
5. Direct paraseptal approach with sphenoidotomy
6. Transethmoidal approach
7. Transethmoidal approach with middle turbinate resection
8. Transethmoidal-transpterygosphenoidal approach

Direct Paraseptal Approach

In this approach, the mucosa is incised all around the bony defect and dissected. If an encephaloceles is associated, the pedicle is dissected free off the mucosa, coagulated and cut. Lateralization of middle turbinate may be performed to increase the exposure. A mucosal graft may be harvested from the turbinate and the defect is closed. Alternatively, a composite graft consisting of bone and mucosa from the turbinate may be harvested for closure of the defect. This approach permits exposure of the olfactory groove area with preservation of anatomical structures (Figure 5).

Direct Paraseptal Approach with Sphenoidotomy

The sphenoid ostia are located along a line posterior to the superior turbinate and sphenoidotomy is done to enter the sphenoid sinus. The mucosa lining the sinus is excised to prevent postoperative mucocele formation. The mucosa may also be used as a component of grafting material. This approach is used to locate and repair defects in posterior wall and roof of sphenoid sinus. This approach also allows preservation of ethmoidal anatomical structures and is used in leaks following transsphenoidal surgery (Figure 6)

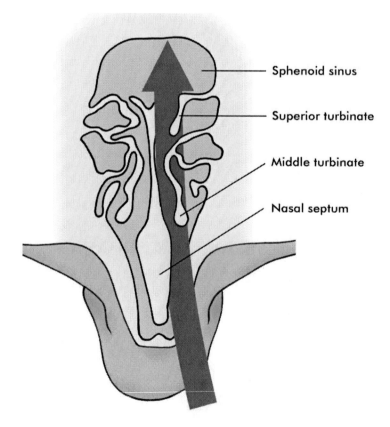

Figure 6. Direct paraseptal approach with sphenoidotomy.

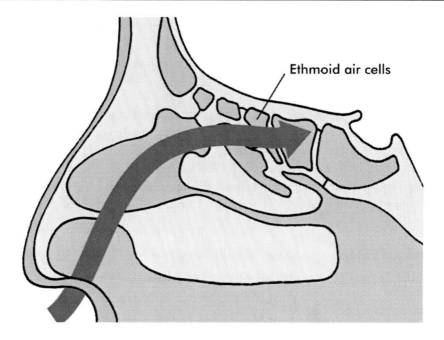

Figure 7. Transethmoidal approach.

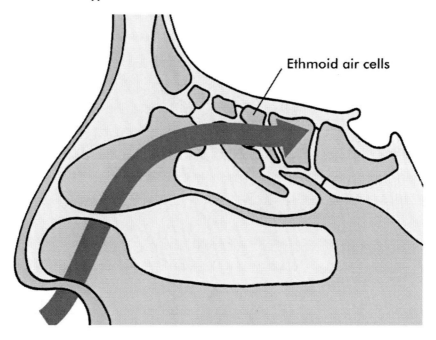

Figure 8. Transethmoidal approach with middle turbinate resection.

Transethmoidal Approach

Defects located lateral to the ethmoidal basal lamella may be exposed with this approach. Removal of the basal lamella and middle turbinate is not done. The ethmoidal roof is dissected to identify the site of leak (Figure 7)

Transethmoidal Approach with Middle Turbinate Resection

When the defect is located medially in the ethmoidal roof, the basal lamella of ethmoid and the middle turbinate are resected to obtain the exposure. Adequate exposure of the defect has to be achieved to enable the graft to adhere to the bone and seal the defect (Figure 8)

Transethmoidal-Transpterygosphenoidal Approach

This approach involves ethmoidectomy followed by sphenoidotomy. Then, a middle antrostomy is done and the posterior wall of maxillary sinus is exposed. The anterior wall of sphenoid sinus and pterygoid process are drilled laterally till the lateral wall of sphenoid sinus. Pterygopalatine artery may be encountered during this process and may need to be coagulated. This approach is used for defects in the lateral wall of sphenoid sinus and middle fossa (Figure 9)

Repair Using a Vascularized Flap

The vascularized flap may be harvested from intranasal tissue or from surrounding regions. Intranasal flaps may be based on sphenopalatine artery, inferior turbinate or middle turbinate artery. Extranasal regional flaps may be based on supraorbital/supratrochlear artery, superficial temporal artery or greater palatine artery. The sphenopalatine artery based nasoseptal flap is the most versatile and most commonly used to repair the anterior cranial base (Figure 10).

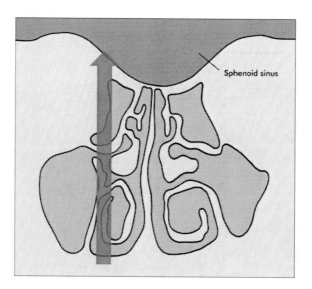

Figure 9. Transethmoidal-transpterygosphenoidal approach.

Post-operative Management

In the post-operative period, patients are advised bed rest for 2-5 days. There are advised not to strain or perform maneuvers that may increase intracranial tension. A slight head end elevation of 30° is advisable so that the dura is pressed against the graft and bone. Lumbar drain may be removed in the immediate post operative period or continued for up to one week depending based on surgeon's preference. We usually remove the lumbar drain by postoperative day 2. Patients are mobilized after 2-3 days and discharged by day 5. They are advised not to resort to strenuous activities for the first month after surgery. MRI evaluation is done at follow-up at 3 months and 6 months and endoscopic evaluation is done if there is suspicion of recurrence of leak.

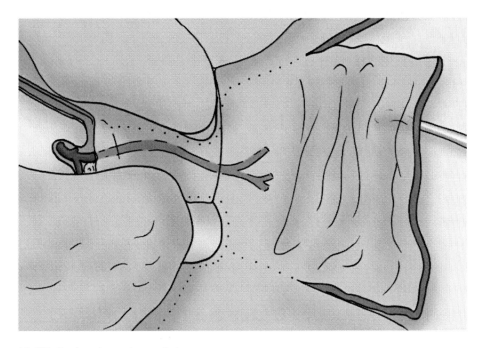

Figure 10. HB flap based on sphenopalatine artery.

Complication Avoidance

The advantages of endoscopic repair over open repair are minimally invasive approach with avoidance of an external scar, excellent visualization of anterior cranial base, ability to harvest locally available tissue for grafting, possibility of a second attempt at repair, shorter duration of hospital stay and success rates comparable to open repair. Anosmia is sometimes associated with open repair. It occurs during retraction of the frontal lobes when the olfactory tract gets sheared off at the cribriform plate. This complication is not encountered with endoscopic repair. Potential complications of frontal lobe retraction such as hemorrhage and cerebral edema are avoided with endoscopic repair.

During endoscopic repair, it is imperative to visualize the entire exposed cranial base for other sites of leak to avoid recurrence of leak in the postoperative period. It is also important to keep in mind that failure to visualize fluoroscein stained CSF does not rule out the presence of a leak. In case of a doubt, it is advised to repair the defect.

Multilayer and combined underlay and overlay repair are preferred to single layer and underlay or overlay repair respectively. Use of a composite graft consisting of a combination of mucosa, fascia lata, fat and bone decreases the likelihood of recurrence of leak.

Recurrence of CSF leak following endoscopic repair can be classified into two categories (1) early recurrence and (2) delayed recurrence. Early recurrence of CSF leak typically occurs within one month of repair and is caused by failure to recognize other sites of leak, slippage of the graft, failure to recognize and treat underlying etiology and excessive straining. Delayed recurrence occurs after months of repair and is usually not directly related to the procedure in most cases. It may be due to dehiscence of adjacent part of bone or failure to treat the underlying etiology.

In case of recurrence of leak in the early post-operative period, the patients are advised absolute bed rest. CT or MRI imaging may be done to visualize the site of leak. Fat suppressed MR imaging may be used to visualize the site of repair and other sites of leak. Endoscopic visualization of cranial base should be done to visualize the site of leak. Small leaks may be managed conservatively with bed rest and avoiding maneuvers that increase the intracranial pressure. Persistent leaks and large quantities of leak are repaired without expectant management. Delayed CSF leaks are managed in a similar manner as fresh CSF leaks.

Another pitfall in management of CSF rRhinorrhea is failure to recognize paradoxical rhinorrhea. One should be careful not to neglect the possibility of CSF tracking along the Eustachian tube and presenting as otorrhea. Mucosal grafts, if placed in an underlay fashion, may lead to the development of intracranial mucocele. Hence, placement of mucosal grafts in the epidural space should be avoided.

Outcome

In a meta-analysis of endoscopic repair of CSF rhinorrhea, the authors observed a success rate of 90.6% and 96.6% following primary repair and second attempt respectively [17]. The highest rate of failures occurred in leaks involving sphenoid sinus (48%) and ethmoidal roof or cribriform plate (41%). The overall complication rate was 0.03%, post-operative meningitis being the most common. In another systematic review of endoscopic reconstruction of skull base defects in which 54% patients underwent free graft repair and 46 underwent vascularized graft repair, the overall rate of CSF leak was 11.5%. The failure rate in patients undergoing free and vascularized graft repair was 15.6% and 6.7% respectively [18].

Two studies have looked into the factors predicting success of CSF leak and the factors that are associated with failure of endoscopic repair [19, 20]. They found that leak location, defect size, presence of meningocele, repair material and method, steroid usage, lumbar drain usage and exposure to post-operative radiation did not have any influence on success of endoscopic repair. The only factors that were found significant were the absence of a discretely identifiable dural defect at the time of endoscopic repair and clinical obesity.

Role of Antibiotics

The use of prophylactic antibiotics remains a controversial topic. With an increased risk of developing meningitis in anterior skull base CSF leaks, the use of prophylactic antibiotics is to decrease the incidence of meningitis. The incidence rate of meningitis ranges between 9% and 50%, while the mortality rate is ranges between 20% - 70% [21]. With such a severe co-morbidity, many implement the use of prophylactic antibiotics in their practice. In a traumatic setting, an anterior skull base CSF leak repair may not be possible due to refractory ICP or cerebral edema. If this requires a delay in surgery, the incidence of meningitis increases at 1 week [22, 23]. A recent retrospective study on traumatic anterior skull base CSF leaks by Sherif et al. found that in patients with either delayed surgical intervention or conservative management for CSF leak, the risk of developing meningitis was approximately 3%, which is much lower than other published data [24]. All patients in this particular study were placed on prophylactic antibiotics.

Opponents to the use of prophylactic antibiotics believe that prophylactic use can lead to gram negative infections by altering the nasopharyngeal microflora [25].

Role of Lumbar Drain

The placement of a lumbar drain to divert CSF from leaking is a conservative method practiced by many. Lumbar drains have been used as a tactic to avoid surgery and allow the CSF leak to heal. They have also been used in the post-operative period to divert CSF from the area of surgical repair. Although a lumbar drain is considered a conservative method of treatment for anterior skull base CSF leaks, it has associated risks. The placement can lead to nerve injury, post-procedure CSF leaks, infection, and there are reports of portions of the catheter severed within the patient. A recent retrospective review by Ransom et al. [26] looked at the risks and benefits of lumbar drain in endoscopic skull base surgery. Out of the 65 patients that had lumbar drains placed prior to surgery in the perioperative period, there were 9 complications associated with the lumbar drain. They included epidural blood patches for post-procedural symptoms, repeat imaging, open retrieval of a catheter fragment, and infectious workup. Another retrospective study by Sherif et al. [24] looked at traumatic CSF leaks. For patients that had CSF leak greater than 48 hrs, a lumbar drain was placed. If the leak resolved within 5 days, no surgery was performed. Of the 34 patients that were treated by this method, only 1 patient developed meningitis. It is presumed that because of the lumbar drain, patient that may have required surgical repair of the defect avoided surgery by diverting CSF from the area of trauma and allowed proper healing. A different retrospective study looked at the effectiveness of lumbar drains in recurrent CSF leaks in patients that underwent endoscopic repair. The results showed that there was no difference in CSF recurrence after repair in patients with or without lumbar drains [27]. In conclusion, the utility of lumbar drains remains controversial and may be deferred to each surgeon's experience.

Case Illustrations

Case 1

A 56 year-old man presented one month after a road traffic accident with watery nasal discharge on the right side. On examination, he had no neurological deficits other than anosmia on the right side. CT cisternogram showed a defect in the cribriform plate on the right side with CSF leaking. T2-weighted magnetic resonance imaging showed hyperintense signal in the ethmoids and across the cribriform plate on the right side. He underwent endoscopic direct paraseptal approach. At surgery, there was a defect in the cribriform plate on the right side. It was closed using abdominal fat graft and fibrin glue. Post-operatively, the patient did well and his rhinorrhea subsided (Figures 11a, 11b, 11c)

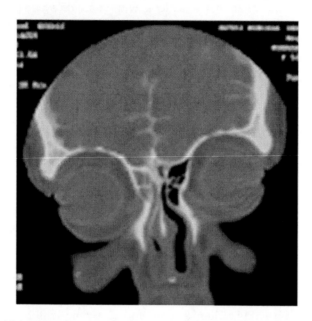

Figure 11a. CT scan with intrathecal contrast showing extravasation of the dye through cribriform plate.

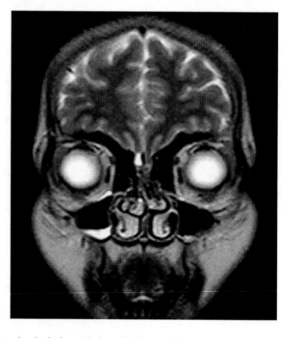

Figure 11b. MR T2WI showing leak through the cribriform plate.

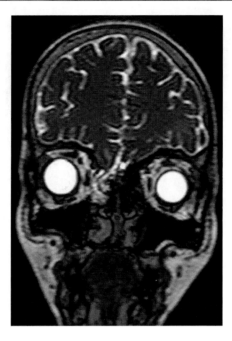

Figure 11c. MR T2WI showing herniation of gliotic brain tissue through the defect in cribriform plate.

Case 2

An 80 year-old lady presented with clear watery CSF discharge from the left nostril of 3 months duration. On examination there were no neurological deficits. Multiplanar CT and MRI showed a defect in the ethmoid on the left side and CSF pooling in the maxillary sinus. She underwent endoscopic transethmoidal approach. At surgery, the defect was visualized in the posterior ethmoids. She underwent repair of the defect with fat graft and fibrin glue. The patient recovered well and her CSF rhinorrhea ceased (figure 12a, 12b).

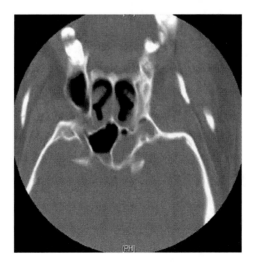

Figure 12a. Coronal CT scan with bone window showing opacification of sphenoid sinus indication fluid collection.

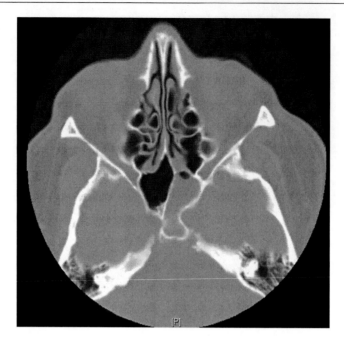

Figure 12b. Axial CT scan with bone window showing opacification of sphenoid sinus indication fluid collection.

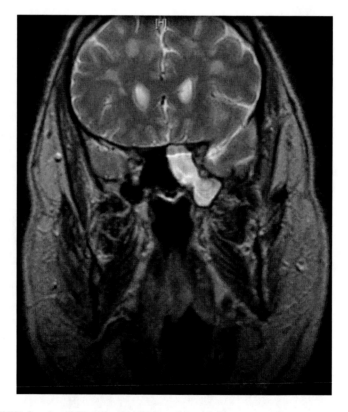

Figure 12c. MR T2WI showing CSF collection in the sphenoid sinus.

Management Algorithm

Based on our experience, we propose a management algorithm for a patient with newly diagnosed CSF rhinorrhea.

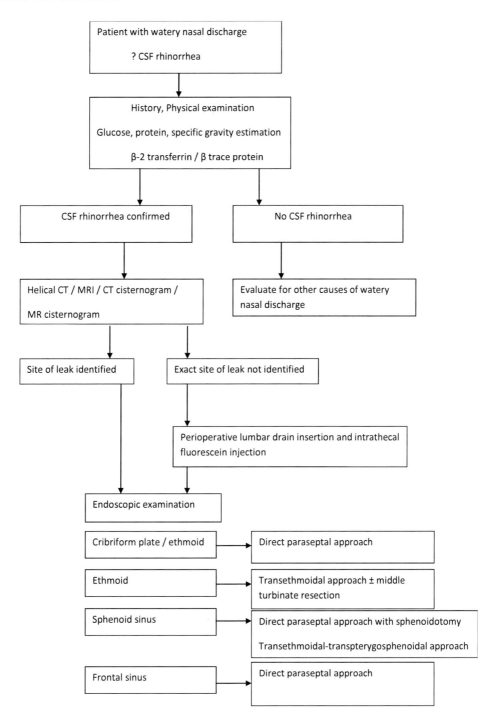

Conclusion

Endoscopic repair of CSF rhinorrhea is a minimally invasive technique with low morbidity and mortality with very good outcome. It is the current standard of treatment and should be the first line of treatment in patients in which surgical repair is contemplated.

References

[1] Zlab MK, Moore GF, Daly DT, Yonkers AJ. Cerebrospinal fluid rhinorrhea: a review of the literature. *Ear, nose, & throat journal.* 1992 Jul;71(7):314-7. PubMed PMID: 1505380. Epub 1992/07/01. eng.

[2] Beckhardt RN, Setzen M, Carras R. Primary spontaneous cerebrospinal fluid rhinorrhea. Otolaryngology--head and neck surgery: *Official journal of American Academy of Otolaryngology-Head and Neck Surgery.* 1991 Apr;104(4):425-32. PubMed PMID: 1903854. Epub 1991/04/01. eng.

[3] Dandy WE. Pneumocephalus (Intracranial Penumatocele or Aerocele. *Archives of Surgery.* 1926;12(5):949-82.

[4] Wigand ME, Haid T, Berg M. The enlarged middle cranial fossa approach for surgery of the temporal bone and of the cerebellopontine angle. *Archives of oto-rhino-laryngology.* 1989;246(5):299-302. PubMed PMID: 2590041. Epub 1989/01/01. eng.

[5] Mattox DE, Kennedy DW. Endoscopic management of cerebrospinal fluid leaks and cephaloceles. *The Laryngoscope.* 1990 Aug;100(8):857-62. PubMed PMID: 2381261. Epub 1990/08/01. eng.

[6] Ommaya AK, Di Chiro G, Baldwin M, Pennybacker JB. Non-traumatic cerebrospinal fluid rhinorrhoea. *Journal of neurology, neurosurgery, and psychiatry.* 1968 Jun;31(3):214-25. PubMed PMID: 5303103. Pubmed Central PMCID: 496347. Epub 1968/06/01. eng.

[7] Ommaya AK. Spinal fluid fistulae. *Clinical neurosurgery.* 1976;23:363-92. PubMed PMID: 975689.

[8] Skedros DG, Cass SP, Hirsch BE, Kelly RH. Sources of error in use of beta-2 transferrin analysis for diagnosing perilymphatic and cerebral spinal fluid leaks. Otolaryngology--head and neck surgery : *Official journal of American Academy of Otolaryngology-Head and Neck Surgery.* 1993 Nov;109(5):861-4. PubMed PMID: 8247566. Epub 1993/11/01. eng.

[9] Lloyd KM, DelGaudio JM, Hudgins PA. *Imaging of skull base cerebrospinal fluid leaks in adults. Radiology.* 2008 Sep;248(3):725-36. PubMed PMID: 18710972. Epub 2008/08/20. eng.

[10] Cui S, Han D, Zhou B, Zhang L, Li Y, Ge W, et al. Endoscopic endonasal surgery for recurrent cerebrospinal fluid rhinorrhea. *Acta oto-laryngologica.* 2010 Oct;130(10):1169-74. PubMed PMID: 20735181. Epub 2010/08/26. eng.

[11] Tomaras CR, Blacklock JB, Parker WD, Harper RL. Outpatient surgical treatment of cervical radiculopathy. *Journal of neurosurgery.* 1997 Jul;87(1):41-3. PubMed PMID: 9202263. Epub 1997/07/01. eng.

[12] Gjuric M, Goede U, Keimer H, Wigand ME. Endonasal endoscopic closure of cerebrospinal fluid fistulas at the anterior cranial base. *The Annals of otology, rhinology, and laryngology*. 1996 Aug;105(8):620-3. PubMed PMID: 8712632. Epub 1996/08/01. eng.

[13] Dunn CJ, Goa KL. *Fibrin sealant: a review of its use in surgery and endoscopy. Drugs.* 1999 Nov;58(5):863-86. PubMed PMID: 10595866. Epub 1999/12/14. eng.

[14] Kassam AB, Thomas A, Carrau RL, Snyderman CH, Vescan A, Prevedello D, et al. Endoscopic reconstruction of the cranial base using a pedicled nasoseptal flap. *Neurosurgery*. 2008 Jul;63(1 Suppl 1):ONS44-52; discussion ONS-3. PubMed PMID: 18728603. Epub 2008/09/09. eng.

[15] Seth R, Rajasekaran K, Benninger MS, Batra PS. The utility of intrathecal fluorescein in cerebrospinal fluid leak repair. Otolaryngology--head and neck surgery : *Official journal of American Academy of Otolaryngology-Head and Neck Surgery*. 2010 Nov;143(5):626-32. PubMed PMID: 20974330. Epub 2010/10/27. eng.

[16] Wormald PJ, McDonogh M. 'Bath-plug' technique for the endoscopic management of cerebrospinal fluid leaks. *The Journal of laryngology and otology*. 1997 Nov;111(11):1042-6. PubMed PMID: 9472573.

[17] Psaltis AJ, Schlosser RJ, Banks CA, Yawn J, Soler ZM. A systematic review of the endoscopic repair of cerebrospinal fluid leaks. Otolaryngology--head and neck surgery: *Official journal of American Academy of Otolaryngology-Head and Neck Surgery*. 2012 Aug;147(2):196-203. PubMed PMID: 22706995.

[18] Harvey RJ, Parmar P, Sacks R, Zanation AM. Endoscopic skull base reconstruction of large dural defects: a systematic review of published evidence. *The Laryngoscope*. 2012 Feb;122(2):452-9. PubMed PMID: 22253060.

[19] Zweig JL, Carrau RL, Celin SE, Schaitkin BM, Police PA, Snyderman CH, et al. Endoscopic repair of cerebrospinal fluid leaks to the sinonasal tract: predictors of success. Otolaryngology--head and neck surgery : *Official journal of American Academy of Otolaryngology-Head and Neck Surgery*. 2000 Sep;123(3):195-201. PubMed PMID: 10964290.

[20] Basu D, Haughey BH, Hartman JM. Determinants of success in endoscopic cerebrospinal fluid leak repair. Otolaryngology--head and neck surgery: *Official journal of American Academy of Otolaryngology-Head and Neck Surgery*. 2006 Nov;135(5):769-73. PubMed PMID: 17071310.

[21] Eljamel MS, Foy PM. Post-traumatic CSF fistulae, the case for surgical repair. *British journal of neurosurgery*. 1990;4(6):479-83. PubMed PMID: 2076209.

[22] Schaller B. Subcranial approach in the surgical treatment of anterior skull base trauma. Acta neurochirurgica. 2005 Apr;147(4):355-66; discussion 66. PubMed PMID: 15726278.

[23] Scholsem M, Scholtes F, Collignon F, Robe P, Dubuisson A, Kaschten B, et al. Surgical management of anterior cranial base fractures with cerebrospinal fluid fistulae: a single-institution experience. *Neurosurgery*. 2008 Feb;62(2):463-9; discussion 9-71. PubMed PMID: 18382325.

[24] Sherif C, Di Ieva A, Gibson D, Pakrah-Bodingbauer B, Widhalm G, Krusche-Mandl I, et al. A management algorithm for cerebrospinal fluid leak associated with anterior skull base fractures: detailed clinical and radiological follow-up. *Neurosurgical review*. 2012 Apr;35(2):227-37; discussion 37-8. PubMed PMID: 21947554.

[25] Bibas AG, Skia B, Hickey SA. Transnasal endoscopic repair of cerebrospinal fluid rhinorrhoea. *British journal of neurosurgery.* 2000 Feb;14(1):49-52. PubMed PMID: 10884886.
[26] Ransom ER, Palmer JN, Kennedy DW, Chiu AG. Assessing risk/benefit of lumbar drain use for endoscopic skull-base surgery. *International forum of allergy & rhinology.* 2011 May-Jun;1(3):173-7. PubMed PMID: 22287368.
[27] Caballero N, Bhalla V, Stankiewicz JA, Welch KC. Effect of lumbar drain placement on recurrence of cerebrospinal rhinorrhea after endoscopic repair. *International forum of allergy & rhinology.* 2012 May-Jun;2(3):222-6. PubMed PMID: 22344940.

Chapter V

Neurosurgical Overview of Minimally Invasive Resection of Anterior Skull Base Malignancy: A Comparison to Traditional Craniofacial Resection

Burak Sade[1], Pete S. Batra[2], Martin J. Citardi[3] and Joung H. Lee[4]

[1]Department of Neurosurgery, Dokuz Eylül University, İzmir, Turkey
[2]Department of Otolaryngology-Head and Neck Surgery and Comprehensive Skull Base Program, University of Texas Southwestern Medical Center, Dallas, TX, US
[3]Department of Otolaryngology- Head and Neck Surgery, University of Texas at Houston, Houston, TX, US
[4]Brain Tumor and Neuro-oncology Center, Cleveland Clinic, Cleveland, OH, US

Abstract

Surgical management of anterior skull base malignancy remains a significant surgical challenge. This chapter aims to serve as an overview of minimally invasive resection of this relatively rare entity in reference to traditional craniofacial resection.

Various aspects of the topic will be discussed using case examples and pertinent literature will be reviewed. Operative management of anterior skull base malignancy is still associated with relatively high morbidity, mainly due to infectious and wound reconstruction-related complications, and minimally invasive endoscopic surgery may serve as a viable option in carefully selected patients.

Introduction

Management of the malignant tumors of the anterior skull base is fraught with significant challenges, mainly due to their rarity, the wide variation of the histological subtypes and advanced stage at presentation with frequent involvement of critical skull base structures, including the brain, cranial nerves, carotid arteries and the orbit. One can add the lack of standardization in staging classification of these tumors (i.e. Kadish, Bilier, TNM systems for olfactory neuroblastoma) as another confounding factor in managing these patients, and analyzing the available literature.

Malignant tumors of the anterior skull base are mainly comprised of tumors of the sinonasal cavity. The incidence of histological subtypes varies in the published series, but the most common ones are olfactory neuroblastoma (13-42%), squamous cell carcinoma (7-28%), and adenocarcinoma (6-29%) [4, 10, 16, 18]. Other histological subtypes include malignant melanoma, carcinomas (i.e. basal cell, undifferentiated, adenoid cystic, transitional cell) as well as sarcomas (fibrous histiocytoma, osteogenic sarcoma, rhabdomyosarcoma, leiomyosarcoma).

Ethmoid cells are the most common site of tumor occurrence (93%), followed by the nasal cavity (88%), cribriform plate (58%), maxillary sinus (53%) and the orbit (51%) [17]. Intradural invasion is encountered in 13-14% of the patients [7, 17].

Although the aim of this chapter is to analyze the role of minimally invasive endoscopic resection (MIER) in the management of these entities, it would be crucial to have a good understanding of the important aspects of the more traditional approaches, namely, craniofacial resection (CFR).

Craniofacial Resection

Craniofacial resection may offer long-term palliation and a potential for cure in selected patients harboring malignant sinonasal tumors. In this context, it satisfies two main goals of the surgery from the neurosurgical perspective:

1. Achieving "cancer surgery", namely en bloc resection of the tumor with negative margin (This concept will be discussed further in the following pages of this chapter).
2. Prevention of neurological complications (i.e. cerebrospinal fluid leak, brain injury, infection, pneumocephalus).

Indications and Contra-Indications

The indications for CFR are involvement of the frontal fossa floor, dura, olfactory nerve and the brain parenchyma. Contra-indications include invasion of the cavernous sinus, internal carotid artery (ICA) and the orbital apex by the tumor (Figure 1). Progressive systemic disease may also be added to this list. In a large series from MD Anderson, it was reported that even in the group of patients in whom the tumor was deemed resectable initially based on the preoperative assessment, obvious tumor was detected in such locations and had to be left behind in as many as 21% of the patients [7].

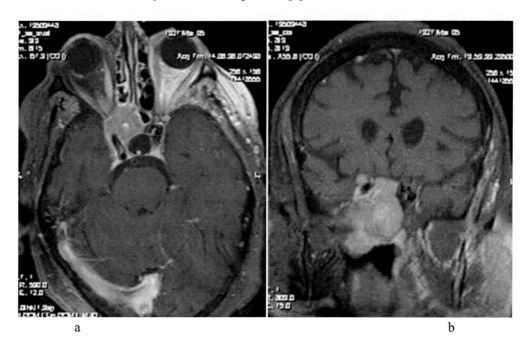

Figure 1. a) Axial and b) coronal post-contrast MRI images of a patient with invasion of the cavernous sinus, ICA, and the orbital apex.

Outcome, Complications and Prognosis

The largest multi-institutional review on this topic has been published by Ganly and colleagues, which consisted of 21 institutions and 1307 patients [9]. In this review, 59% of the patients had various extents of skull base and intracranial involvement (22% bone, 27% dura, 9% brain parenchyma). Negative resection margin was achievable in 70% of the cases.

Postoperative mortality was 5% and overall morbidity was 36%. Age and medical co-morbidity were risk factors for mortality where as medical co-morbidity, history of previous radiotherapy and dural invasion were risk factors for morbidity [9, 10]. Wound complications were seen in as much as 20% of the patients, for which medical co-morbidity and prior history of radiotherapy were risk factors [8] Neurological complications were seen in 16% of the patients, for which previous radiotherapy and dural invasion were risk factors.

In this multi-institutional review, overall poor prognostic factors were identified as positive surgical margin, invasion of brain parenchyma and a histopathological diagnosis of melanoma [9].

Technique for Craniofacial Resection

Craniofacial resection consists of transcranial and transfacial components. The workhorse for the transcranial component is bifrontal craniotomy. In addition to its relative simplicity, this approach allows for direct exposure of the anterior fossa to protect its contents, requires minimal brain retraction (with the aid of lumbar drain), at times facilitates en bloc resection of the tumor with a negative margin and provides access for a reliable reconstruction of the skull base. The workhorse for the transfacial component has traditionally been the lateral rhinotomy approach, with midface degloving approach being utilized as an alternative in some select cases. This approach is described elsewhere in detail and will not be discussed here, as it is beyond the main scope of this chapter.

Reconstruction is a critical aspect of CFR. As appropriately stated by Goel, a successful radical resection can be jeopardized by a poor reconstruction [12]. There are a number of options available for this purpose. For instance, one can use pericranium, fascia latae or temporalis fascia for dural reconstruction; autologous bone, split calvarium, or alloplastic material for reconstruction of the bony floor or may simply prefer not to reconstruct it. For more advanced reconstruction options, one may use non-vascularized (fascia latae, pericranium, fat), pedicled (pericranial flap, temporalis myofacial flap, trapezius myocutaneous flap) or free microvascular flaps (rectus abdominis, lattismus dorsi etc.) [3, 15]. In our experience, combination of a layer of collagen matrix, a layer of vascularized pericranial flap, autologous bony support and abdominal fat graft has produced satisfactory results (Figure 2).

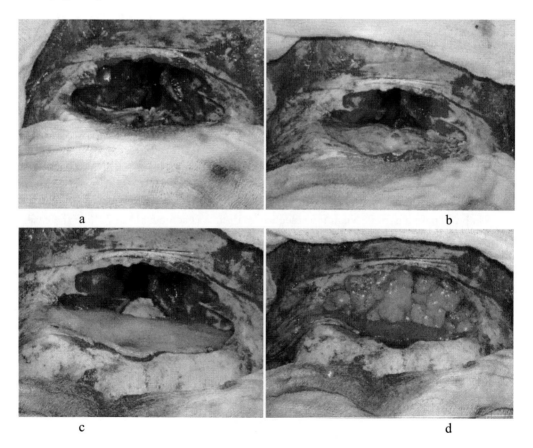

Figure 2. Reconstruction of skull base in traditional CFR. a) Dural and skull base defect after the resection b) Vascularized pericranial flap for dural reconstruction c) Collagen matrix to enhance the dural reconstruction and posterior wall of the frontal sinus for bony reconstruction d) Abdominal fat.

Extent of the Approach

There has been an ongoing debate in the skull base literature as to how extensive the approach should be, when one deals with the malignant tumors of the anterior skull base. In carefully selected cases, even a unilateral frontal craniotomy may suffice, whereas in others, individual preferences may prompt more extended variations.

The main aspects and advantages of bifrontal craniotomy have been discussed above in this chapter. To the proponents of more extensive approaches (i.e. transbasal / orbitonasofrontal), the additional osteotomies enhance the exposure and reduce the risk of retraction injury to the brain. Another advantage of these approaches has been proposed as the lack of need for facial incisions and therefore superior cosmesis. However, more bone removal from the skull base does not always guarantee less brain injury. Besides, with each additional procedure in the skull base, comes the added risk of new set of complications [6] Also, it has to be kept in mind that these are malignant lesions, which will require radiotherapy at some point. It has been reported that osteoradionecrosis of the bone flaps can be in the range of 20%, and when it involves the naso-orbito-frontal bone, it can actually result in inferior cosmesis [11].

Such concerns and debates lead us to another very important question that is relevant to CFR and the application of MIER in this area: Is en bloc resection really necessary or not?

En Bloc Resection versus Piecemeal Resection

The classic oncological principle is that monobloc excision of malignancy is required to minimize the risk of recurrence and overall survival with the belief that surgical violation of the tumor would predispose to spread of the cancer through lymphatic or vascular channels. However, this has not been systematically reviewed, especially in the skull base literature.

To the best of our knowledge, there has been only one study which compared the results of en bloc versus piecemeal resection in skull base malignancy [20]. In this study, Wellman and colleagues compared the outcomes of 16 patients in whom en bloc resection was performed with those of 14 for whom piecemeal resection was carried out. In all patients, the cranial exposure was achieved via a bifrontal craniotomy, whereas the approach used for resection of the facial portion was dictated by the size and location of the tumor below the anterior fossa. When the piecemeal approach was used, the most dangerous parts were left for last, and the margins were assessed with frozen and permanent sections once gross-total resection was completed. Interestingly, the outcome of the piecemeal group was quite comparable to the en bloc group. The incidence of positive surgical margin was 19% for the en bloc group as compared to 36% for the piecemeal counterpart. This was attributed to the discrepancy of tumor size and extension between the two groups. However, despite this difference, 1- year and 3-year survival rates were quite comparable: 94% and 56% for the en bloc group, and 83% and 70% for the piecemeal group. As expected, major complications were higher in the en bloc group (31%), as compared to the piecemeal group (21%). In the former group, 4 additional surgeries were required to manage the complications.

At this point, the issue of safety is also worth emphasizing. Wellman and colleagues stated that it might be safer to remove tumors that approach vital structures in piecemeal fashion [20]. This would not only improve control of these structures, it may also improve visualization of tumor margins. Others have also supported similar management strategies, suggesting that although radical resection is the aim, it can be compromised in favor of safety [12]. We would also agree with this opinion.

The debate on en bloc vs. piecemeal resection brings us to the use of endoscope and the use of minimally invasive techniques in the surgical management of anterior skull base malignancy.

Minimally Invasive Endoscopic Resection As an Alternative to Traditional Craniofacial Resection

From early on, our group was among the supporters of the idea that MIER could be a viable alternative to CFR in the management of anterior skull base malignancies [1]. In one of our earlier papers, we compared 16 patients, who were treated via traditional CFR with 9, who were operated via MIER. Tumors from both groups had comparable extension into contiguous structures. There was no statistically significant difference between the operative time, blood loss, or hospital stay between the two groups. Intensive care unit was shorter (3 days to 1 day) in the MIER group. Negative surgical margin was achieved in 78% of the patients in this group, which was also comparable to the traditional counterpart. Major complications (pneumocephalus, seizure, cerebrospinal fluid leak, hemorrhage, meningitis, frontal lobe infarction) were seen in 44% in the traditional CFR and 22% of the MIER groups. Recurrence was 36% in the traditional CFR group with a mean follow-up was 2.6 years, as compared to 33% in the MIER group with a mean follow-up of 2 years.

In their series consisting of 48 patients treated with traditional CFR and 18 treated with MIER, Eloy and colleagues showed a decreased hospital stay and faster recovery for the latter group [5], whereas there was no difference in survival, or complications. Having reviewed the literature on 226 patients, Higgins and colleagues found no statistically different outcome between the two approaches when dealing with low-stage tumors (T1-T2, Kadish A-B) [13] They could not perform a meaningful comparison on high-stage tumors due to the paucity of data on MIER on this subgroup. Komotar and colleagues, in their review of olfactory neuroblastomas, also reported a higher incidence of Kadish A tumors in the MIER group [14].

In a more recent study, we have reported on 25 patients with anterior skull base malignancies, whom all but one were treated with MIER alone [2]. One patient was treated with combined MIER and staged craniotomy. All patients reported in that series were stage T3 or T4. Surgery was performed with curative intent in all but 3, and negative margin was achieved in 74% of the patients, which was comparable to traditional series. Cavernous sinus, vidian nerve, infratemporal fossa, and clivus were sites of residual tumor. Major complications were seen in 19%. Overall survival at 2 and 5 years was 86% and 59% for the overall cohort, respectively. We believe that this study was important to show that even higher stage malignancies of the anterior skull base may be candidates for MIER. That said, we believe that there has to be a good understanding of the limitations of this technique (Table 1).

One has to keep in mind though that MIER does not render the traditional approaches obsolete. Rather, these techniques serve as complementary options in the armamentarium of the skull base surgeon. The advantages and disadvantages of both approaches are summarized in Table 2.

Table 1. Limitations of minimally invasive endoscopic approach

Limitations
Extensive involvement of the dura / brain parenchyma
Lateral extension of the tumor over the orbit(s)
Involvement of the facial soft tissue, lacrimal pathway, anterior table of frontal sinus
Vascular tumor
Extensive bilateral disease
Involvement of the orbita / infratemporal fossa

Table 2. Comparison traditional craniofacial resection (CFR) and minimally invasiave endoscopic resection (MIER)

CFR	MIER
Advantages	
Facilitates en bloc resection	Better assessment of tumor margin
	Better preservation of uninolved structures
	No facial incisions
Disadvantages	
Higher morbidity and mortality	En bloc resection typically not possible
Limited visualization of certain areas*	
Inferior cosmesis	
Criticism	
Feasibility and necessity of en bloc resection	Concern for tumor seeding during piecemeal resection

*Sphenoid sinus, orbital apex, frontal recess.

Given the rarity, and histological diversity of these lesions, along with the range of techniques are applied by surgeons with different backgrounds, definitive prospective studies comparing both techniques are unlikely to be available. Therefore, the goal of comparing both techniques should probably be limited to demonstrate that MIER is at least as safe and effective as the traditional CFR in carefully selected patients. In this context, as stated by others as well, completeness of the surgical resection is more important than the specific surgical technique used to achieve negative margin, and therefore, the best practice would be to use the technique that is more likely to result in complete tumor removal [19]. Depending on the extent of the disease, and the preference of the surgical team, there are more options in the contemporary management of anterior skull base malignancies (traditional CFR alone, MIER alone, combination of the two in the form of MIER with craniotomy) as compared to the past. Figures 3 and 4 show the preoperative MRI and intraoperative images of a patient with olfactory neuroblastoma, who was operated via a combined approach. The sinonasal component of the tumor was removed via MIER, and the cranial component through a craniotomy. On the other hand, a patient with limited intracranial disease with limited involvement of the dura (Figure 5) can be managed via MIER alone.

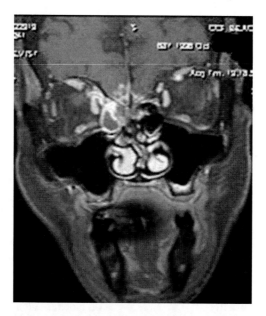

Figure 3. Preoperative coronal post-contrast MR image of a patient with olfactory neuroblastoma.

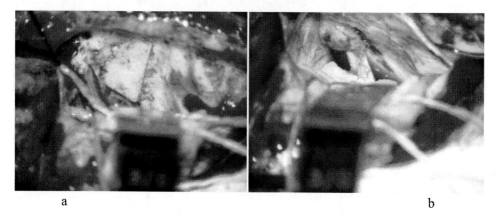

Figure 4. Intraoperative pictures of the same patient depicted in Figure 3. a) Visualization of the skull base from above and identification of the reconstruction performed during the endoscopic endonasal aspect of the surgery. b) Intradural exposure and identification of the olfactory nevre invaded by the tumor.

Technique for Minimally Invasive Endoscopic Resection

The advent of rigid endoscopes, the availability of sophisticated skull base instrumentation and surgical navigation, and advances in endoscopic techniques has facilitated the consideration of minimally invasive endoscopic resection of anterior skull base malignancy. The rigid endoscope provides brilliant illumination and magnification of the paranasal sinus confines to achieve tumor removal (Figure 6). Initially, the tumor is extirpated from the sinonasal region to create an optical cavity. Next, the skull base and orbital interfaces area addressed based on tumor extension.

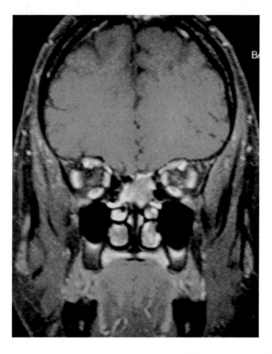

Figure 5. Coronal MR image demonstrates clear cell carcinoma of the anterior skull base with limited dural involvement.

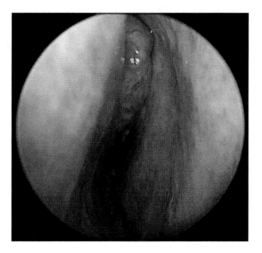

Figure 6. Endoscopic view demonstrates left sinonasal adenocarcinoma.

The endoscopic approach allows for removal of the tumor adjacent to the lacrimal system, periorbita, orbital apex, and optic nerve region ifn clinically required. Furthermore, the endoscopic approach allows for removal of the cribriform plate/ethmoid roof complex, dura, and/or olfactory nerves if needed (Figure 7). Though brain parenchyma may also be addressed endoscopically, this is better managed via a bifrontal craniotomy. Further, if there is extensive brain invasion by malignancy, one may need to question the necessity of the surgery, as the patient may be better served by a non-surgical approach, such as concurrent chemoradiation.

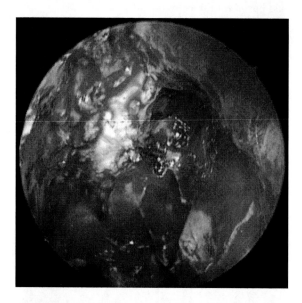

Figure 7. Intraoperative endoscopic view demonstrates anterior skull base view after minimally invasiave resection. Intraoperative dural biopsies were negative.

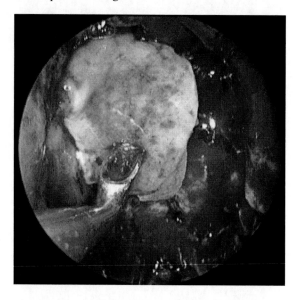

Figure 8. Endoscopic view illustrates mucosal graft used for skull base reconstruction.

Reconstruction of the skull base is an essential aspect of the endoscopic approach. Careful separation of the sinonasal space from the intracranial space decreases the risk of CSF leak, pneumocephalus, and ascending meningitis. A variety of tissue arrays have been employed to achieve this surgical objective.

Typically, multilayered closure is required in most patients, especially in cases with extensive dural defect. The initial layers are placed as an underlay in the subdural or epidural space. This could be comprised ofn dural substitutes, acellular dermis, temporalis fascia, or fascia lata. Rigid bone grafts are not required and may increase risk of infection and extrusion as most of these patients require postoperative radiation therapy. The final layer is usually comprised of vascularized pedicled flaps to facilitate robust skull base reconstruction. The pedicled nasoseptal flap based on the posterior septal artery is the workhorse for skull base reconstruction; alternate options include inferior turbinate flap, middle turbinate flap,or transglabellar pericranial flap. Free mucosal grafts from the septum or floor of nose may also provide a viable option for the final layer of reconstruction (Figure 8).

Conclusion

Traditional CFR and MIER serve as complementary techniques in the management of anterior skull base malignancies and should be available in the armamentarium of any skull base team set out to deal with these pathologies. The evidence shows us that completeness of the surgical resection remains to be one of the most important indicators of outcome, and therefore should have more priority over which surgical technique is used to achieve this goal.

The decision to choose the surgical approach should be made with this priority in mind, and therefore this sense of priority should give the surgeon the flexibility for decision-making on a case-by-case basis.

References

[1] Batra PS, Citardi MJ, Worley S, Lee J, Lanza DC. Resection of anterior skull base tumors: comparison of combined traditional and endoscopic techniques. *Am. J. Rhinol* 19:521-528, 2005.

[2] Batra PS, Luong A, Kanowitz SJ, Sade B, Lee J, Lanza DC, Citardi MJ. Outcomes of minimally invasive endoscopic resection of anterior skull base neoplasms. *Laryngoscope* 120:9-16, 2010.

[3] Boyle JO, Shah KC, Shah JP. Craniofacial resection for malignant neoplasms of the skull base. *J. Surg. Oncol.* 69: 2275-284, 1998.

[4] Catalano PJ, Hecht CS, Biller HF, Lawson W, Post KD, Sachdev V, Sen C, Urken ML. *Otolaryngol Head Neck Surg* 120:1203-1208, 1994.

[5] Eloy JA, Vivero RJ, Hoang K, Civantos FJ, Weed DT, Morcos JJ, Casiano RR. Comparison of transnasal endoscopic and open craniofacial resection for malignant tumors of the anterior skull base. *Laryngoscope* 119: 834-839, 2009.

[6] Feiz-Erfan I, Han PP, Spetzler RF, Porter RW, Klopfenstein JD, Ferreira MAT, Beals SP, Joganic EF. Exposure of midline cranial base without a facial incision through a combined craniofacial-transfacial procedure. *Neurosurgery* 56 (ONS Suppl 1):28-35, 2005.

[7] Feiz-Erfan I, Suki D, Hanna E, deMonte F. Prognostic significance of transdural invasion of cranial base malignancies in patients undergoing craniofacial resection. Neurosurgery 61:1178-1185, 2007.

[8] Ganly I, Patel SG, Singh B, et al. Complications of craniofacial resection for malignant tumors of the skull base: report of an international collaborative study. *Head Neck* 27:445-451, 2005.

[9] Ganly I, Patel SG, Singh B, et al. Craniofacial resection for malignant paranasal sinus tumors: report of an international collaborative study. *Head Neck* 27:575-584, 2005.

[10] Ganly I, Patel SG, Singh B, Kraus DH, Cantu G, Fliss DM, Kowalski L, Snyderman C, Shah JP. Craniofacial resection in the elderly. *Cancer* 117:563-571, 2011.

[11] Gil Z, Fliss DM. Pericranial wrapping of the frontal bone after anterior skull base tumor resection. *Plastic Reconstr Surg* 116:395-398, 2005.

[12] Goel A, Molyadi A. Evolved craniofacial surgery. *World Neurosurgery* 78:62-63, 2011 (Commentary).

[13] Higgins TS, Thorp B, Rawlings BA, Han JK. Outcome results of endoscopic vs craniofacial resection of sinonasal malignancies: a systematic review and pooled-data analysis. *Int Forum Allergy Rhinol* 1:255-261, 2011.

[14] Komotar RJ, Starke RM, Raper DM, Anand VK, Schwartz TH. Endoscopic endonasal compared with anterior craniofacial resection of esthesioneuroblastoma. *World Neurosurg* (Epub).

[15] Liu JK, Niazi Z, Couldwell WT. Reconstruction of the skull base after tumor resection. *Neurosurg Focus* 12(5):e9, 2002.

[16] Lund VJ, Howard DJ, Wel WI, Cheesman A. Craniofacial resection for tumors of the nasal cavity and paranasal sinuses. *Head Neck* 20:97-105, 1998.

[17] McCutcheon IE, Blacklock JB, Weber RS, deMonte F, Moser RP, Byers M, Goepfert H. Anterior transcranial resection of tumors of the paranasal sinuses: surgical techniques and results. *Neurosurgery* 38:471-479, 1996.

[18] Raza S, Muvdi TG, Gallia G, Tamargo RJ. Craniofacial resection of midline anterior skull base malignancies: a reassessment of outcomes in the modern era. *World Neurosurgery* 78:128-136, 2011.

[19] Soler Z, Smith TL. Endoscopic versus open craniofacial resection of esthesioneuroblastoma: what is the evidence? *Laryngoscope* 122:244-245, 2012.

[20] Wellman BJ, Traynelis VC, McCulloch TM, Funk GF, Menezes AH, Hoffman HT. Midline anterior craniofacial approach for malignancy: results of en bloc versus piecemeal resections. *Skull Base Surg.* 9:41-46, 1999.

Chapter VI

Expanded Endoscopic Endonasal Approaches to the Middle Cranial Fossa

Danielle de Lara[1,2], Daniel M. Prevedello[1,*], Ricardo L. Carrau[2], Leo F. S. Ditzel Filho[1], Jun Muto[1], Pornthep Kasemsiri[2], Bradley A. Otto[2], and Amin B. Kassam[3]

[1]Department of Neurosurgical Surgery, Wexner Medical Center at The Ohio State University, Columbus, Ohio, US
[2]Department of Otolaryngology- Head and Neck Surgery, Wexner Medical Center at The Ohio State University, Columbus, Ohio, US
[3]Department of Neurological Surgery, University of Ottawa, Ottawa, ON, Canada

Abstract

Objective: Endoscopic endonasal approaches (EEAs) offer a safe and effective option for the surgical treatment of ventral skull base lesions. These EEAs expanded in a paramedian (i.e., lateral to the ICA) direction, being referred as the coronal plane approaches to the anterior, middle and posterior cranial fossae. In this chapter we describe the technical and anatomical nuances related to EEAs to the middle cranial fossa.

Methods: Surgical indications, limitations and technical aspects pertaining to the EEA to the middle cranial fossa are systematically reviewed. Anatomical key landmarks regarding the approach and case examples are discussed with special attention to caveats, pitfalls, common complications and how to avoid them.

Conclusions: Approaches to the paramedian skull base are the most challenging and complex of all endoscopic endonasal techniques. A transpterygoid corridor typically precedes EEAs to the middle and posterior paramedian approaches. EEAs to the middle paramedian region provide wide exposure of the petrous apex, middle cranial fossa (including cavernous sinus and Meckel's cave), and the infratemporal and

---

[*] Corresponding author: Daniel M. Prevedello, M.D., The Ohio State University, Department of Neurological Surgery, N-1011 Doan Hall, 410 W. 10th Avenue, Columbus, OH, 43210. Phone: 614-293-7190; E-mail: Daniel.Prevedello@osumc.edu.

pterygopalatine fossae. Meningiomas, chondrosarcomas, schwannomas and cholesterol granulomas are some of the pathologies presenting in this area that can be treated through and endoscopic endonasal approach. Team approach with a complete understanding of the internal carotid artery anatomy and its relationships is paramount during these approaches to avoid and deal with unexpected complications.

Introduction

Endoscopic Endonasal skull base surgery, originally targeting the sella turcica, has progressively evolved to more complex technique and, currently, Endoscopic Endonasal Approaches (EEA) allow the treatment of benign and malignant pathologies in several different areas of the skull base [1-3]. These different target areas have been organized in anatomy-based modular approaches and divided in those that access the median ventral skull base, known as the "sagittal plane" approaches (between the internal carotid arteries - ICA) and those approaches addressing areas lateral to the ICAs know as paramedian or the "coronal plane" approaches [2, 4, 5]. The Coronal Plane module addresses lesions related to the ICA, systematized from an anterior to posterior direction. As one approaches the skull base in the paramedian plane through an endoscopic approach, the anterior coronal plane correlates to the anterior fossa and orbits, the middle coronal plane to the middle fossa and temporal lobe, and the posterior coronal plane to the posterior fossa. This chapter aims to discuss the endoscopic endonasal approaches to the middle fossa [2, 5, 6].

Traditionally, middle cranial fossa pathologies have been treated through lateral routes, such as pterional, petrosal (anterior and posterior) and retrosigmoid approaches. These approaches are well recognized and effective, especially in those patients whose lesions extend to the lateral aspect of the middle fossa [7, 8]. However, lateral approaches may require undesirable manipulation and retraction of neural tissue when used for lateral lesions that extend medially to the ventral brainstem. Under these circumstances, paramedian EEAs can be a great alternative to avoid neural tissue and vascular retraction [5, 8, 9].

Pathologies arising in the middle fossa can be directly treated through an EEA and are the most common indications for this approach. Meningiomas and trigeminal nerve schwannomas are the most frequent ones. Other tumors can extend or invade the middle fossa, growing from the pterygopalatine fossa (PPF), sella turcica, infratemporal fossa (ITF) and even from the paranasal sinuses. Some examples are juvenile angiofibromas, aggressive pituitary adenomas, chordomas and adenoid cystic carcinomas [2, 5, 10].

Understanding the surrounding anatomy is key to perform a successful endoscopic endonasal approach to the middle fossa. In the coronal plane, the most critical and defining structure is the internal carotid artery [2, 9]. Additional important anatomic landmarks include the pterygoid plates (PTPs), the vidian canal, and foramina rotundum and ovale (with the second and third branches of the trigeminal nerve respectively) [6, 10, 11].

Relevant Anatomy

The location of the sphenoid bone, just in the center of the cranial base, poses it as a major anatomical structure to all endoscopic endonasal approaches. The sphenoid bone is deeply related to the internal carotid arteries, basilar artery, cavernous sinuses and other important arterial, venous and neural structures [10, 12].

Additionally, its pneumatization offers the surgeon several natural corridors, with straightforward access to many neurovascular structures.

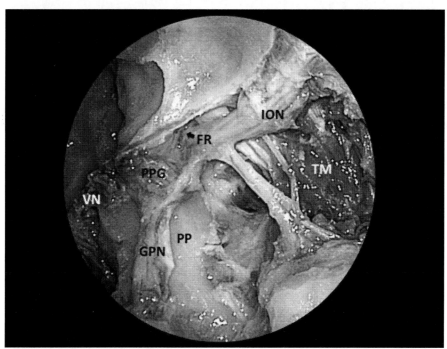

ION: Infraorbital nerve; FR: Foramen Rotundum; GPN: Great Petrosal Palatine nerve; PP: Pterygoid Plate; PPG: Pterygopalatine ganglion; TM: Temporalis Muscle; VN: Vidian nerve.

Figure 1. Endoscopic view of the left pterygopalatine fossa.

Understanding the pneumatization of the sphenoid sinus extending into the greater sphenoid wing is particularly important for EEAs to the middle cranial fossa. It creates the lateral sphenoid recess, projecting under the middle fossa in between the vidian and the maxillary nerves [11]. Removal of the lateral wall of the sphenoid sinus exposes the periosteum of the middle fossa [5, 13]. If the periosteum is transgressed at the anteromedial portion of middle fossa, the correspondent Meckel's cave, containing the Gasserian ganglion, is entered. The greater wing of the sphenoid bone needs to be well understood when addressing the paramedian skull base endonasally. It forms a considerable portion of the lateral wall of the orbit, the floor of the middle fossa, and the roof of the infratemporal fossa [11]. The infratemporal fossa is an anatomic space located under the floor of the middle cranial fossa and posterior to the maxilla. It relates medially with the pterygopalatine fossa through the pterygomaxillary fissure, which is continuous with the inferior orbital fissure superiorly [4, 12-14]. The pterygopalatine fossa includes a segment of the corridor to most of the approaches to the middle cranial fossa. It is also important because its neural compartment includes the pterygopalatine ganglion, the infraorbital nerve, the vidian nerve, and the maxillary nerve (V2), which are vital structures that can be used as landmarks to safely navigate on the parasagittal plane [13, 15]. [Figure 1]

Foramen rotundum and maxillary nerve (V2) can be classically identified following the infraorbital nerve proximally from anterior to posterior [10, 13]. However, retracting the periorbita and the pterygopalatine fossa's periosteum laterally can directly identify it. The vidian nerve, formed by the union of the greater superficial petrosal nerve and the deep petrosal nerve, runs along with the vidian artery through the pterygoid (or vidian) canal in the sphenoid bone. It guides the surgeon to the level of the horizontal ICA in the petrous bone, being considered one of the most important surgical landmarks, as it allows the surgeon to safely identify the ICA [6, 16, 17]. [Figure 2]

Bounded by the petrous ICA inferiorly, by the paraclival ICA medially, by the V2 laterally and by the sixth cranial nerve (VI) superiorly is the quadrangular space (QS). The QS offers direct access to the anteromedial segment of the Meckel's cave, which contains the trigeminal ganglion [5, 9, 10, 18]. [Figure 3] The trigeminal or gasserian ganglion guards the middle cranial fossa. From the ganglion arise the three branches of the trigeminal nerve: the ophthalmic nerve (V1), the maxillary nerve (V2) and the mandibular nerve (V3) [9, 10].

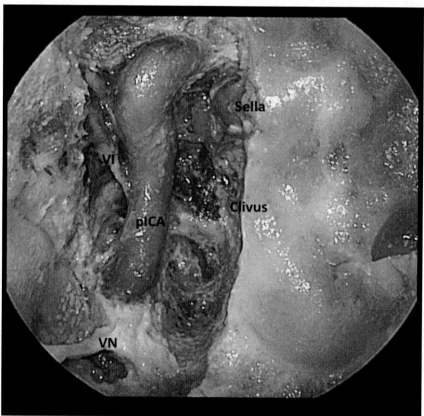

pICA: paraclival internal carotid artery; FL: cartilage of the Foramen Lacerum; VN: Vidian nerve; VI: sixth cranial nerve.

Figure 2. Endoscopic view of a cadaveric dissection of the sphenoid bone.

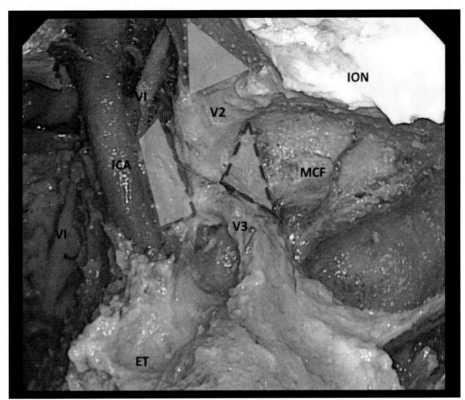

ICA: Internal Carotid Artery; ION: Infraorbital nerve; ET: Eustachian Tube; VI: sixth cranial nerve; V2: maxillary nerve; V3: mandibular nerve; MCF: middle cranial fossa. Quadrangular Space: green dotted line; Anteromedial triangle: orange dotted line; Anterolateral triangle: blue dotted line.

Figure 3. Endoscopic view of Middle fossa and surrounding anatomy.

Surgical Technique

Preoperative Considerations

Patients undergo preoperative computed tomographic angiography (CTA) and magnetic resonance imaging (MRI), which are combined and used for comprehensive intraoperative image guidance [19]. Neurophysiological monitoring (including electromyography) with greater attention to the third, fourth, motor V3 and sixth cranial nerves, and somatosensory evoked potentials are of vital relevance and are monitored during the entire surgical procedure [9, 10].

Patient is positioned supine with the head fixed in a 3-pin head holder. The neck is then gently extended and the head turned slightly to the right and tilted to the left. The facial skin, nasal cavity and periumbilical area (for a possible fat graft) are prepared and draped as standard fashion [2].

Exposure

The procedure begins with the use of a 0°-lens endoscope. Before entering the sphenoid sinus, a nasoseptal flap pedicled on the posterior nasal artery (a branch of the sphenopalatine artery) is elevated on the contralateral side in relation to the disease being accessed. This flap can be stored in the nasopharynx and used for the closure of the anticipated skull base defect. A posterior nasal septectomy in addition to wide bilateral sphenoidotomies and posterior ethmoidectomies completes the nasal corridor [20].

The nasoseptal flap is elevated in all cases of EEA to middle cranial fossa even if there is low chance of CSF exposure. The main goal of the flap in these cases is to cover the ICA at the end of the procedure, which is often exposed.

After exposing the initial nasal corridor, a transpterygoid approach must be performed using the maxillary sinus as the initial working corridor [14].

The transpterygoid approach is initiated with an uncinectomy, anterior ethmoidectomy (with bulla ethmoidalis removal), and enlargement of the maxillary ostium to obtain a wide middle meatus antrostomy. The sphenopalatine and posterior nasal arteries are identified at the sphenopalatine foramen and their branches are coagulated and transected [14]. The infraorbital canal is visualized superiorly and the posterior wall of the maxillary sinus is removed, exposing the soft tissue contents of the pterygopalatine fossa (PPF) protected by its periosteum. Lateralization of the contents of the PPF is achieved once the palatovaginal nerve and artery are identified and transected, allowing identification of the opening of the vidian canal at the base of the pterygoid plates [6, 12].

The vidian foramen and foramen rotundum are identified posteriorly in the sphenoid bone. Once the vidian nerve is identified, the surrounding bone is drilled down in an anterior to posterior direction. The floor of the sphenoid sinus and the base of the pterygoid plates are drilled completely exposing the sphenoid sinus lateral recess. In case of malignancies, a completely lateralization of the pterygopalatine fossa can be achieved after transecting the vidian nerve and the descending palatine artery with greater and lesser palatine nerves. It allows exposure of the lateral pterygoid plate and allows direct access to the medial infratemporal fossa. The vidian nerve can be transposed and preserved in selected cases and followed posteriorly to the anterior genu of the ICA [6, 11, 16, 21].

Middle Fossa Approach

At this moment of the approach, key anatomical landmarks: vidian nerve, V2, infraorbital nerve (ION) and the ICA should be easily identified. Moreover, other important neurovascular structures that limit the boundaries of the approach to the middle fossa/ quadrangular space (QS) must be recognized. The horizontal (petrous) and vertical (paraclival) segments of the ICA represent the inferior and medial limits of the QS. The superior orbital fissure (SOF), located inferior to the lateral optic carotid recess (LOCR) and anterior to the ICA syphon, represents the superior limit of approaches to the middle fossa [10, 16].

The V2 is then followed superiorly until the foramen rotundum is identified, and the bone covering the middle fossa is completely drilled. The periosteum of the middle fossa is exposed, allowing identification of the gasserian impression and trigeminal branches [9, 14].

Lesions relating directly to the gasserian ganglion can be approached with a dural opening at the quadrangular space. The dura is opened in a medial-to-lateral direction, from the ICA genu towards the V2. Staying below the level of the superior border of V2 is crucial to avoid an abducens nerve injury. Neurophysiology should also be used before dural opening to assure safe tumor removal [10].

Additionally, the middle cranial fossa can be approached through the anteromedial triangle, corresponding to the area between the SOF/ophthalmic nerve (V1) and V2, and through the anterolateral triangle, between V2 and V3, when access to the lateral aspect of the Meckel's cave is needed [9, 10, 13]. [Figure 3]

Tumor pathology and behavior will determine the location and extension of dural opening and neurovascular structures manipulation. After initial tumor removal, it can be extended in a posterior and lateral direction to Meckel's cave, petrous bone and into the posterior fossa, if necessary.

Complications

The most feared complication related to EEA to the middle fossa is the injury of the cranial nerves and to the ICA. The abducens, V1, V2 and V3 nerves are the ones in most risk of being harmed [2].

As previously mentioned, neurophysiological monitoring is fundamental to help the surgeon in preventing any nerve injuries. When performing an EEA to the middle fossa, it should be always used before dural opening [2]. Additionally, to avoid abducens nerve injury, the opening should not transgress the level of the superior limit of V2 [9, 18]. V1 and V2 sensory deficits or mastication limitation (due to V3 manipulation) may occur after performing Meckel's cave dissection [10] and functional recovery is related to nerve damage and also to tumor malignancyaggressiveness.

Even though major vascular injuries are not expected with this approach, some patient and tumor characteristics may increase the risk of perioperative bleeding. Prior surgery, prior radiation therapy and ICA tumor encasement are the most significant ones [2, 22].

The risk of CSF leaks after an EEA to the middle cranial fossa is usually lower when compared to other endonasal approaches to the skull base, since cisternal CSF exposure is not commonly performed. Furthermore, the addition of a vascularized nasoseptal flap reconstruction dramatically decreases the incidence of CSF leaks [2, 20].

Conclusion

EEAs have flourished as an important option for the treatment of ventral skull base lesions, as they offer the possibility of tumor resection with minimal brain and cranial nerves manipulation. However, approaches in the coronal plane represent the most challenging and complex of all endoscopic endonasal skull base surgicalery techniques. Detailed understanding of the ventral skull base anatomy, extensive training in endonasal endoscopic surgery and proper microsurgical technique are the key elements to achieve a successful paramedian endoscopic approach. EEAs for middle fossa pathology does not substitute open approaches. They are complementary and every case needs to be analyzed individually. The best approach may be a lateral craniotomy for some situations; an EEA may be the way to go for some others and a combination of both approaches the safest and more effective for others.

References

[1] Bhatki, A. M., Carrau, R. L., Snyderman, C. H., Prevedello, D. M., Gardner, P. A., Kassam, A. B. Endonasal surgery of the ventral skull base--endoscopic transcranial surgery. *Oral and maxillofacial surgery clinics of North America.* Feb. 2010;22(1):157-168.

[2] Kassam, A. B., Prevedello, D. M., Carrau, R. L., et al. Endoscopic endonasal skull base surgery: analysis of complications in the authors' initial 800 patients. *Journal of neurosurgery.* Jun. 2011;114(6):1544-1568.

[3] Kassam, A., Snyderman, C. H., Mintz, A., Gardner, P., Carrau, R. L. Expanded endonasal approach: the rostrocaudal axis. Part II. Posterior clinoids to the foramen magnum. *Neurosurgical focus.* Jul. 15 2005;19(1):E4.

[4] Falcon, R. T., Rivera-Serrano, C. M., Miranda, J. F., et al. Endoscopic endonasal dissection of the infratemporal fossa: Anatomic relationships and importance of eustachian tube in the endoscopic skull base surgery. *The Laryngoscope.* Jan. 2011;121(1):31-41.

[5] Kassam, A. B., Gardner, P., Snyderman, C., Mintz, A., Carrau, R. Expanded endonasal approach: fully endoscopic, completely transnasal approach to the middle third of the clivus, petrous bone, middle cranial fossa, and infratemporal fossa. *Neurosurgical focus.* Jul. 15 2005;19(1):E6.

[6] Kassam, A. B., Vescan, A. D., Carrau, R. L., et al. Expanded endonasal approach: vidian canal as a landmark to the petrous internal carotid artery. *Journal of neurosurgery.* Jan. 2008;108(1):177-183.

[7] Al-Mefty, O., Kadri, P. A., Hasan, D. M., Isolan, G. R., Pravdenkova, S. Anterior clivectomy: surgical technique and clinical applications. *Journal of neurosurgery.* Nov. 2008;109(5):783-793.

[8] Carrabba, G., Dehdashti, A. R., Gentili, F. Surgery for clival lesions: open resection versus the expanded endoscopic endonasal approach. *Neurosurgical focus.* 2008;25(6):E7.

[9] Kassam, A. B., Prevedello, D. M., Carrau, R. L., et al. The front door to meckel's cave: an anteromedial corridor via expanded endoscopic endonasal approach- technical considerations and clinical series. *Neurosurgery.* Mar. 2009;64(3 Suppl.):ons71-82; discussion ons82-73.

[10] Prevedello, D. M., Ditzel Filho, L. F., Solari, D., Carrau, R. L., Kassam, A. B. Expanded endonasal approaches to middle cranial fossa and posterior fossa tumors. *Neurosurgery clinics of North America.* Oct. 2010;21(4):621-635, vi.

[11] Mousa Sadr Hosseini, S., Razfar, A., Carrau, R. L., et al. Endonasal transpterygoid approach to the infratemporal fossa: Correlation of endoscopic and multiplanar CT anatomy. *Head Neck.* Mar. 2012;34(3):313-320.

[12] Theodosopoulos, P. V., Guthikonda, B., Brescia, A., Keller, J. T., Zimmer, L. A. Endoscopic approach to the infratemporal fossa: anatomic study. *Neurosurgery.* Jan. 2010;66(1):196-202; discussion 202-193.

[13] Hofstetter, C. P., Singh, A., Anand, V. K., Kacker, A., Schwartz, T. H. The endoscopic, endonasal, transmaxillary transpterygoid approach to the pterygopalatine fossa, infratemporal fossa, petrous apex, and the Meckel cave. *J. Neurosurg.* Nov. 2010;113(5):967-974.

[14] Rivera-Serrano, C. M., Terre-Falcon, R., Fernandez-Miranda, J., et al. Endoscopic endonasal dissection of the pterygopalatine fossa, infratemporal fossa, and post-styloid compartment. Anatomical relationships and importance of eustachian tube in the endoscopic skull base surgery. *Laryngoscope.* 2010;120 Suppl. 4:S244.

[15] Cavallo, L. M., Messina, A., Gardner, P., et al. Extended endoscopic endonasal approach to the pterygopalatine fossa: anatomical study and clinical considerations. *Neurosurgical focus.* Jul. 15 2005;19(1):E5.

[16] Cai, W. W., Zhang, G. H., Yang, Q. T., et al. [Endoscopic endonasal anatomy of pterygopalatine fossa and infratemporal fossa: comparison of endoscopic and radiological landmarks]. *Zhonghua Er Bi Yan Hou Tou Jing Wai Ke Za Zhi.* Oct. 2010; 45(10):843-848.

[17] Dallan, I., Lenzi, R., Bignami, M., et al. Endoscopic transnasal anatomy of the infratemporal fossa and upper parapharyngeal regions: correlations with traditional perspectives and surgical implications. *Minim. Invasive Neurosurg.* Oct. 2010;53(5-6): 261-269.

[18] Iaconetta, G., Fusco, M., Cavallo, L. M., Cappabianca, P., Samii, M., Tschabitscher, M. The abducens nerve: microanatomic and endoscopic study. *Neurosurgery.* Sep. 2007;61(3 Suppl.):7-14; discussion 14.

[19] Morera, V. A., Fernandez-Miranda, J. C., Prevedello, D. M., et al. "Far-medial" expanded endonasal approach to the inferior third of the clivus: the transcondylar and transjugular tubercle approaches. *Neurosurgery.* Jun. 2010;66(6 Suppl. Operative):211-219; discussion 219-220.

[20] Kassam, A. B., Thomas, A., Carrau, R. L., et al. Endoscopic reconstruction of the cranial base using a pedicled nasoseptal flap. *Neurosurgery.* Jul. 2008;63 (1 Suppl. 1): ONS44-52; discussion ONS52-43.

[21] Prevedello, D. M., Pinheiro-Neto, C. D., Fernandez-Miranda, J. C., et al. Vidian nerve transposition for endoscopic endonasal middle fossa approaches. *Neurosurgery.* Dec. 2010;67(2 Suppl. Operative):478-484.

[22] Kassam, A., Snyderman, C. H., Carrau, R. L., Gardner, P., Mintz, A. Endoneurosurgical hemostasis techniques: lessons learned from 400 cases. *Neurosurgical focus.* Jul. 15 2005;19(1):E7.

Chapter VII

The Endoscopic Endonasal Transtuberculum-Transplanum Approach

Luigi Maria Cavallo*, Domenico Solari, Alessandro Villa, Michelangelo de Angelis, Teresa Somma, Felice Esposito and Paolo Cappabianca

Department of Neurological Sciences, Division of Neurosurgery,
Department of Neurosciences, Reproductive & Odontostomatological Sciences
Università degli Studi di Napoli Federico II
Naples, Italy

Abstract

In the last decades, the widespread use of the endoscope has shed new interest on the transsphenoidal surgery, affording its extension. The wider panoramic view offered by the endoscope has improved the safety of this technique, extending its applicability to the removal of different supradiaphragmatic lesions.

The sphenoid sinus represents the gateway to the skull base: on the posterior wall the sellar floor occupies a central position, the sphenoid planum is above and the clival indentation below; on both sides, laterally to the sellar floor, the two bony prominences of the intracavernous carotid arteries and, slight superiorly, the optic nerves. Immediately above, the tuberculum sellae can be seen as an indent represented by the angle formed by the convergence of the sphenoid planum with the sellar floor; recently, according to the identification of this shape, we renamed this structure as the "suprasellar notch". Anteriorly, the sphenoid planum lies, limited on both sides by the optic nerves protuberances, diverging towards the orbits.

---

* Address correspondence and requests for reprints to: Luigi M. Cavallo, MD, PhD. Division of Neurosurgery - Dept of Neurolosciences, Reproductive & Odontostomatological gical Sciences. Università degli Studi di Napoli Federico II via S. Pansini 5, 80131 Naples, Italy. ph. +39-081-7462582. fax +39-081-7462594. E-mail: lcavallo@unina.it.

Middle turbinectomy on one side, with homolateral ethmoidectomy (in the right nostril for right handed surgeon and vice versa for left handed surgeons) and a wider anterior sphenoidectomy along with the removal of the posterior portion of the nasal septum are common steps required to create a wider surgical room.

The removal of the upper half of the sella, the tuberculum sellae and the posterior portion of the planum sphenoidale offers the possibility to explore the suprasellar region. The entire suprasellar region could be divided in four intradural areas: the supra-chiasmatic region, the sub-chiasmatic region, the retrosellar region and the ventricular region.

The endoscopic endonasal approach (EEA) is a 2-surgeon, 3- or 4-handed technique: one surgeon pilots the endoscope "dynamically" helping the dissection with his free hand, while the other performs bimanual dissections and tumor removal, according to the same paradigms of microsurgery, i.e. internal debulking of the solid component and/or cystic evacuation and fine dissection of the capsule from the main neurovascular structures. In case of meningiomas approached through an EEA an early tumor devascularization can be achieved, so that debulking and extra-arachnoidal dissection from the surrounding microvascular structures result easier. In case of craniopharyngiomas, the surgical removal should proceed according to Kassam's classification, which addresses different lesion types as their relationship with the infundibulum. Finally, even pituitary adenomas, with rubbery and/or fibrous consistency could be approached via this technique, in order to achieve an extracapsular dissection.

After tumor removal, the reconstruction of the osteodural defect is tailored to the entity of CSF leakage, mostly related to the degree of subarachnoid dissection and/or opening of the third ventricle.

This technique provides a wider and multi-angled, close-up view of the surgical field allowing a safer dissection and lesion removal without any brain retraction and nerve manipulation. On one hand it requires a thorough knowledge of the ventral skull base anatomy, on the other it facilitates intraoperative understanding of tumor inner features.

Introduction

Transsphenoidal surgery is a very effective technique for the treatment of pituitary tumors and the other lesions of the sellar area [1-10]. Such technique, developed in the late 1800s thanks to the anatomical studies of the Italian surgeon Davide Giordano [11], was first described by the Viennese surgeon Schloffer [12, 13] and refined, over the years, through the contributions of giants such as Kocher [14], Halstead [15], Cushing [16], Guiot [17] and Hardy [1].

Guiot first proposed the use of the endoscope during a classic transsphenoidal trans-naso-rhinoseptal microsurgical approach in order to explore the sellar contents [18], thus expanding the field of vision of this kind of surgery. After sporadic reports of the use of the endoscope [19-23] – amongst all Apuzzo [24] and Bushe and Halves [9] – the main cornerstone of the evolution process of the endoscopic transsphenoidal surgery could be considered the otorhinolaryngologists experience with Functional Endoscopic Sinus Surgery (FESS) [25-27]. Finally it was the Pittsburgh duo, i.e. an ENT surgeon Carrau and a neurosurgeon Jho [4, 5], that defined and introduced in the clinical practice the pure endoscopic endonasal approach to the sellar area, followed by our group in Naples [28] and several other groups all around the world.

As a matter of facts, the endoscopic transsphenoidal route represents a minimally invasive approach commonly used in many centers throughout the world, under the same indications as the conventional microsurgical technique. Indeed, several advantages arise from the use of the endoscope itself and from the absence of the transsphenoidal retractor could be achieved; the endoscope provides the surgeon with an increased panoramic close-up view of the surgical area [2, 29].

In the last decades, the widespread use of the endoscope along with the evolution of the surgical techniques and the technological progress, have led to the progressive reduction of invasiveness together with decreased morbidity. The interest for the transsphenoidal surgery has been renewed so that this technique has been seldomly adopted for the management of different lesions, involving anatomical areas around the sella [30-40].

It was the definition of the extended transsphenoidal approach – first described by Weiss [41] – that disclosed new horizons in transsphenoidal surgery, opening a new corridor to the suprasellar space. This approach, initially performed with microsurgical technique, provides a midline access and a direct visualization of the suprasellar space without brain manipulation, thus offering the possibility to treat small/medium midline suprasellar lesions traditionally approached transcranially. Indeed, thanks to an additional bone removal from the midline anterior cranial base, i.e. the tuberculum sellae and the posterior portion of the planum sphenoidale, it provides an excellent, direct, access to the suprasellar/supradiaphragmatic area [6, 39, 42-49].

More recently we further refined the approach [31, 50, 51] to this area and carefully analyzed the anatomical details to be faced when accessing the skull base via a ventral-extracranial corridor; in a recent publication, though we named the "tuberculum sellae" as the "suprasellar notch" (SSN): it depicts a sort of "wide shallow indentation", observed in the majority of cases just between the superior aspect of the sella turcica and the declining part of the planum sphenoidale [52].

We herein describe the endoscopic endonasal transtuberculum transplanum approach for the management of skull base lesions involving the suprasellar space, detailing pros and cons of the this technique also in regards of the anatomical landmarks as seen from the endoscopic endonasal standpoint and the pathology dealt with.

Endoscopic Endonasal Transtuberculum-Transplanum Approach

Surgical Procedure

The endoscopic endonasal approach is a 2-surgeon, 3-or 4-handed technique procedure performed using a rigid endoscope (0 degrees), 18 cm in length 4 mm in diameter (Karl Storz® & Co, Tuttlingen, Germany), as the sole visualizing instrument of the surgical field. At the beginning of the procedure, the primary surgeon is placed at the right of patients' bed regardless he-she is right or left-handed, with the assistant surgeon on the left side of the patient and the scrub nurse at the level of the patient's legs. Each surgeon looks into a dedicated monitor, adjusted in front of him-her at personalized height and distance.

A detailed, complete preoperative planning, even integrated by tridimensional computerized reconstruction of MRI and/or CT scans, has to be the backbone of the surgical procedure.

An image-guided system (neuronavigator) is also useful – especially when the classic landmarks are not easily identifiable -; it provides information for midline and trajectory and offers more precision in defining the bony boundaries and the neurovascular spatial relationships.

Finally, it is extremely important to use dedicated instruments: different endonasal bipolar forceps have been designed, in order to be easily introduced and maneuvered in the nasal cavitiesy, with various diameters and lengths. High-speed low-profile drills, long enough but not too bulky, may be very helpful for the opening the bony structures to gain access to the dural space. Finally, it is of utmost importance to use the micro Doppler probe to insonate the major arteries.

To increase the working space and the maneuverability of the instruments it is necessary: I) to remove the middle turbinate and the homolateral ethmoid cells on one side; II) to lateralize the middle turbinate in the other nostril; and III) to remove the posterior portion of the nasal septum. A wider anterior sphenoidotomy, as compared to the standard approach, is performed, especially in lateral and superior directions where bony spurs are flattened in order to create an adequate space for the endoscope during the deeper steps of the procedure. All the septa inside the sphenoid sinus are removed including those attached on the bony protuberances and depressions on the posterior wall of the sphenoid sinus cavity. These surgical maneuvers allow the use of both nostrils, so that two or three instruments plus the endoscope can be inserted.

Patient Positioning

The patient is placed supine or in slight Trendelenburg's position with the head turned 10°-15° on the horizontal plane, towards the surgeon, fixed in a rigid three-pin Mayfield-Kees skeletal fixation device, if necessary for the use of the neuronavigation systems or otherwise is fixed with a strip. On the sagittal plane, the head is extended for about 10-15 degrees to achieve a more anterior trajectory avoiding that either the endoscope and the surgical instruments hit on the thorax of the patient.

Nasal Phase

The procedure begins with the removal of middle turbinate in one nostril and the lateralization in the other; the side of turbinectomy is chosen according to primary surgeons' need, namely whether he/she is right or left handed.

The head of the middle turbinate is cut with nasal scissors and pushed downward, up to expose its tail; after haemostasiis has been performed, the turbinate is removed and stored in order to realize with it a mucopericondrium flap for the reconstruction of the osteo-dural defect after lesion removal. It is not mandatory to remove the tail of the middle turbinate because it does not affect the surgical trajectory and furthermore since its removal could increase the risk of sphenopalatine artery bleeding [53-55]. Whether a vascularized pedicled naso-septal flap should be used, it has to be harvested at this time according to the Hadad-Carrau [56, 57] technique and stored in the choana or in the maxillary sinus up to the reconstruction phase. Then, the posterior nasal septum is detached from the sphenoid prow with a microdrill and removed with a retrograde bone punch; the mucosal edges are accurately coagulated with the bipolar forceps.

The Sphenoid Phase

The anterior sphenoidotomy required is wider as compared to that performed in the standard approach; therefore, it is useful the removal of the posterior ethmoid air cells (bilaterally) and eventually superior turbinates is useful, to create enough space. All the septa inside the sphenoid sinus are removed including those attached on the bony protuberances and depressions on the posterior wall of the sphenoid sinus cavity. It has to be said that each single irregularity of the bone and/or either of the mucosa covering the limit between sphenoidal and ethmoidal planum, should be flattened to allow a better maneuverability of the endoscope and of the surgical instruments, while working above the sella. Once the complete exposure of the posterior wall of the sphenoid sinus is achieved, a series of protuberances and depressions (according to the grade of pneumatization) becomes visible; the precise knowledge of such anatomical landmark is of great importance for the correct orientation and the bone opening.

From now over, the surgeon has the possibility to perform a bimanual dissection while a co-worker holds the endoscope and, if needed another surgical instrument, according to the so-called "3-4 hands technique" [58]; it requires a good collaboration between of two surgeons, one holding the endoscope as a "navigator" and the other, the "pilot", handling two surgical instruments inside the surgical field, as in the traditional microsurgical technique. The "pilot" surgeon moves actively the instruments whereas the "navigator" "dynamically" follows such in and out movements with the endoscope, being responsible for having them under direct endoscopic view. The "pilot" and the "navigator" continuously pass between the close-up view, as during the dissecting maneuvers, and a panoramic view of the neurovascular structures.

The Bone Resection over the Sella

The bone removal starts with the drilling of the upper half of the sella in order to expose and isolate the superior intercavernous sinus. In case of accidental damaging during the bone resection, the bleeding can be controlled with temporary gentle compression with cottonoids and with different haemostatic agents, such as the Floseal® (Baxter BioSciences, Vienna, Austria).

The tuberculum sellae, observed from below by the endoscope through the sphenoid sinus cavity, corresponds to the angle formed by the planum sphenoidale with the anterior surface of the sella, namely what we call the "suprasellar notch": it is thinned with an high speed microdrill moving from the center towards both the medial opto-carotid recesses, which represent the lateral limits of the approach at this level.

Once the tuberculum sellae is drilled out from the two medial opto-carotid recesses and from the planum sphenoidale, it is gently dissected from the dura and periosteum and then removed; by using a Kerrison's rongeur the bone removal can be extended anteriorly up to the falciform ligament, or more, according to the anterior extension of the lesion. To better define such limit, the role of the neuronavigator is quite useful. In the upper part, just above the medial opto-carotid recesses, the bone removal can be extended more laterally, so that the opening resembles a "chef's hat" (Figure 1). This particular shape is due to the fact that the inferior part of the osteo-dural opening is narrower since it is limited by the parasellar portion of both the intracavernous carotid arteries and the optic nerves at their entrance in the optic canals, whereas it can be extended laterally in its superior half since the optic nerves diverges towards the optic canals (Figure 1 b).

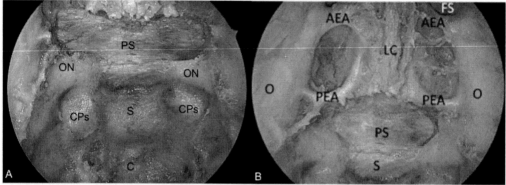

PS: planum sphenoidale; S: sella turcica; C: clivus; CPs: parasellar segment of the carotid protuberance; ON: optic nerve; AEA: anterior ethmoidal artery; PEA: posterior ethmoidal artery; O: Lamina papyracea of the orbit; LC: lamina cribriformis; PS: planum sphenoidale; FS: frontal sinus.

Figure 1. Endoscopic endonasal anatomic view of the anterior skull base. The superior part of the nasal septum has been removed and the sphenoid sinus widely opened to expose a) the posterior wall of the sphenoid sinus and the bone opening so-called "chef's hat" and b) the most anterior part of the planum sphenoidale and ethmoidale with the anterior and posterior ethmoidal arteries well visible enclosing the olfactory groove.

Dural Opening

After the bone has been removed, bleeding from the superior intercavernous sinus, especially in case of suprasellar craniopharyngioma, is not uncommon and therefore should be managed. Despite the use of haemostatic agents, such as the Floseal® (Baxter BioSciences, Vienna, Austria) results effective, the sinus should be closed with the bipolar forceps instead of using the hemoclips, which narrows the dural opening. The dura is incised horizontally few millimeters above and below the superior intercavernous sinus in order to coagulate the sinus between the two tips of the bipolar forceps in its median portion. It is then incised with microscissors, and the two resulting dural flaps are coagulated thus achieving their retraction and the consequential enlargement of the dural opening, which over the planum can be easily cut using the Kerrison punch (Figure 2).

The Intradural Phase

The dissection and removal maneuvers are tailored to each lesion following the same principles of transcranial microsurgery; nevertheless, before entering the intradural space it has to be controlled any bleeding coming from the sphenoidotomy into the surgical field.

In the case of prechiasmatic lesions, i.e. a tuberculum sellae meningioma, the tumor immediately becomes visible after the dural opening so that its base could be easily devascularized, whereas in case of retrosellar lesions, the tumor appears behind the pituitary stalk and therefore should be first dissected using a surgical corridor lateral to such structure.

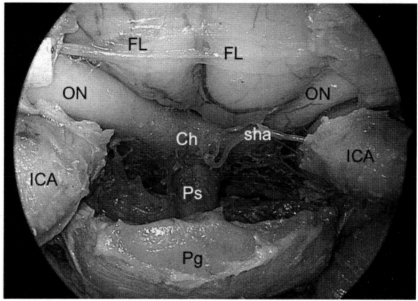

ICA: internal carotid artery; ON: optic nerve; Ch: optic chiasm; Ps: pituitary stalk; sha: superior hypophyseal artery; FL: frontal lobe.

Figure 2. Endoscopic endonasal anatomic intradural close-up view. After dural opening the chiasm, the pituitary stalk with the superior hypophyseal arteries appear at the center of the view in the suprasellar area.

Thereafter the internal debulking of the solid part and/or cystic evacuation is performed either with suction and radiofrequency monopolar wire electrodes (SurgiMax, Elliquence©, LLC., Baldiwin, NY) and/or CUSA devices. Once the internal decompression has been made, the surrounding arachnoid is dissected away from the tumor's capsule; during such maneuvers the pituitary stalk, the dorsal aspect of the pituitary gland, the chiasm and both the superior hypophyseal arteries need to be identified and respected (Figure 2). Continuous irrigation and blunt dissection with the instruments under a close-up view are required in order to avoid injuries of neurovascular structures and moreover the sacrifices of the small arterial perforators. Dissections thus goes on continuously taking care to respect arachnoidal plane; once it has been freed from any adherence the tumor is removed in a piecemeal fashion or "en block" (Figures 3 and 4). Finally, an accurate haemostasis is performed, since a clear, bloodless operating field is achieved.

In the final step of the procedure the operative field is inspected even with angled scopes and irrigated.

Closure

Due to frequent intraoperative cerebrospinal fluid (CSF) leakage resulting from wider dural opening, an accurate reconstruction of the skull base defect is mandatory after lesion removal. The reconstruction should ideally be watertight to prevent postoperative CSF leak, whose risk is higher in the endoscopic endonasal approach for craniopharyngiomas because of the large opening of the arachnoid cisterns and/or of the third ventricle. Furthermore, if dissection of the arachnoid is extended into the retrosellar area the Liliequist's membrane could be disrupted, with a large amount of CSF flowing off the basal cisterns.

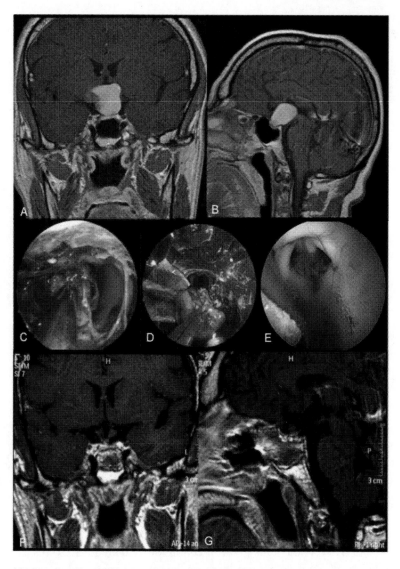

Figure 3. a) and b) Pre-operative MRI coronal and sagittal scans showing a case an intra-suprasellar craniopharyngioma, extending from the infundibular area into the third ventricle. Intraoperative pictures. c) and d) the tumor is dissected and removed in piecemeal fashion. e) After the removal of the lesion, the close-up endoscopic view (45° angled endoscope) allows the visualization inside the third ventricle. f) and g) Post-operative MRI coronal and sagittal scans showing the total removal of the lesion.

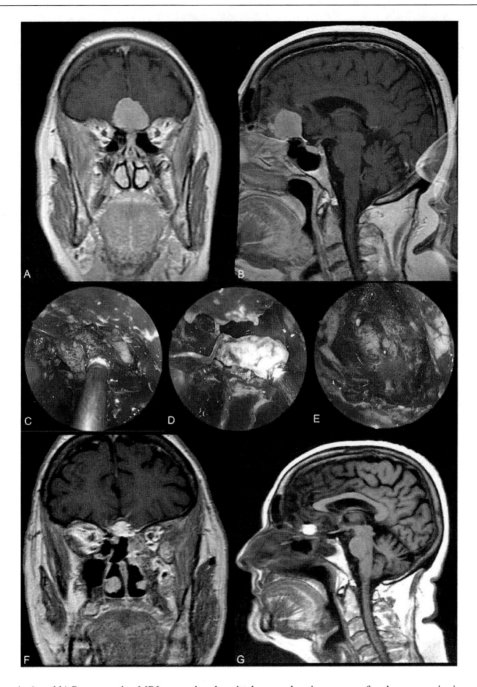

Figure 4. a) and b) Pre-operative MRI coronal and sagittal scans showing a case of a planum meningioma. c) and d) the tumor is dissected and removed in piecemeal fashion. e) The endoscopic exploration of the surgical cavity does not reveal any tumor remnant. f) and g) Post-operative MRI coronal and sagittal scans showing the total removal of the lesion with the autologous fat in the surgical cavity (hyperintensity) used for the reconstruction.

Conventional reconstruction techniques yield inadequate results for several reasons: i] the size of the defect; ii] the irregular shape of the defect, due to the close distance of the osteo-dural defect to the optic nerves and carotid arteries and, finally, iii] the wide intradural empty space lacking arachnoidal barriers.

Thus, the repair should proceed as for a Grade 3 of Kelly's classification [59], in order to achieve:

1. intradural sealing of the arachnoid;
2. watertight closure of the osteo-dural skull base defect;
3. packing of the sphenoid.

The closure is achieved with several layers of dural substitute positioned in the intradural space as first barrier; thereafter, autologous abdominal fat is introduced together with a small amount of fibrin glue (Tisseel®, Baxter, Vienna, Austria) to fill the dead space [60]. Subsequently, a further layer of dural substitute is positioned in the extradural space to cover the dural opening, being wedged under the bony margins.

Once a watertight barrier has been achieved, a mucosal flap – usually a free muchoperichondrium flap harvested from the middle turbinate or from the nasal septum or a vascularized Haddad pedicled flap [56, 57] – is used to cover the posterior wall of the sphenoid sinus. Fibrin glue is used to to hold the material in place. Lumbar drainage is positioned at the beginning of the procedure and left in place for 5-7 days.

At the end of the procedure, bipolar haemostasis is obtained over the border of sphenoidotomy and final irrigation of the nasal cavities is performed.

Advantages and Drawbacks of the Extended Transsphenoidal Approach

The endoscopic transtuberculum-transplanum approach provides a direct view of the neurovascular structures of the suprasellar region from below, avoiding any brain retraction and optic pathways manipulation; furthermore, this exposure is median and bilateral. This route minimizes the risk of postoperative visual loss, which is strictly related to the integrity of the vascularization of the optic chiasm. In fact, at this level, most of the blood supply to the optic system comes from the its superior surface, from the branches of the anterior cerebral and anterior communicating arteries, thus rendering the inferior approaches less dangerous than the transcranial routes [50, 51, 55, 61, 62].

Dividing the suprasellar region by two ideal planes, one passing through the inferior surface of the chiasm and the mamillary bodies and one, passing through the posterior edge of the chiasm and the dorsum sellae, four anatomical areas are attained: suprachiasmatic, infrachiasmatic, retrosellar and intraventricular [31]. In both anatomic dissections and clinical experiences we noticed that the endoscopic endonasal route allows a wider and better view of all these area except the suprachiasmatic, if compared to the offered by the transcranial approach. The endoscopic endonasal approach to the supradiaphragmatic compartment offers some advantages due to the properties of the endoscope itself. Such approach provides a wider, close-up view of the surgical field that permits a better identification of many surgical landmarks, either in the sphenoid sinus and even intradurally, in the suprasellar area, thus allowing a safer dissection of the tumor from the neurovascular structures without any brain retraction and optic apparatus manipulation. In these regards it is advisable, during lesion removal maneuvers, to not extend the dissection too laterally thus crossing the nerves and increasing the risk of their injuries; indeed, according to Kassam, "never cross the nerves" is one of the basic rule of this kind of surgery [63].

This direct approach from below offers also the possibility in lesions such as meningiomas, to perform an early devascularization, thus achieving an easier and almost bloodless tumor debulking. This technique indeed carries better cosmetic results associated with an increased patient comfort.

Despite these advantages, some limitations should be considered, when approaching a skull base lesion through such an approach: this procedure is more technically demanding and requires some additional skills because of the opposite anatomical point of view. Furthermore, it should be mentioned that some factors can affect the lesion dissection and removal, either related to the lesion and to the anatomy: while a well pneumatized sphenoid sinus allows a better visualization of all important landmarks of its posterior wall that have to be identified before starting the drilling of the medial optocarotid recesses and the removal of the tuberculum sellae, a conchal-type sphenoid sinus could be considered as an obstacle for the extended approach. Besides this, it has to be mentioned that a small sella could determine a narrower approach, due to closeness of the two intracavernous carotids [51, 64, 65].

Concerning lesion-related conditions, it has to be said that a size exceeding 2–2,5 cm and/or an eccentric shape of the tumor, the extension with involvement of the optic canal[s] and/or encasement of one or both ICAs and/or of the AcomA complex, could represent an obstacle for the effectiveness of the extended transsphenoidal approach.

Moreover, in case of lesion displacing posteriorly the optic chiasm, or since it is already post-fixed, tumor removal can be easier: indeed, the internal debulking of the mass gives enough space for the dissection maneuvers from the chiasm and from the optic nerves. Conversely, the transsphenoidal approach provides a more limited access to the suprasellar and supradiaphragmatic areas if the chiasm is pre-fixed or anteriorly displaced [50, 51].

Finally, it should be reminded that the problems concerning the bleeding control from main vessels in such a narrow space [53-55] and the higher risk of postoperative CSF leak, as compared with the transcranial approaches, are still a challenging matter. Nevertheless, improvements in closure techniques and the use of new materials and dedicated instruments seem to significantly reduce such risks [56, 57, 66, 67].

The endoscopic endonasal trans-tuberculum approach seems to be the less invasive and more direct route to the suprasellar area; therefore, it could be adopted in case of small and medium size suprasellar midline lesions, without or with a limited parasellar extension and without involvement of vascular structures, such as craniopharyngiomas [68-71], tuberculum sellae and planum meningiomas [40, 72-74], suprasellar Rathke's cleft cyst or large fibrotic pituitary adenomas, in which extracapsualar dissection is required [75]. This procedure can be proposed, as well, for the treatment of recurrent lesions, already operated on transcranially and/or of suprasellar lesions with an unclear preoperative diagnosis, with bioptic or partial removal finalities [76].

However, it has to be kept in mind that adequate endoscopic equipment, image-guidance systems, dedicated instruments and above all an increased experience with the endoscopic transsphenoidal techniques are essential skill for the realization of the extended transsphenoidal approach to the suprasellar area.

Conclusion

The endoscopic endonasal approach to the suprasellar area, via a trastuberculum transplanum route, offers an excellent, wide exposure of this area, requiring a wide bone removal of the upper half of the sella turcica, along with the tuberculum sellae and the posterior portion of the planum sphenoidale. The main limits of this technique are the inner features of lesions dealt with (relatively big size, asymmetry, encasement of main vessels, invasion of one or both the optic canals), the anatomical configuration (pneumatization of sphenoid sinus, sellar size), and the need of proper endoscopic technical skills. Furthermore, the reconstruction of the osteodural defect can be particularly troublesome. Pathologies arising or extending in these regions – once approachable only with the more invasive transcranial surgery - such as craniopharyngiomas, tuberculum sellae meningiomas, macroadenomas involving the suprasellar space are candidate to be removed via the endonasal route; however, patients should be carefully evaluated and indications are to be tailored to the case, the surgeon's confidence and the anatomy.

References

[1] Hardy J. Transphenoidal microsurgery of the normal and pathological pituitary. *Clin. Neurosurg.* 1969;16:185-217.

[2] Cappabianca P, Cavallo LM, de Divitiis E. Endoscopic endonasal transsphenoidal surgery. *Neurosurgery.* 2004 Oct;55(4):933-40; discussion 40-41.

[3] de Divitiis E, Cappabianca P, Cavallo LM. Endoscopic endonasal transsphenoidal approach to the sellar region. In: de Divitiis E, Cappabianca P, editors. *Endoscopic Endonasal Transsphenoidal Surgery.* Wien - New York: Springer; 2003. p. 91-130.

[4] Jho HD, Carrau RL, Ko Y. Endoscopic pituitary surgery. In: Wilkins H, Rengachary S, editors. *Neurosurgical Operative Atlas.* Park Ridge, IL: American Association of *Neurological Surgeons*; 1996. p. 1-12.

[5] Carrau RL, Jho HD, Ko Y. Transnasal-transsphenoidal endoscopic surgery of the pituitary gland. *Laryngoscope.* 1996 Jul;106(7):914-8.

[6] Laws ER, Lanzino G, editors. *Transsphenoidal surgery.* Philadelphia: Saunders - Elsevier; 2010.

[7] Laws ER, Jr., Thapar K. Pituitary surgery. *Endocrinol Metab Clin North Am.* 1999 Mar;28(1):119-31.

[8] Laws ERJ. Transsphenoidal surgery. In: Apuzzo MLJ, editor. *Brain surgery: Complications Avoidance and Management.* New York: Churchill Livingstone; 1993. p. 357-62.

[9] Elias WJ, Laws ER. Transsphenoidal approach to lesion of the sella. In: Schmideck HH, editor. *Schmideck & Sweet Operative neurosurgical technique Indications, methods and results.* 4° ed. Philadelphia: WB Saunders; 2000. p. 373-84.

[10] Kennedy DW, Cohn ES, Papel ID, Holliday MJ. Transsphenoidal approach to the sella: the Johns Hopkins experience. *Laryngoscope.* 1984 Aug;94(8):1066-74.

[11] Artico M, Pastore FS, Fraioli B, Giuffre R. The contribution of Davide Giordano (1864-1954) to pituitary surgery: the transglabellar-nasal approach. *Neurosurgery.* 1998 Apr;42(4):909-11; discussion 11-12.

[12] Liu JK, Das K, Weiss MH, Laws ER, Jr., Couldwell WT. The history and evolution of transsphenoidal surgery. *J. Neurosurg.* 2001 Dec;95(6):1083-96.

[13] Kanter AS, Dumont AS, Asthagiri AR, Oskouian RJ, Jane JA, Jr., Laws ER, Jr. The transsphenoidal approach. A historical perspective. *Neurosurg Focus.* 2005 Apr 15;18(4):e6.

[14] Lanzino G, Laws ER, Jr. Pioneers in the development of transsphenoidal surgery: Theodor Kocher, Oskar Hirsch, and Norman Dott. *J. Neurosurg.* 2001 Dec;95(6):1097-103.

[15] Halstead AE. Remarks on the operative treatment of tumors of the hypophysis, with report of two cases operated on by an oronasal method. *Trans Am. Surg Assoc.* 1910;28:73-93.

[16] Cushing H. The Pituitary body and its disorders: clinical states produced by disorders of the Hypophysis Cerebri. Philadelphia: *J.B. Lippincott*; 1912. p. 296-305.

[17] Guiot G. Transsphenoidal approach in surgical treatment of pituitary adenomas: general principles and indications in non-functioning adenomas. In: Kohler PO, Ross GT, editors. *Diagnosis and treatment of pituitary adenomas.* Amsterdam: Excerpta Medica; 1973. p. 159-78.

[18] Guiot G, Rougerie J, Fourestier M, Fournier A, Comoy C, Voulmiere J, et al. *intracranial endoscopic explorations* [in French]. Presse Med. 1963;71:1225-8.

[19] Dandie GDC, Pell MF, Atias MD. Endoscopic transsphenoidal approach to the pituitary fossa: technical note. *J. Clin. Neuroscience.* 1996;3(1):65-8.

[20] Gamea A, Fathi M, el-Guindy A. The use of the rigid endoscope in trans-sphenoidal pituitary surgery. *J. Laryngol Otol.* 1994 Jan;108(1):19-22.

[21] Jankowski R, Auque J, Simon C, Marchal JC, Hepner H, Wayoff M. Endoscopic pituitary tumor surgery. *Laryngoscope.* 1992 Feb;102(2):198-202.

[22] Heilman CB, Shucart WA, Rebeiz EE. Endoscopic sphenoidotomy approach to the sella. *Neurosurgery.* 1997 Sep;41(3):602-7.

[23] Yaniv E, Rappaport ZH. Endoscopic transseptal transsphenoidal surgery for pituitary tumors. *Neurosurgery.* 1997 May;40(5):944-6.

[24] Apuzzo ML, Heifetz MD, Weiss MH, Kurze T. Neurosurgical endoscopy using the side-viewing telescope. *J. Neurosurg.* 1977 Mar;46(3):398-400.

[25] Kennedy DW. Functional endoscopic sinus surgery. Technique. *Arch Otolaryngol.* 1985 Oct;111(10):643-9.

[26] Stammberger H, Posawetz W. Functional endoscopic sinus surgery. Concept, indications and results of the Messerklinger technique. *Eur. Arch. Otorhinolaryngol.* 1990;247(2):63-76.

[27] Stubbs WK. Functional endoscopic sinus surgery. *J. Fla Med Assoc.* 1989 Feb;76(2):245-8.

[28] Cappabianca P, Alfieri A, de Divitiis E. Endoscopic endonasal transsphenoidal approach to the sella: towards functional endoscopic pituitary surgery (FEPS). *Minim Invasive Neurosurg.* 1998 Jun;41(2):66-73.

[29] Tabaee A, Anand VK, Barron Y, Hiltzik DH, Brown SM, Kacker A, et al. Endoscopic pituitary surgery: a systematic review and meta-analysis. *J. Neurosurg.* 2009 Sep;111(3):545-54.
[30] de Divitiis E, Cappabianca P, Cavallo LM. Endoscopic transsphenoidal approach: adaptability of the procedure to different sellar lesions. *Neurosurgery.* 2002 Sep;51(3):699-705; discussion -7.
[31] Cavallo LM, de Divitiis O, Aydin S, Messina A, Esposito F, Iaconetta G, et al. Extended endoscopic endonasal transsphenoidal approach to the suprasellar area: anatomic considerations - part 1. *Neurosurgery.* 2007;61:ONS-24-ONS-34.
[32] Frank G, Pasquini E. Approach to the cavernous sinus. In: de Divitiis E, Cappabianca P, editors. *Endoscopic endonasal transsphenoidal surgery.* Wien NewYork: Springer-Verlag; 2003. p. 159-75.
[33] Jho HD, Ha HG. Endoscopic endonasal skull base surgery: Part 1--The midline anterior fossa skull base. *Minim Invasive Neurosurg.* 2004 Feb;47(1):1-8.
[34] Jho HD, Ha HG. Endoscopic endonasal skull base surgery: Part 2--The cavernous sinus. *Minim Invasive Neurosurg.* 2004 Feb;47(1):9-15.
[35] Jho HD, Ha HG. Endoscopic endonasal skull base surgery: Part 3--The clivus and posterior fossa. *Minim Invasive Neurosurg.* 2004 Feb;47(1):16-23.
[36] Kassam A, Snyderman CH, Mintz A, Gardner P, Carrau RL. Expanded endonasal approach: the rostrocaudal axis. Part II. Posterior clinoids to the foramen magnum. *Neurosurg Focus.* 2005 Jul 15;19(1):E4.
[37] Kassam A, Snyderman CH, Mintz A, Gardner P, Carrau RL. Expanded endonasal approach: the rostrocaudal axis. Part I. Crista galli to the sella turcica. *Neurosurg Focus.* 2005 Jul 15;19(1):E3:1-12.
[38] Cappabianca P, Frank G, Pasquini E, de Divitiis O, Calbucci F. Extended endoscopic endonasal transsphenoidal approaches to the suprasellar region, planum sphenoidale and clivus. In: de Divitiis E, Cappabianca P, editors. *Endoscopic endonasal transsphenoidal surgery.* Wien - New York: Springer; 2003. p. 176-87.
[39] Couldwell WT, Weiss MH, Rabb C, Liu JK, Apfelbaum RI, Fukushima T. Variations on the standard transsphenoidal approach to the sellar region, with emphasis on the extended approaches and parasellar approaches: surgical experience in 105 cases. *Neurosurgery.* 2004 Sep;55(3):539-50.
[40] Laufer I, Anand VK, Schwartz TH. Endoscopic, endonasal extended transsphenoidal, transplanum transtuberculum approach for resection of suprasellar lesions. *J. Neurosurg.* 2007 Mar;106(3):400-6.
[41] Weiss MH. The transnasal transsphenoidal approach. In: Apuzzo MLJ, editor. *Surgery of the third ventricle.* Baltimore: Williams & Wilkins; 1987. p. 476-94.
[42] Mason RB, Nieman LK, Doppman JL, Oldfield EH. Selective excision of adenomas originating in or extending into the pituitary stalk with preservation of pituitary function. *J. Neurosurg.* 1997 Sep;87(3):343-51.
[43] Kouri JG, Chen MY, Watson JC, Oldfield EH. Resection of suprasellar tumors by using a modified transsphenoidal approach. Report of four cases. *J. Neurosurg.* 2000 Jun;92(6):1028-35.
[44] Kim J, Choe I, Bak K, Kim C, Kim N, Jang Y. Transsphenoidal supradiaphragmatic intradural approach: technical note. Minim Invasive Neurosurg. 2000 Mar;43(1):33-7.

[45] Kitano M, Taneda M. Extended transsphenoidal approach with submucosal posterior ethmoidectomy for parasellar tumors. Technical note. *J. Neurosurg.* 2001 Jun;94(6):999-1004.

[46] Kato T, Sawamura Y, Abe H, Nagashima M. Transsphenoidal-transtuberculum sellae approach for supradiaphragmatic tumours: technical note. *Acta Neurochir* (Wien). 1998;140(7):715-9.

[47] Laws ER, Kanter AS, Jane JA, Jr., Dumont AS. Extended transsphenoidal approach. *J. Neurosurg.* 2005 May;102(5):825-7; discussion 7-8.

[48] Cook SW, Smith Z, Kelly DF. Endonasal transsphenoidal removal of tuberculum sellae meningiomas: technical note. Neurosurgery. 2004 Jul;55(1):239-44.

[49] Dusick JR, Esposito F, Kelly DF, Cohan P, DeSalles A, Becker DP, et al. The extended direct endonasal transsphenoidal approach for nonadenomatous suprasellar tumors. *J. Neurosurg.* 2005 May;102(5):832-41.

[50] Cappabianca P, Cavallo LM, Esposito I, Solari D. Sellar-Tuberculum approach. In: Kassam AB, Gardner P, editors. *Endoscopic Approaches to the skull base*. Pittsburgh: Karger; 2012. p. 41-59.

[51] de Divitiis E, Cavallo LM, Cappabianca P, Esposito F. *Extended endoscopic endonasal transsphenoidal approach for the removal of suprasellar tumors*: Part 2. Neurosurgery. 2007 Jan;60(1):46-58; discussion -9.

[52] de Notaris M, Solari D, Cavallo LM, D'Enza AI, Ensenat J, Berenguer J, et al. The "suprasellar notch," or the tuberculum sellae as seen from below: definition, features, and clinical implications from an endoscopic endonasal perspective. *J. Neurosurg.* 2012 Mar;116(3):622-9.

[53] Laws ER, Kern EB. Complications of transsphenoidal surgery. In: Laws ER, Randall RV, Kern EB, Abboud CF, editors. *Management of pituitary adenomas and related lesions with emphasis on transsphenoidal microsurgery*. New York: Appleton-Century-Crofts; 1982. p. 329-46.

[54] Cappabianca P, Cavallo LM, Colao A, de Divitiis E. Surgical complications associated with the endoscopic endonasal transsphenoidal approach for pituitary adenomas. *J. Neurosurg.* 2002 Aug;97(2):293-8.

[55] Kassam AB, Prevedello DM, Carrau RL, Snyderman CH, Thomas A, Gardner P, et al. Endoscopic endonasal skull base surgery: analysis of complications in the authors' initial 800 patients. *J. Neurosurg.* 2011 Jun;114(6):1544-68.

[56] Hadad G, Bassagasteguy L, Carrau RL, Mataza JC, Kassam A, Snyderman CH, et al. A novel reconstructive technique after endoscopic expanded endonasal approaches: vascular pedicle nasoseptal flap. *Laryngoscope.* 2006 Oct;116(10):1882-6.

[57] Kassam AB, Thomas A, Carrau RL, Snyderman CH, Vescan A, Prevedello D, et al. Endoscopic reconstruction of the cranial base using a pedicled nasoseptal flap. *Neurosurgery.* 2008 Jul;63(1 Suppl 1):ONS44-ONS52; discussion ONS-ONS3.

[58] Castelnuovo P, Pistochini A, Locatelli D. Different surgical approaches to the sellar region: focusing on the "two nostrils four hands technique". *Rhinology.* 2006 Mar;44(1):2-7.

[59] Esposito F, Dusick JR, Fatemi N, Kelly DF. Graded repair of cranial base defects and cerebrospinal fluid leaks in transsphenoidal surgery. *Neurosurgery.* 2007 Apr;60(4 Suppl 2):295-303; discussion -4.

[60] Cappabianca P, Esposito F, Magro F, Cavallo LM, Solari D, Stella L, et al. Natura abhorret a vacuo--use of fibrin glue as a filler and sealant in neurosurgical "dead spaces". Technical note. *Acta Neurochir* (Wien). 2010 May;152(5):897-904.

[61] Frank G, Pasquini E, Mazzatenta D. Extended transsphenoidal approach. *J. Neurosurg.* 2001 Nov;95(5):917-8.

[62] Dehdashti AR, Ganna A, Witterick I, Gentili F. Expanded endoscopic endonasal approach for anterior cranial base and suprasellar lesions: indications and limitations. *Neurosurgery.* 2009 Apr;64(4):677-87; discussion 87-89.

[63] Kassam AB, Vescan AD, Carrau RL, Prevedello DM, Gardner P, Mintz AH, et al. Expanded endonasal approach: vidian canal as a landmark to the petrous internal carotid artery. *J. Neurosurg.* 2008 Jan;108(1):177-83.

[64] Fernandez-Miranda JC, Prevedello DM, Madhok R, Morera V, Barges-Coll J, Reineman K, et al. Sphenoid septations and their relationship with internal carotid arteries: anatomical and radiological study. *Laryngoscope.* 2009 Oct;119(10):1893-6.

[65] Gardner PA, Kassam AB, Rothfus WE, Snyderman CH, Carrau RL. Preoperative and intraoperative imaging for endoscopic endonasal approaches to the skull base. *Otolaryngol Clin North Am.* 2008 Feb;41(1):215-30, vii.

[66] Cavallo LM, Messina A, Esposito F, de Divitiis O, Dal Fabbro M, de Divitiis E, et al. Skull base reconstruction in the extended endoscopic transsphenoidal approach for suprasellar lesions. *J. Neurosurg.* 2007 Oct;107(4):713-20.

[67] Leng LZ, Brown S, Anand VK, Schwartz TH. "Gasket-seal" watertight closure in minimal-access endoscopic cranial base surgery. *Neurosurgery.* 2008 May;62(5 Suppl 2):ONSE342-3; discussion ONSE3.

[68] Kassam AB, Gardner PA, Snyderman CH, Carrau RL, Mintz AH, Prevedello DM. Expanded endonasal approach, a fully endoscopic transnasal approach for the resection of midline suprasellar craniopharyngiomas: a new classification based on the infundibulum. *J. Neurosurg.* 2008 Apr;108(4):715-28.

[69] Cavallo LM, Prevedello DM, Solari D, Gardner PA, Esposito F, Snyderman CH, et al. Extended endoscopic endonasal transsphenoidal approach for residual or recurrent craniopharyngiomas. *J. Neurosurg.* 2009 Sep;111(3):578-89.

[70] Gardner PA, Kassam AB, Snyderman CH, Carrau RL, Mintz AH, Grahovac S, et al. Outcomes following endoscopic, expanded endonasal resection of suprasellar craniopharyngiomas: a case series. *J. Neurosurg.* 2008 Jul;109(1):6-16.

[71] de Divitiis E, Cappabianca P, Cavallo LM, Esposito F, de Divitiis O, Messina A. Extended endoscopic transsphenoidal approach for extrasellar craniopharyngiomas. *Neurosurgery.* 2007 Nov;61(5 Suppl 2):219-27; discussion 28.

[72] Gardner PA, Kassam AB, Thomas A, Snyderman CH, Carrau RL, Mintz AH, et al. Endoscopic endonasal resection of anterior cranial base meningiomas. *Neurosurgery.* 2008 Jul;63(1):36-52; discussion -4.

[73] Ceylan S, Koc K, Anik I. Extended endoscopic approaches for midline skull-base lesions. *Neurosurg Rev.* 2009 Jul;32(3):309-19; discussion 18-9.

[74] de Divitiis E, Cavallo LM, Esposito F, Stella L, Messina A. Extended endoscopic transsphenoidal approach for tuberculum sellae meningiomas. *Neurosurgery.* 2007 Nov;61(5 Suppl 2):229-37; discussion 37-8.

[75] Di Maio S, Cavallo LM, Esposito F, Stagno V, Corriero OV, Cappabianca P. Extended endoscopic endonasal approach for selected pituitary adenomas: early experience. *J. Neurosurg.* 2011 Feb;114(2):345-53.

[76] Tabaee A, Nyquist G, Anand VK, Singh A, Kacker A, Schwartz TH. Palliative endoscopic surgery in advanced sinonasal and anterior skull base neoplasms. *Otolaryngol Head Neck Surg.* Jan;142(1):126-8.

In: Minimally Invasive Skull Base Surgery
Editor: Moncef Berhouma

ISBN: 978-1-62808-567-9
© 2013 Nova Science Publishers, Inc.

Chapter VIII

Craniopharyngiomas: The Endonasal Endoscopic Approach

Daniel M. S. Raper, Robert M. Starke, John A. Jane and Ricardo J. Komotar[*]

University of Virginia, Charlottesville, US

Abstract

The endoscopic, endonasal approach for craniopharyngiomas represents an alternative to more traditional open transcranial approaches. Even for lesions with significant suprasellar extension, an extended endonasal approach may offer the potential for complete resection with reduction in operative morbidity. In this chapter, the various surgical approaches for craniopharyngiomas are reviewed. The endonasal endoscopic approach is outlined, with particular attention to the extended intracranial endoscopic techniques necessary for resection of suprasellar craniopharyngiomas. Newer techniques of dural closure to reduce the incidence of postoperative CSF leak are also reviewed. A higher rate of gross total resection and improved visual outcomes may also be achievable utilizing an endonasal approach for carefully selected patients; however, the risk of postoperative CSF leak remains a significant drawback of the endonasal extended approaches. The endonasal endoscopic approach is a safe and effective alternative for the treatment of select craniopharyngiomas. Larger lesions with more lateral extension may be more suitable for an open approach. In all cases, an individualized treatment plan devised through experienced multidisciplinary collaboration is essential for the success of a minimally invasive, endonasal endoscopic approach.

---

[*] Corresponding author: E-mail address: rms6bx@virginia.edu.

Introduction

Craniopharyngiomas are benign epithelial neoplasms that originate from the embryonic craniopharyngeal duct, or Rathke's pouch [1]. Craniopharyngiomas are often associated with cysts, which may be large and project in either a superior-inferior or medial-lateral direction and may grow to be larger than the solid component of the tumor. Despite their benign histological appearance, craniopharyngiomas arise in close proximity to vital CNS structures such as the optic nerves, pituitary stalk and chiasm, and major blood vessels such as the internal carotid artery. As such, they may present with signs and symptoms of local compression, mass effect, or global increase in intracranial pressure. Endocrinopathies are a common finding and are due to compression of normal pituitary tissue. The optimal management paradigm for craniopharyngiomas incorporates maximal safe surgical resection often followed by adjuvant radiotherapy [2, 3]. Initial gross total resection has been correlated with increased long-term survival [4].

A variety of surgical approaches have been utilized in the treatment of craniopharyngiomas. These lesions are primarily midline, often project through the suprasellar cisterns towards the hypothalamus, and may even extend into the third ventricle. Open skull base techniques, most commonly accessed via a pterional craniotomy, require tumor resection through small windows between the optic nerves and carotid artery, and do not allow good visualization of contralateral structures which may be involved by the pathology. A ventral, midline approach is logical, because it provides direct access to the tumor without brain retraction. For intrasellar lesions, this may account for the observation that transsphenoidal surgery can result in lower morbidity than open transcranial approaches [5]. Craniopharyngiomas, however, often present with significant suprasellar extension, and gaining complete resection of these lesions via a transsphenoidal microscopic approach can be difficult, due to the fall-off in illumination from the operative microscope at extended working distances. The endoscope provides a potential solution for this problem by transposing the light source closer to the point of interest and enlarging the field of view via angled lenses which allows dissection of critical structures from the tumor capsule under direct visualization [6].

Extended Endonasal Endoscopic Technique

The extended endonasal endoscopic technique proceeds from the philosophy that 'in approaching a midline skull base tumor, no cranial nerve or vessel should be traversed' – theoretically, this provides the safest possible approach for preservation of these vital neurovascular structures. The technique is performed under general anesthesia often after placement of a lumbar drain. Many surgeons prefer the patient to be supine with slight Trendelenburg positioning. For craniopharyngiomas, a bimanual, bilateral technique is utilized.

The first stage of the procedure involves ipsilateral lateralization of the middle turbinate and partial posterior septectomy, and lateralization of the contralateral middle turbinate. If the middle turbinate is removed, particularly in pediatric patients with small nasal cavities, it is first removed using an incision near the head of the turbinate. Complete removal of the tail is not needed as it does not interfere with the trajectory, and removal increases risk of bleeding from the sphenopalatine artery. For smaller tumors that are primarily within the sella with only moderate suprasellar extension, the posterior part of the nasal septum is then removed, depending on the desired trajectory and working space. Removal of the posterior septum creates a larger cavity through which the binarinal technique with greater ease of instrument manipulation. For larger tumors, or those that require removal of a portion of the tuberculum sellae and planum sphenoidale, a vascularized nasoseptal flap is harvested with care to preserve the branches of the sphenopalatine artery. Then, the contralateral middle turbinate is lateralized to allow greater bilateral exposure.

The wide anterior sphenoidectomy is performed with the aim of exposing the sella, opticocarotid recesses, and planum sphenoidale. This part of the procedure is slightly modified from a standard sellar exposure in order to provide access to the suprasellar cistern and pathology in this region. The inferior one-third of the superior turbinate is removed to expose the posterior ethmoid air cells either bilaterally. This maximizes the lateral exposure and provides ready visualization of the optico-carotid recesses. Then the anterior sphenoid sinus wall is enlarged, all intersphenoid septae are removed, and the planum flattened with a drill. Tuberculum sellae is thinned with a 2mm diamond burr, and the superior part of the anterior wall is removed, exposing the intercavernous sinus. The tuberculum is then mobilized from the medial optico-carotid recesses and from the planum sphenoidale and removed. The planum sphenoidale is removed using Kerrison's rongeur up to the falciform ligament, which forms a landmark for the anterior extent of the bony and dural opening [7]. Prior to bone removal, it is essential to develop a precise understanding of the three-dimensional anatomy of the areas deep to the surgical field. Micro-Doppler may be useful in mapping out the location of the carotid arteries.

The dura is opened in linear fashion horizontally on either side of the superior intercavernous sinus. After adequate bipolar cauterization, the dura may be opened across the sinus. This sinus may consist of a plexus of fine veins, rather than a single channel, and bleed profusely. A combination of hemostatic agents, Floseal, bipolar cautery and cottonoids may be required. The resultant dural flaps are reflected laterally, before enlarging the dural opening to the required size. The edges of the dura are coagulated with bipolar cautery to cause retraction, but care must be taken not to coagulate too far lateral due to the proximity of the optic nerves and carotid arteries.

Once intradural access has been achieved, dissection through the Liliequist membrane is performed with care to preserve the superior hypophyseal arteries bilaterally. These arteries often must be sharply dissected free from the tumor capsule. Dissection of the tumor capsule may be proceeded either below the optic nerves and chiasm. If proceeding below the optic nerves, further extension upwards into the third ventricle or downward towards the interpeduncular cistern is possible, and either or both may be needed in the complete resection of some craniopharyngiomas. Dissection of tumor from the optic chiasm, pituitary stalk, and carotid branches requires similar microsurgical techniques to those required for open microneurosurgery. Following complete or subtotal tumor removal, the surgical field is inspected with a combination of angled or a multi-angled rotational endoscopes, and meticulous hemostasis is achieved.

Reconstruction techniques continue to evolve, but are based on the principles of establishing an ultimate watertight closure with multiple layers of inlay and/or onlay graft, often bolstered by rigid struts and fascia lata. A complete discussion of the various closure techniques following endoscopic skull base surgery is beyond the scope of this chapter, but an important and relatively recent development has been the incorporation of vascularized nasoseptal flaps, which have been quite successful in more recent series of skull base lesions treated endoscopically (see below).

The extended transsphenoidal technique was first introduced in 1987 by Martin Weiss who described the transspheoidal transplanar approach. Although surgeons had described the transsphenoidal approach for craniopharyngiomas initially as a tool for creating a drainage tract for craniopharyngioma cysts into the sphenoid sinus. As experience with the endoscopic technique increased, it began to be used in select cases as a primary approach, with gross total resection reported in a majority of patients in small series [9]. An endonasal approach was found to be especially useful for retroinfundibular lesions as it allows transposition of the pituitary gland and stalk [10]. The endonasal technique may be associated with lower rates of seizure, bone flap infection and procedure-related visual deterioration. However, the major limitation of early extended endonasal techniques remained postoperative CSF leak. Early series experienced CSF leak in 20-58% of cases [11-14]. Intraoperative entry into the third ventricle is one risk factor for postoperative CSF leak [12], but it may occur after any extended ventral approach for craniopharyngioma. In our review of the current literature the rate of this complication was significantly higher in the endoscopic group then the open transcranial group (18.4% vs. 2.6%, p<0.003) [15]. As a way of addressing this risk, reconstruction techniques have evolved from simple fascial or bone overlays to more sophisticated, multilayer closures incorporating vascularized septal flaps. Newer series, which incorporate these multilayer closures, report fewer CSF leaks; as low as 0-4% have been observed [16-20]. The major development, which should be incorporated in every closure of an extended endoscopic craniopharyngioma resection, has been the inclusion of a vascularized nasoseptal flap. A consistent postoperative CSF leak rate of less than 5%, though achievable, requires considerable experience, as well as a constant evolution of closure technique.

A steep learning curve is observed in initial experience with the endoscopic technique. Effective control of bleeding using this technique can be difficult, and proved a limitation in early experience with the technique. Due to the technical difficult of extended endoscopic approaches and the risk of CSF leak, lesions that encase vascular structures or extend far into the ventricles are still not thought to be primary candidates for a minimally invasive approach [7, 17, 21]. Pre-operative assessment includes determination of the goal of surgery (gross total vs. subtotal resection followed by adjuvant radiation therapy). It is beneficial to have two surgeons with familiarity with the skull base and/or endoscopic anatomy of the nasal cavity and sphenoid sinus to be involved in case [11].

Comparison with Transsphenoidal Microscopic and Open Transcranial Approaches

The choice of approach for resection of craniopharyngiomas must incorporate assessment of the anatomical location, relationship to neurovascular structures, projection of the lesion, location/extent of any cystic component, and considerations related to previous surgical resection. Structures that particularly influence the choice of approach include the sella turcica, optic chiasm, and third ventricle.

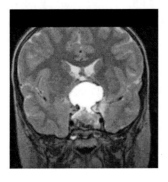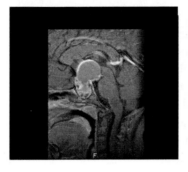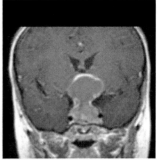

Figure 1. A (Coronal T2), 1B (Sagital T1), and 1C (Coronal T1) MRI and of a 10 year old boy presenting with short stature. Laboratory examination demonstrated panhypopituitary and nero-ophthalmic examination demonstrated a left APD, a superior temporal defect (OS) and a temporal defect (OD).

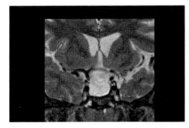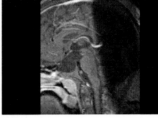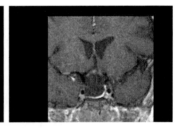

Figure 2. A (Coronal T2), 2B (Sagital T1), and 2C (Sagital T1) post-operative MRI following endoscopic endonasal resection of the craniopharyngioma. The patient experienced improvement in neuro-ophthalmic examination.

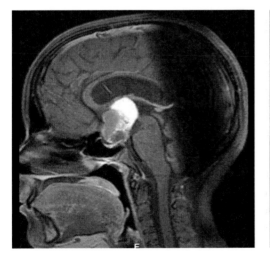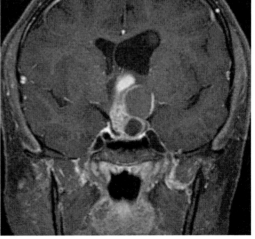

Figure 3. A (Sagital) and 3B (Coronal) MRI of a craniopharyngioma without an expanded sella turcica. This indicated a supradiaphragmatic origin. The craniopharyngioma is more likely adherent and invading the third ventricular floor. There is also often a smaller interval between optic chiasm and pituitary. This requires a transsphenoidal transplanum sphenoidale approach.

Transsphenoidal Microscopic Approach

The transsphenoidal microscopic approach for craniopharyngiomas offers many of the theoretical anatomical advantages of the endoscopic endonasal approach, with some additional benefits and limitations. The operative microscope is a familiar tool to cranial and skull base neurosurgeons, and the transsphenoidal approach is routinely used in pituitary tumor resection. Prior to the development of endoscopic techniques, however, these approaches were limited by the technical difficulty of operating far from the light source (microscope). Traditionally, the transsphenoidal microscopic approach was primarily used for intrasellar craniopharyngiomas or lesions with sellar enlargement [22-25]. Compared with open approaches, transsphenoidal approaches allow resection of subdiaphragmatic lesions with better preservation of visual function, and may be associated with better preservation of anterior pituitary function [26]. Primarily cystic lesions are also able to be addressed with transsphenoidal removal. For suprasellar lesions, the technique may be modified by takedown of the posterior planum sphenoidale [27], though there is a higher risk of CSF leak. Overall, the transsphenoidal microscopic approach is most appropriate for intrasellar and subdiaphragmatic lesions.

Subfrontal Approach

The subfrontal approach incorporates a low frontal (may be minimal) craniotomy, with direct access to tumor at the anterior and middle skull base and straightforward access to the third ventricle. It allows visualization of the neurovascular structures on both sides, and is useful for dissection of tumor anterior to the chiasm. This technique has been shown to be effective for a majority of craniopharyngiomas, but may be a more technically challenging skull base approach, particularly in cases of pre-fixed chiasm. There is a risk of damage to the olfactory tract, optic nerves and breach of the frontal sinus.

Anterolateral Approaches

The pterional craniotomy has been traditionally used to access a wide variety of skull base pathology. It has the advantage of being a widely known and familiar surgical route for all neurosurgeons, and provides satisfactory access to the sellar region. However, this approach requires tumor resection in and around the optic nerves, chiasm and carotid artery via the interoptic space, opticocarotid triangle or caroticosylvian aperture. Surgical manipulation in and around these areas carries significant surgical risk. The approach also provides poor visualization of contralateral neurovascular structures, which may be involved in the tumor, and posterior third ventricle, which lies behind the hypothalamus when approached from the side. Additionally, depending on their size, craniopharyngiomas may distort the apposing vessels and nerves, distorting the expected surgical anatomy when approaching from a pterional or transcallosal route. Variations of the traditional pterional approach have been used for lesions with extension to the hypothalamus [28], but higher rates of CSF leak and other operative morbidity have been used as justification for more conservative management for these difficult lesions [29]. An orbitozygomatic or modified orbitozygomatic approach, with removal of the supraorbital rim and/or zygomatic arch, may be useful in some suprasellar lesions, as it allows greater inferior-to-superior working space. Anterior petrosal approaches provide additional options for large retrochiasmatic lesions and in select lesions the sigmoid sinus may be mobilized medially for wide visualization. However, the increased operative morbidity associated with this approach warrants careful consideration.

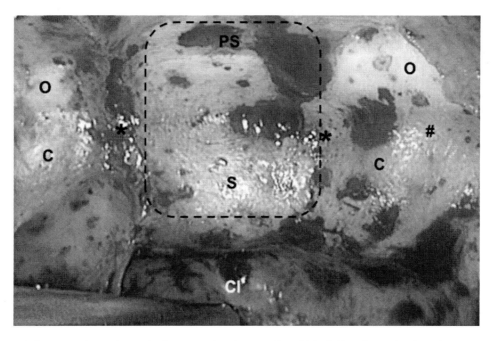

Figure 4. Intraoperative endoscopic view notes the planum sphenoidale (PS), sella turcica (S), clinoid (Cl), carotid arteries (C), and optic nerves (O).

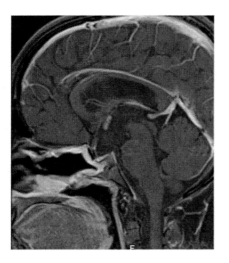 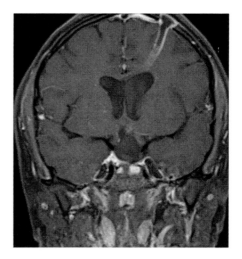

Figure 5. A (Sagital) and 5B (Coronal) MRI following endoscopic endonasal transsphenoidal transplanum sphenoidale resection.

Transcallosal Approach

A transcallosal approach may be used (or combined with other approaches) for the resection of intraventricular craniopharyngiomas. Though primarily intraventricular craniopharyngiomas are relatively rare, this approach provides a direct route to the pathology with minimal brain retraction. There is, however, a risk of callosal and forniceal injury, as well as postoperative seizures. Alternatives for these craniopharyngiomas include pterional or subfrontal, trans-lamina terminalis approaches, which risk optic pathway damage and carry the same risk as other anterolateral approaches.

Other and Combined Approaches

Most combined approaches incorporate pterional craniotomy with adjunctive transcranial or transsphenoidal approaches for resection of large or extensive tumors with significant superior, lateral or intraventricular extension. One major predictor of recurrence following craniopharyngioma resection is the presence of intrasellar residual tumor. Ventral midline approaches may be a useful alternative or adjunct to open transcranial approaches in cases of extensive tumor with superior and inferior extension. Up to 50% of recurrences following open resection occur in the pituitary fossa, which is not exposed in basal transcranial approaches [30]. Essentially, combined approaches are most often used for larger lesions that would not be appropriate for a primary endoscopic or other ventral midline approach.

Systematic Review

In our review of 88 published studies in the modern literature (1995-2010) that reported outcomes for craniopharyngioma resection [15], gross total resection was achieved in 48.3%, 69.1% and 66.9% of patients treated with open transcranial, transsphenoidal microscopic and endonasal endoscopic approaches respectively. Average lesion size was smaller in the endoscopic and transsphenoidal groups than the open transcranial group. Complications encountered in the open group included hydrocephalus (10%), seizure (8.5%), stroke (2.9%), CSF leak (2.6%) and meningitis (2.3%). Postoperative CSF leak was observed following transsphenoidal microscopic surgery in 9% of cases, with meningitis occurring in 1.8%. For the endonasal endoscopic cohort, CSF leak occurred in 18.4%, with meningitis in 5.1% of cases. There was a higher rate of improved visual outcome postoperatively in patients treated endoscopically vs. those with an open approach, but also a higher rate of postoperative CSF leak. Though interpretation of these results is limited by potential biases inherent in analysis of retrospective data, the literature overall supports the use of endoscopic endonasal technique as a safe and effective alternative for appropriate craniopharyngiomas.

There is likely a bias, on the part of early adopters of the endoscopic technique for craniopharyngiomas, to apply it first to lesions in which it would be teleologically ideal – that is, for small, midline lesions with limited superior extension. Though this may have contributed to the favorable results observed in case series of endoscopic craniopharyngioma surgery, it is an instructive in identifying those lesions most appropriate for endoscopic surgery. For lesions that do not extend down to the floor of the pituitary fossa, a transsphenoidal or endonasal approach risks further damage to potentially functional pituitary tissue during the approach. For lesions that extend far lateral to the lamina papyracea, endoscopic surgery may not be able to achieve satisfactory rates of complete tumor removal at the lateral extent of the lesion. One important determinant of outcome in craniopharyngioma surgery is extent of resection at initial surgery [31]. As a newer surgical technique, the endonasal endoscopic approach has been employed often in patients with residual tumor after a previous subtotal open resection [15]. This may affect the observed tumor-free and overall survival and should be taken into account in evaluating reported outcomes, until greater experience and longer follow-up after primary endoscopic treatment is gathered.

A slightly different calculus regarding the selection of surgical approach may be necessary when considering craniopharyngioma in children. Although the general goal of complete resection remains, the implications of subtotal resection followed by radiotherapy on neurocognitive development, as well as the risk of secondary malignancy, is more of a concern in children [2]. There is, therefore, greater debate regarding the role of aggressive total resection vs. subtotal resection with adjuvant radiotherapy [32-34]. Anatomical differences between children and adults, such as the presence of a non-pneumatized sphenoid sinus in younger patients, may make ventral minimally invasive approaches more technically challenging. However, endoscopic endonasal approaches have been used in patients as young as 2 years [9].

Conclusion

Craniopharyngiomas, particularly those with suprasellar extension, with or without significant cystic component, and those occurring in children, require careful preoperative evaluation. This includes CT to identify calcifications, MRI to characterize tumor extent and cystic components, consideration of CSF diversion if hydrocephalus is present, ophthalmologic evaluation and full endocrinological workup and treatment. These components of the initial evaluation of the patient with suspected craniopharyngiomas are essential in the selection of surgical approach, which may be influenced by tumor size and imaging characteristics, lateral and inferior extension, patient preference and surgical experience. The endonasal endoscopic route appears to be comparable to open transcranial and transsphenoidal microscopic approaches for suitable lesions, i.e. smaller lesions without significant lateral extension past the lamina papyracea. Indeed, in the published literature endonasal approaches appear to demonstrate higher rates of gross total resection, improved outcome, and a trend towards fewer recurrences. The endonasal endoscopic approach is limited by a higher rate of postoperative CSF leak, though other operative morbidity may be lower than with transcranial techniques. Endonasal endoscopic techniques, if used in an extended fashion for suprasellar lesions such as craniopharyngiomas, require a surgical team with expertise in this area. There is a steep learning curve in endoscopic skull base surgery and close collaboration between the neurosurgeon and otorhinolaryngologists is essential to the successful performance of minimally invasive skull base surgery. As more such teams become established, a more detailed and specific understanding of the precise role of endonasal endoscopic techniques as a primary approach for craniopharyngiomas, both suprasellar and intrasellar, will continue to be refined.

References

[1] Karavitaki, N., Cudlip, S., Adams, B. T. & Wass, J. A. H. (2006). *Endocrine Rev.*, *27*, 371-397.

[2] Komotar, R. J., Roguski, M. & Bruce, J. N. (2009). *J. Neurooncol.* *92*, 283-296.

[3] Yang, I., Sughrue, M. E., Rutkowski, J. M., Kaur, R., Ivan, M. E., Aranda, D., Barani, I. J. & Parsa, A. T. (2010). *Neurosurg. Focus*, *28*, E5.

[4] Yasargil, M. G., Curcic, M., Kis, M., Siegenthaler, G., Teddy, P. J. & Roth, P. (1990). *J. Neurosurg. 73*, 3-11.

[5] Gardner, P. A., Prevedello, D. M., Kasam, A. B., Snyderman, C. H., Carrau, R. L. & Mintz, A. H. (2008). *J. Neurosurg.* 108, 1043-1047.

[6] Schwartz, T. H., Fraser, J. F., Brown, S., Tabaee, A., Kacker, A. & Anand, V. K. (2008). *Neurosurgery*, 62, 991-1005.

[7] De Divitiis, E., Cavallo, L. M., Cappabianca, P. & Esposito, F. (2007). *Neurosurgery*, *60*, 46-59.

[8] Mason, R. B., Nieman, L. K., Doppman, J. L. & Oldfield, E. H. (1997). Selective excision of adenomas originating in or extending into the pituitary stalk with preservation of pituitary function. *J Neurosurg.* *87*, 343-351.

[9] Locatelli, D., Levi, D., Rampa, F., Pezzotta, S. & Castelnuovo, P. (2004). *Childs Nerv. Syst. 20*, 863-867.

[10] Kassam, A. B., Prevedello, D. M., Thomas, A., Gardner, P., Mintz, A., Snyderman, C. & Carrau, R. (2008). *Neurosurgery*, *62* (ONS Suppl 1), 57-74.

[11] de Divitiis, E., Cappabianca, P., Cavallo, L. M., Esposito, F., de Divitiis, O. & Messina, A. (2007). *Neurosurgery*, 61 (ONS Suppl 2), 219-228.

[12] Frank, G., Pasquini, E., Doglietto, F., Mazzatenta, D., Sciarretta, V., Farneti, G. & Calbucci, F. (2006). *Neurosurgery*, 59 (ONS Suppl 1), 75-83.

[13] Gardner, P. A., Kassam, A. B., Snyderman, C. H., Carrau, R. L., Mintz, A. H., Grahovac, S. & Stefko, S. (2008). *J. Neurosurg.*, *109*, 6-16.

[14] Stamm, A. C., Vellutini, E., Harvey, R. J., Nogeir Jr, J. F. & Herman, D. R. (2008). *Laryngoscope*, *118*, 1142-1148.

[15] Komotar, R. J., Starke, R. M., Raper, D. M., Anand, V. K. & Schwartz, T. H. (2012). *World Neurosurg.* 77, 329-341.

[16] Leng, L. Z., Greenfield, J. P., Souweidane, M. M., Anand, V. K. & Schwartz, T. H. (2012). *Neurosurgery*, *70*, 110-23.

[17] Dehdashti, A. R., Ganna, A., Witterick, I. & Gentili, F. (2009). *Neurosurgery*, *64*, 677-689.

[18] Jane Jr, J. A., Kiehna, E., Payne, S. C., Early, S. V., Laws Jr, E. R. (2010). *Neurosurg. Focus*, *28*, E9.

[19] Leng, L. Z., Brown, S., Anand, V. K. & Schwartz, T. H. (2008). *Neurosurgery*, *62* (ONS Suppl 2), E342-E343.

[20] Cavallo, L. M., Prevedello, D., Esposito, F., Laws Jr, E. R., Dusick, J. R., Messina, A., Jane Jr, J. A., Kelly, D. F. & Cappabianca, P. (2008). *Neurosurg. Rev. 31*, 55-64.
[21] Samii, M. & Tatagiba, M. (1997). *Neurol. Med. Chir. (Tokyo), 37*, 141-149.
[22] Gsponer, J., de Tribolet, N., Deruaz, J. P., Janzer, R., Uske, A., Mirimanoff, R. O., Reymond, M. J., Rey, F., Temler, E., Gaillard, R. C. & Gomez, F. (1999). *Medicine, 78*, 236-269.
[23] Maira, G., Anile, C., Rossi, G. F. & Colosimo, C. (1995). *Neurosurgery, 36*, 715-724.
[24] Abe, T. & Ludecke, D. (1999). *Neurosurgery, 44*, 957-964.
[25] Norris, J. S., Pavaresh, M. & Afshar, F. (1998). *Br. J. Neurosurg. 12*, 305-312.
[26] Honegger, J., Buchfelder, M. & Fahlbusch, R. (1999). *J. Neurosurg. 90*, 251-257.
[27] Khouri, J. G., Chen, M. Y., Watson, J. C. & Oldfield, E. H. (2000). *J. Neurosurg. 92*, 1028-1035.
[28] Inoue, H. K., Fujimaki, H., Kohga, H., Ono, N., Hirato, M. & Ohye, C. (1997). *Childs Nerv. Syst. 13*, 250-256.
[29] Caldarelli, M., Massimi, L., Tamburrini, G., Cappa, M. & Di Rocco, C. (2005). *Childs Nerv. Syst. 21*, 747-757.
[30] Fahlbusch, R., Honegger, J., Paulus, W., Huk, W. & Buchfelder, M. (1999). *J. Neurosurg. 90*, 237-250.
[31] Yasargil, M. G., Curcic, M., Kis, M., Siegenthaler, G., Teddy, P. J. & Roth, P. (1990). *J. Neurosurg., 73*, 3-11.
[32] Sosa, I. J., Krieger, M. D. & McComb, J. G. (2005). *Childs Nerv. Syst. 21*, 785-789.
[33] Villani, R. M., Tomei, G., Bello, L., Sganzerla, E., Ambrosi, B., Re, T. & Barilari, M. G. (1997). *Childs Nerv. Syst. 13*, 397-405.
[34] Zuccaro, G., Jaimovich, R., Mantese, B. & Monges, J. (1996). *Childs Nerv. Syst. 12*, 385-391.

In: Minimally Invasive Skull Base Surgery
Editor: Moncef Berhouma

ISBN: 978-1-62808-567-9
© 2013 Nova Science Publishers, Inc.

Chapter IX

Endoscopic Optic Nerve Decompression

D. D. Sommer[1], M. S. Gill[1], K. Reddy[2] and A. Vescan[3]*

[1]Division of Otolaryngology and Head and Neck Surgery, McMaster University,
[2]Division of Neurosurgery, McMaster University
[3]Department of Otolaryngology and Head and Neck Surgery,
University of Toronto

Abstract

Introduction: There have been significant advancements in minimally invasive endoscopic surgery over the past 2 decades. Optic nerve decompression (OND) is particularly suited to an endoscopic approach as the optic nerve is well accessed trans-nasally.

Indications: Although initially thought to be effective in traumatic optic neuropathy, some evidence suggests the potential for spontaneous improvement leaving the role of decompression unclear in the traumatic setting. This controversy will be discussed and an evidence based approach for optic nerve decompression in selected individuals will be reviewed.

Non-traumatic indications for optic nerve decompression are variable and include optic nerve neoplasms (eg. Selected meningioma causing optic nerve compression), fibrous dysplasia, inflammatory optic neuropathy, idiopathic intracranial hypertension, mucocele, hemangioma, endocrine orbitopathy and osteopetrosis. The use of this procedure as an adjunct to tumour removal adjacent to the optic nerve (eg. meningioma) will also be discussed and our own experience with this approach presented.

Technique: A bilateral ethmoidectomy and sphenoidotomy are performed exposing the lamina papyracea and optic nerves. Next, a posterior septectomy allows bilateral access. The posterior lamina papyracea is carefully outfractured preserving the lateral periosteum to avoid orbital fat herniation. The 2 mm diamond drill is inserted from the contralateral side to remove remaining bone with the endoscope and irrigation introduced from the ipsilateral side.

---

[*] Corresponding author email: mandeep.gill@medportal.ca.

Discussion: Some of the advantages of the endoscopic approach include better cosmesis and lack of an external scar, preservation of olfaction, shorter recovery times and excellent visualization of the inferomedial aspect of the nerve. The major risks associated with endoscopic decompression of the optic nerve will also be discussed. Technical pearls and pitfalls including the potential for thermal damage will also be highlighted.

Conclusion: Endoscopic nerve decompression is a promising technique with the potential for low morbidity. The indications and limitations of the technique are still evolving and further studies will continue to elucidate these.

Introduction

The literature on optic nerve decompression (OND) dates back to the early 1900s. Early descriptions focused on traumatic indications and this remains the most common indication today. Historically, the intracranial approach was the initially described [1-3]. With advances in anatomical and surgical experience of the region, other approaches were defined. In the 1920's, Sewall described the transethmoidal optic canal decompression method. Other external approaches followed and include the transorbital, transnasal and transantral routes [3]. Over the past two decades, with progressive advancements in technology and expertise in endoscopic sinus surgery, the endoscopic technique has gained prominence. This was first described by Kennedy and Michel in the early 1990s [1, 2]. Its popularity has risen since that time given the decreased morbidity associated with this approach. Kennedy and colleagues described both endoscopic optic nerve decompression as well as endoscopic transnasal orbital decompression [1]. This chapter will only focus on the former. The two aforementioned papers have helped shift the current state of practice. Kennedy's initial report demonstrated comparable results of decompression when comparing the traditional external ethmoidectomy or Caldwell-Luc antrotomy to that of the endoscopic transnasal approach [1]. The main benefits of an endoscopic approach includes the absence of an external scar, the excellent exposure-allowing for decompression of at least 180 degrees, superb visualization of structures and preservation of olfaction. The approach also avoids injury to the dental structures and their innervations [1, 2, 4]. As with other endoscopic approaches, a more rapid postoperative recovery is achieved compared to the more traditional approaches [4].

Relevant Anatomy

Orbit

The orbits lie underneath the anterior cranial fossa and are in close relation to the middle cranial fossa. Each orbit is surrounded by an osseous framework of seven bones, which gives it its characteristic pyramidal shape. The base of the pyramid is anterior whereas the apex is located posterior-medially. The optic foramen at the apex of orbit forms the origin of the optic canal. The optic nerve, ophthalmic artery (OA) and branches of the autonomic nervous system travel through this canal to enter the orbital cavity. At its entry-point at the optic foramen, the optic nerve is surrounded by the fibrous annulus of Zinn at its inferior, superior and medial aspects. The optic canal penetrates the lesser wing of the sphenoid bone and is separated laterally from the superior orbital fissure by the optic strut (inferior root of the anterior clinoid process). The optic strut, whose base forms the lateral optico-carotid recess (OCR), connects the body to the lesser wing of the sphenoid bone and forms the infero-lateral wall of the optic canal [5-8]. The medial wall of the optic canal, the optic tubercle, separates it from the lateral junction of the sphenoid and posterior ethmoid sinuses (Figure 1). The roof of the optic canal is formed by the superior root of the anterior clinoid process of the sphenoid bone and this separates it from the anterior cranial fossa. As the optic nerve travels in a posteromedial direction it enters into the chiasmatic groove at a 35 degree angle. The average length of the optic canal is 8-12mm and its diameter is approximately 6.5mm [5, 9, 10].

There are four segments of the optic nerve, the intraocular segment, the intraorbital segment, the intracanalicular segment and the intracranial segment. The average lengths of these segments are ~ 1mm, 20-30mm, 5-11mm and 3-16mm respectively [5, 11]. The optic ring is the most distal aspect of the canal and is significant because of its narrow diameter, often less than 5mm. This is postulated to be the area most susceptible to compression [5, 9, 11]. A notable aspect that makes the optic nerve unique is the similarities it shares with the white matter of the brain. It has been described as a nerve fiber tract as opposed to a true nerve. As such, all three layers of meninges surround it and its fibers are surrounded by glial cells [9].

Sino-Nasal

The optic nerve is endoscopically accessed after ethmoidectomy and sphenoidotomy have been performed. In a significant proportion of patients, the optic nerve travels through an Onodi cell, a posterior ethmoid cell which lies lateral and/or superior to the sphenoid. There is significant variation in the literature about the proportion of patients who have an Onodi cell (12-60%) [4, 8, 12]. Depending on pneumatization, a variable proportion of the circumference (up to 360 degrees) of the optic nerve may be exposed in the ethmoid or sphenoid sinuses (Figure 1).

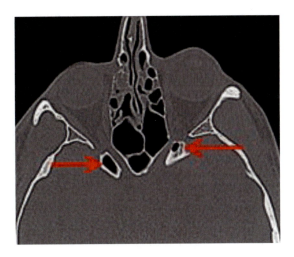

Figure 1. Axial CT showing pneumatization (arrow) of the sphenoid lateral to the optic nerve.

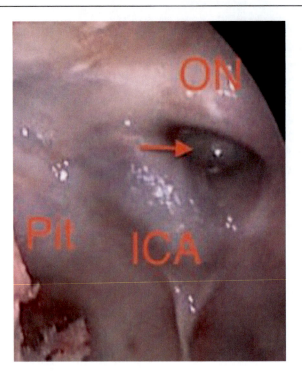

Figure 2. Left wall of sphenoid sinus demonstrating optic nerve (ON) and internal carotid (ICA) protuberances, pituitary (Pit) and lateral OCR (arrow).

In the majority of patients, the optic nerve is seen as a protuberance in the supero-lateral surface of the sphenoid sinus. This is located just superior to the bulge created by the internal carotid artery on the lateral surface of the sphenoid sinus. The optico-carotid recesses (OCR) are located between these two prominences. Both lateral and medial OCR are described. In reality, the lateral OCR is a much more reliable landmark than either the optic nerve protuberance or the medial OCR [6, 7, 13] (Figure 2). The lateral OCR is a result of pneumatization of the base of the optic strut [5]

Ophthalmic Artery

Knowledge of the course of the ophthalmic artery (OA) is also important in surgical approaches to the optic nerve. The OA arises from the internal carotid artery after it exits the cavernous sinus along the medial side of the anterior clinoid process. In the majority of patients, the artery runs infero-medially to the nerve as it originates. As it runs anteriorly, it moves infero-lateral to the nerve as it enters the optic canal. It is however noteworthy that there are significant variations in the medial/lateral location of the OA as it enters the orbit [5, 11, 14-16]. Locatelli and colleagues demonstrated that the OA emerged from the ICA infero-medial to the ON in 50% of cadaveric decompressions, inferior to the ON in 30% of cases and infero-lateral to the nerve in 20% of cases [11]. Wang and colleagues examined the relationship between the ON and OA in two regions, the orbital aperture of the ON canal and the intracranial interior opening of the ON canal. They determined that the OA was located lateral to the ON 85% of the time at the orbital aperture of the ON canal. When examining the relationship from the intracranial interior opening of the ON canal, the OA was directly inferior to the ON in 80% of cases [14]. These studies demonstrate that there are significant variations in the relationship between the OA and ON. Understanding the possible variations will help one navigate appropriately during endoscopic decompression of the optic nerve.

Technique

Optic nerve decompression is potentially difficult and not without risks and should only be performed by those with sufficient experience and training. Pre-operative imagining should be carefully reviewed and evaluated for the course/position of the optic nerve, the presence of Onodi cells and the pneumatization patterns and septations of the ethmoid and sphenoid sinuses. Image guidance is generally indicated and should be calibrated with a high degree of accuracy especially in patients with anatomic variation or alteration.

Once general anesthesia and nasal preparation is completed, bilateral maxillary antrostomies, ethmoidectomies and wide sphenoidotomies and undertaken. The middle and superior turbinates are resected on the ipsilateral side of the decompression and a posterior septectomy is performed to allow for a bi-nasal technique. As dissection proceeds posteriorly, septations must be removed especially laterally towards the orbit and superiorly towards the skull base. The lateral wall of the sphenoid/posterior ethmoids is evaluated for the presence and locations of the carotid protuberance, optic nerve protuberance and lateral optico-carotid recess.

Next, the medial orbital wall/lamina papyracea is identified and then outfractured medially in its posterior portion using a spoon curette or Freer elevator 1-2 cm anterior to the sphenoid sinus (Zone 1 in Figure 3). This must be performed delicately to preserve the lateral periosteum and prevent orbital fat herniation.

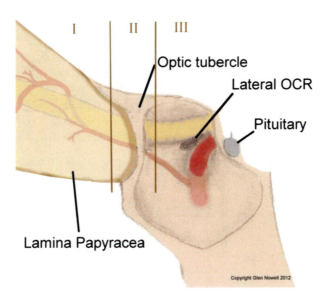

Figure 3. Sagittal view of region for optic nerve decompression in the lateral nasal wall. Note thin bone covering lamina papyracea, thicker bone posteriorly (optic tubercle, lateral sphenoid wall).

If this does occur it will impact the visualization of the next steps in the procedure. This may be ameliorated by placing one or two 1x3 cm saline soaked patties over the fat to try to keep it laterally away from the surgical field. Fat herniation may also hamper later stages of the procedure as it may get caught up in an instrument or the drill.

Next at the junction of the posterior ethmoids and the sphenoid, the optic tubercle and annulus of Zinn are exposed. This is approached by carefully removing the lamina papyracea posteriorly. The optic tubercle may containrelatively thick bone (Zone 2 in Figure 3), and often requires a drill to thin it out before removal. This is accomplished by use of a 2mm diamond drill from the contralateral side while the endoscope and irrigation is provided from the ipsilateral side (Figure 4). Figure 5 depicts an intra-operative picture of the drill in close proximity to the ON. The drill should not be used on any one spot for prolonged periods to reduce the risk of thermal injury. As well, irrigation is important to minimize this risk. Once the bone is adequately thinned, a low profile microcurette, ear instrument or other thin instrument can be used to fracture the bone of the optic canal in a medial direction. It is important not to try to insert a thick instrument into this constricted region to remove the tubercle as this may increase pressure on the nerve and result in injury.

As dissection proceeds postero-medially, the depth of bone overlying the optic nerve in the supero-lateral sphenoid is variable (Zone 3 in Figure 3). In some patients, the bony layer is essentially absent, while in others, a drill may be required to thin the bone before removal (Figure 6).

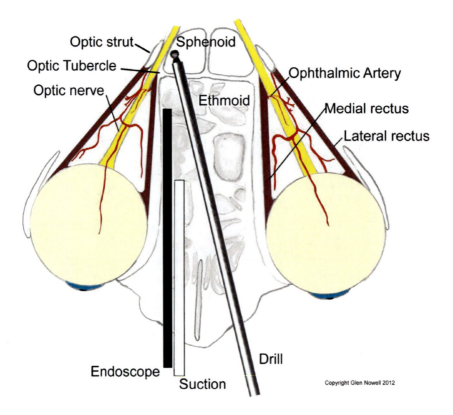

Figure 4. Illustrating how to place the instruments when completing endoscopic decompression of the optic nerve.

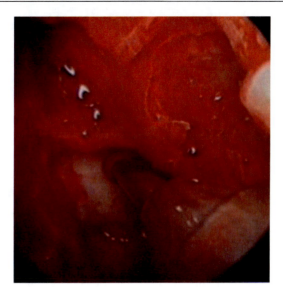

Figure 5. Intra-operative picture of the 2mm diamond drill in close proximity to the ON.

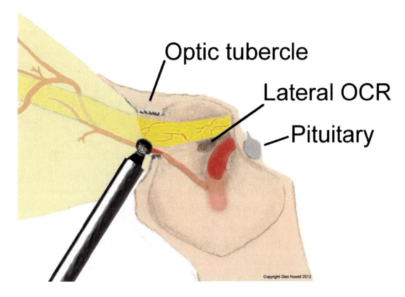

Figure 6. Sagittal schematic of the course of the optic nerve in the supero-lateral ethmoids and sphenoid wall. Bone removal may require drilling at optic tubercle / sphenoid regions.

A variety of opinions regarding the length of decompression are found in the literature. The most consistent is to complete the decompression up to approximately 1 cm posterior to the face of the sphenoid sinus for cases of TON. For other cases it is likely sufficient to decompress only the area involved. If the nerve is to be decompressed prior to tumor resection, the decompression is performed posteriorly up to the region of the tumor.

Another significant area of controversy surrounds the incision of the optic nerve sheath. Normally this is completed with a sickle knife. Currently, there are no consistent guidelines to help guide surgeons, and we do not feel there is sufficient evidence to perform this routinely. An important anatomical variation that must be noted for any surgeon opening up the optic nerve sheath is the OA. Normally it passes inferio-laterally. However, in a significant proportion of cases it can run inferio-medially. It is thus recommended that if one is making the incision that it should be carried out in a supero-medial location to avoid injuring the artery. To repair a cerebrospinal fluid fistula after the sheath is opened, tissue adhesives and possibly other methods of repairs can be used depending on the severity of the leak.

Indications

There are a variety of indications for decompression of the optic nerve, with traumatic optic neuropathy being one of the more common [17-27]. Although surgery was initially considered to be effective in the treatment of traumatic optic neuropathy, further evidence suggests the potential for spontaneous improvement leaving the role of decompression unclear in this setting. Recent evidence suggests that a significant proportion of patients will undergo spontaneous recovery without intervention (surgery, high dose steroids). In a meta-analysis conducted by Cook and colleagues it was shown that treatment with either corticosteroids, decompression or both were better than no treatment in cases of TON [24]. Levis and colleagues addressed this again in 2003, and found that corticosteroids and surgery do not add any benefit over observation. A review of the literature by Yu-Wai-Man and colleagues in 2011 showed insufficient quality of studies to reach a firm conclusion regarding the role of decompression for TON. They stated that there was no convincing evidence of additional benefit from surgical decompression [19, 20]. However, there are several articles that do support either corticosteroid therapy or endoscopic decompression or both. Additional evidence suggests that in select patients, a more active intervention algorithm may be appropriate. A study by Li et al. reviewed 176 patients who were actively treated and found that a combination treatment regime of steroids and endoscopic decompression can be an effective and safe treatment option for patients with TON. In their study, 55% of patients had an improvement in total vision when treated with a combination of steroids and endoscopic decompression, while 51% of patients had improvements if treated with endoscopic decompression alone. The differences between these two groups were not significant. This group also reported that none of the patients involved in the study had worsened vision as a result of OND. Notably, the success rate of surgery was lower in those who had immediate loss of vision compared to a gradual loss of vision - 38%, 67% respectively [25]. Moreover, Yang and colleagues were able to identify a subgroup of patients who had little benefit from endoscopic decompression. The three most significant variables that they found to be associated with poor recovery of visual acuity comprised of absence of light perception immediately after the injury, undergoing endoscopic decompression more than three days after the injury and evidence of blood in the ethmoid and/or sphenoid sinus [23]. An animal model study conducted by Mabuchi and colleagues illustrated that even though nerve swelling, capillary dilation and nerve necrosis can occur within 24 hours after injury to the optic nerve, it takes three days for the degeneration of the myelin sheath and for optic nerve infarction to occur in their mouse model [28]. Moreover, additional factors that were associated with a worse prognosis were fractures involving the lateral optic canal wall or multiple fractures. Patients who suffered a single fracture in the medial wall of the optic canal had better overall prognosis after endoscopic decompression of the optic nerve.

In the pediatric literature, the degree of injury was demonstrated to have a direct correlation to the success of surgery. Peng and colleagues recommend a combination of therapy (endoscopic decompression and steroids) when managing TON in children. They reviewed 43 cases of TON in 41 children at their facility. With combined therapy, 36 out of the 41 children had improved their vision postoperatively [29].

Controversies

There are considerable controversies surrounding optic nerve decompression especially in patients with traumatic optic neuropathy. There are no randomized controlled trials comparing close observation vs steroids vs surgical decompression. The precise role of surgical decompression remains unclear. Often, surgical decompression is utilized in those individuals who do not respond to steroid therapy or get worse on more conservative therapies. This may bias the true effectiveness of endoscopic surgical decompression of the optic nerve.

Another significant area of debate is the timing of surgical intervention. It appears that the consensus among experts is to not to perform surgical decompression right away but to try conservative measures first. If high dose steroid therapy fails to improve symptoms after 72 hours, it may be reasonable to proceed with surgical decompression. Some studies state that it is important to conduct surgical decompression earlier while others state that there is no difference. The timing is a yet unresolved issue in TON, with some reports advocating for surgical intervention up to seven days after injury while others state better response occur if surgical intervention occurs within three days [18-25].

Other Indications/Local Experience

There are several other less common relative indications for optic nerve decompression involving compressive optic neuropathy present in the literature.Many ofthe reports are from the pre-endoscopic era. Pathological indications include fibro-osseous lesions (eg. fibrous dyplasia), mucocele, inflammatory optic neuropathy, pseudotumour cerebrii, endocrine orbitopathy and a variety of neoplasms [30-32].

A retrospective review of our early experience in patients with benign soft tissue lesions abutting the optic nerve was conducted at our centre to review the outcomes [33]. There were a total of 12 patients who underwent the procedure with the age range spanning from 22 to 91. Among these, 8 were diagnosed with meningioma (Figure 7), one with Rathke's Cleft cyst and 3 with craniopharyngioma (Figure 8). The surgical indications used were acute deterioration in visual acuity in 7 of the patients, acute visual field deterioration in 4 patients and significant compression of the optic nerve based on MRI in 1 patient. Post-operatively, 7 patients had improved visual acuity of at least 2 lines on the Snellen chart and among these patients 5 had normal visual acuity post operatively. Three of the patients improved by 1 line on the Snellen chart, 1 patient demonstrated improvement only in visual fields and 1 patient's visual acuity decreased post operatively. All 12 patients had intra-operative CSF leaks (which were repaired at the time of surgery), and 5 had minor bleeding intra-operatively.

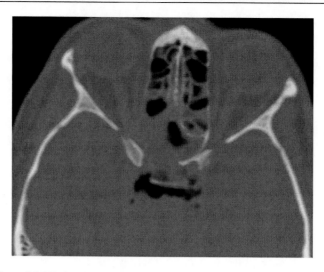

Figure 7. Postoperative axial CT showing right optic nerve decompression in a meningioma patient.

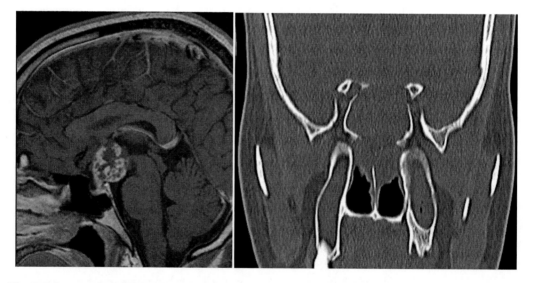

Figure 8. Preoperative MRI and Postoperative CT scan in a craniopharyngioma patient demonstrating left optic nerve decompression.

There were no other intra-operative complications. Our more recent experiences have been similar. Thus we believe that addressing the affected optic nerve prior to tumor resection may be beneficial in selected patients as it relieves pressure on the nerve and may be advantageous in the mobilization of the lesion from the nerve and thus merits further evaluation. As well, the nerve is identified early in a consistent location and is more easily followed posteriorly to the area of the tumor where it may be somewhat obscured by pathology and more difficult to identify safely. Although promising, due to the relative scarcity of these conditions, the data on the subject is not sufficient to make definitive conclusions regarding the efficacy and indications of treatment [33, 34].

Conclusion

Endoscopic optic nerve decompression can be a safe and effective method to treat both traumatic and non-traumatic pathologies affecting the optic nerve. The preciserole of surgery is still being debated for a variety of applications. As experience grows and technology and instrumentation continue to advance, both the limitations and indications of this approach will likely evolve.

Acknowledgments

We would like to thank Glen Nowell for his medical illustrations.

References

[1] Kenndy DW, Goodstein ML, Miller NR, Zinreich SJ. Endoscopic Transnasal Orbital Decompression. (1990). *Arch Otolaryngol Head Neck Surg.* 116 275-280.

[2] Michel O, Bresgen K, Rüssmann W, Thumfart WF, Stennert E. Endoscopically-controlled endonasal orbital decompression in malignant exophthalmos. *Laryngorhinootologie* 1991. Dec;70(12):656-662.

[3] Maliawan S, Mahadewa, TGB, Putra AAM. Lateral orbitotomy for traumatic optic neuropathy and traumatic opthalmoplegia: Is it beneficial? (2009). *Neurology Asia,* 14: 35-39.

[4] Luxenberger, W. Stammberger, H. Jebeles, Walch, C. (2009). Endoscopic Optic Nerve Decompression: *The Graz Experience.* 108(6) 873-882.

[5] Janfaza P, Nadol JB, Galla R, Fabian R, Montogomery W. *Surgical Anatomy of the Head and Neck.* (2001). Lippincott Williams & Wilkins.

[6] Ozcan T, Yilmazlar S, Aker, S, Ender K. Surgical Limits in Transnasal Approach to Opticocarotid Region and PlanumSphenoidale: an Anatomic Cadveric Study. (2010). *World Neurosurg.* 73 (4): 326-333.

[7] Yilmazlar S, Saraydaroglu O, Korfali. Anatomical aspects in transsphenoidal – transethmoidal approach to the optic canal: An anatomic–cadaveric study. (2012). *Journal of Carnio-Maxillo-facial Surgery.* 40: 198-205.

[8] Thanaviratananich, S, Chaisiwamongkol K, Kraitrakul S, Tangsawad A. The prevalence of an Onodi cell in adult Thai cadavers. (2003) *Ear, Nose & Throat Journal.* 82 (3): 200-204.

[9] Mafee MF. The eye, orbit: embryology, anatomy and pathology. In: Som PM, Curtin HD, eds. *Head and Neck Imaging.* St. Louis: Mosby: 2003: 441-654.

[10] Daniels DL, Mark, LP, Mafee, MF, Massaro B, Hendrix LE, Shaffer KA, Morrissey D, Horner CW. Osseous Anatomy of the Orbital Apex. (1995), *AJNR.* 16: 1929-1935.

[11] Locatelli, M. Et al. Endoscopic transsphenoidal optic nerve decompression: an anatomical study. (2011). *SurgRadiol Anat.* 33:257-262.

[12] Weinberger DG, Anand VK, Al-Rawi, M, Cheng HJ, Messina AV. Surgical Anatomy and Variations of the Onodi Cell. (1996). *American Journal of Rhinology.* 10 (60): 365-370.

[13] Li J, Wang J, Jing X, Zhang W, Zhang X, Qiu Y. Transsphenoidal optic nerve decompression: an endoscopic anatomic study (2008). *J. Craniofac Surg.* 19(6):1670-4.

[14] Wang T, Kang Z, Li P, Liu X, Zhang G, Li Y. Anatomical and Imagining Studies of Endoscopic Optic Nerve Decompression. (2012) *The Journal of Bioscience and Medicine* 2: 1-6.

[15] Perri P, Cardia A, Fraser K, Lanzino G. A microsurgical study of the anatomy and course of the ophthalmic artery and its possible dangerous anastomoses. (2007). *J Neurosurg* 106: 142-150.

[16] Kocabiyik, N, Yazar F, Ozan H. The intraorbital course of ophthalmic artery and its relationship with the Optic nerve. (2009). *Neuroanatomy* 8: 36-38.

[17] Kountakis S, Maillard AAJ, El-Harazi SM, Longhini L, Urso RG. Endoscopic Optic Nerve Decompression for Traumatic Blindness. (2000) Otolaryngology) *Head and Neck Surgery* 123(1) 34-37.

[18] Jiang RS, Hsu CY, Shen BH. Endoscopic optic nerve decompression for treatment of traumatic optic neuropathy. (2001). *Rhinology,* 39, 71-74.

[19] Yu Wai Man P, Griffiths PG. Surgery for traumatic optic neuropathy. *Cochrane Database Syst Rev.* 2005;(4):CD005024.

[20] Yu-Wai-Man P, Griffiths P. Steroids for traumatic optic neuropathy. *Cochrane Database Syst Rev.* 2007;(4):CD006032.

[21] Pletcher SD, Sindwani R, Metson R. Endoscopic and optic nerve Decompression. (2006) *Otolaryngology Clinics of North America.* 39(5): 943- 958.

[22] Robinson D, Wilcsek G, Sacks R. Orbit and Orbital Apex. (2011). *Otolaryngol Clin N Am.* 44: 903-922.

[23] Yang Q-T, Zhang G-H, Liu X, Ye J, Li Y. The therapeutic efficacy of endoscopic optic nerve decompression and its effects on the prognoses of 96 cases of traumatic optic neuropathy (2012). *Journal of Trauma.* 72(5): 1350-1355.

[24] Cook MW, Levin LA, Joseph MP, Pinczower EF. Traumatic Optic Neuropathy. (1996). *Arch Otolaryngol Head Neck Surg.* 122: 389-392.

[25] Li H, Zhou B, Shi J, Cheng L, Wen W, Xu G. Treatment of traumatic optic neuropathy: our experience of endoscopic optic nerve decompression. (2008) *The Journal of Laryngology & Otology.* 122: 1325- 1329.

[26] Sofferman RA. Transnasal Approach to Optic Nerve Decompression. (1991) *Operative Techniques in Otolaryngology- Head and Neck Surgery.* 2(3): 150-156.

[27] Lai, D. Stankiewica, JA. Endoscopic optic nerve decompression. (2009) *Operative Techniques in Otolaryngology- Head and neck surgery.* 20(2) 96-100

[28] Mabuchi F, Aihara M, Mackey MR, Lindsey JD, Weinreb RN. (2004) Optic nerve damage in experimental mouse ocular hypertension. *Invest Opthtalmol Vis Sci.* 44: 4321-4330.

[29] Peng A, Li Y, Hu P, Wang Q. Endoscopic optic nerve decompression for traumatic optic neuropathy in children. (2011). *International Journal of Paediatric Otorhinolaryngology* 75: 992-998.

[30] Sleep, TJ, Hodgkins PR, Honeybul S, Neil-Dwyer G, Lang D, Evans B. Visual function following neurosurgical optic nerve decompression for compressive optic neuropathy (2003). *Eye.* 17 (5): 571-8.
[31] Kaufman D, Dickerson K, Kelman S, Langenberg P, Newman N, Wilson PD. Ischemic Optic Neuropathy Decompression Trial. (2000). *Arch Opthalmol* 118: 793-798.
[32] Pletcher SD, Metson R. Endoscopic Optic Nerve Decompression for Nontraumatic Optic Neuropathy. (2007) *Arch Otolaryngol Head Neck Surg.* 133 (8): 780-783.
[33] Kamian K, Reddy K, Sommer DD. Trans-nasal endoscopic decompression of the optic nerves. *44th Congress of the Canadian Neurological Sciences Federations.* June 2009. Halifax, NS, Canada.
[34] Maurer J, Hinni M, Mann W, Pfeiffer N. Optic nerve decompression in trauma and tumor patients. (1999) *Eur. Arch. Otorhinolaryngol.* 256(7):341-5.

Chapter X

Endoscopic Approach and Management of Parasellar Lesions

Mahmoud Messerer[1], Moncef Berhouma[2,3], Marc Sindou[2,3] and Emmanuel Jouanneau[2,3]

[1]Department of Clinical Neurosciences, Department of Neurosurgery, Centre Hospitalier Universitaire Vaudois, Lausanne, Switzerland
[2]University of Medicine, Claude Bernard University Lyon, France
[3]Department of Neurosurgery A, Pierre Wertheimer Neurological and Neurosurgical Hospital, Hospices Civils de Lyon, Lyon, France

Abstract

The parasellar region is the location of a wide variety of inflammatory and benign or malignant lesions. A pathological diagnostic strategy may be difficult to establish relying solely on imaging data. Percutaneous biopsy through the foramen ovale using the Hartel technique has been developed for decision-making process. It is an accurate diagnostic tool allowing pathological diagnosis to determine the best treatment strategy. However, in some cases, this procedure may fail or may be inappropriate particularly for anterior parasellar lesions.

Over these past decades, endoscopy has been widely developed and promoted in many indications. It represents an interesting alternative approach to parasellar lesions with low morbidity when compared to the classic microscopic sub-temporal extradural approach with or without orbito-zygomatic removal.

In this chapter, we describe our experience with the endoscopic approach to parasellar lesions. We propose a complete overview of surgical anatomy and describe methods and results of the technique. We also suggest a model of a decision-making tree for the diagnosis and treatment of parasellar lesions.

Introduction

A large variety of lesions either benign or malignant [1] may be located in the parasellar region. The most common are meningiomas and neurinomas but there are also chondrosarcomas, chordomas, epidermoid and dermoid cysts, inflammatory diseases, carcinomas, metastases, lymphoma and many others [2-4]. A specific management is required for each type of lesion. Neuroimaging including MR spectroscopy and diffusion sequences may approach the diagnosis with more or less accuracy, but does not replace pathological and/or cytological assessment. To allow exact diagnosis and adapted management, the percutaneous biopsy through the foramen ovale has been developed. It uses the same approach as that performed for the percutaneous treatment of trigeminal neuralgia, which was first described by Hartel [5]. It is an accurate diagnostic tool avoiding many unnecessary open surgeries. However it may fail for calcified or dense tumors. Another issue is represented by the important frequency of meningeal cells due to meningeal hyperplasia or presence of cartilaginous tissue of skull base inducing pitfalls in diagnosis. Lastly, the trajectories through the foramen ovale limit targeting to the Meckel's cave, the posterior part of the cavernous sinus and the upper part of the petroclival region. The access to the anterior part of the cavernous sinus is difficult. Endoscopy has been developed since 1990's and is more and more used for skull base surgery and therefore may represent a good alternative to percutaneous biopsy for such locations. In this chapter, we present endonasal endoscopic steps to reach parasellar lesions and propose an algorithm to manage such lesions.

Anatomy

Nasal Fossa (Figure 1)

Medially, the septum is a rigid structure, which is the association of the perpendicular plate of the ethmoid on superior plane, the vomer on the inferior plane and the quadrangular cartilage on the anterior plane. Sometimes, there is septal deviation that may complicate surgery. Posteriorly, there are the choanae giving access to the cavum. Laterally, there are three longitudinal folds i.e. the turbinates. The first seen turbinate is named the inferior turbinate. It is attached to the ethmoid. Posterior and above of it, there is the middle turbinate that is a part of the ethmoid. This middle turbinate has its head free and possibly pneumatized (concha bullosa) so may narrow surgical corridor. The superior turbinate is only seen after the mobilization of the middle turbinate as well as the sphenoid ostium. This is the entry point to the sphenoid sinus, located approximately 15 mm above the choanae. Passing just below the sphenoid ostium into the mucosae to reach the posterior septum is the nasal posterior nasal artery. Branch of the spheno-palatine artery, it may be source of postoperative epistaxis and have to be coagulated most of the time.

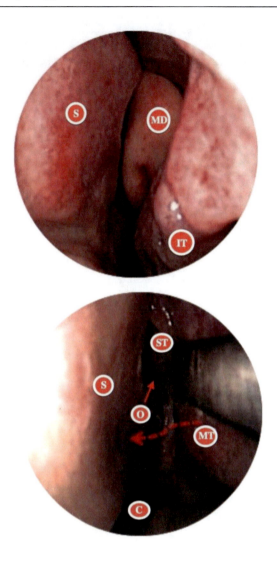

Figure 1. Operative view, nasal anatomy; S: septum; IT: inferior turbinate; MT: middle turbinate; ST: superior turbinate; O: ostium; C: choanae; Red arrow: Nasal posterior artery.

Sphenoid Sinus and Parasellar Anatomy

The sphenoid sinus has a different degree of pneumatization among individuals, as well as the numerous septa variations. The sphenoid sinus is often pneumatized (sellar type) with one or two simple septa. Some individual especially young do not have any sinus (conchal type) [6]. Key landmarks have to be located and understand before doing any surgery in this area (Figure 2). On the midline, from anterior to posterior are the planum, the tuberculum of the sella, the sella turcica and the clivus that is in front of the brainstem. Laterally are the optic nerve, the lateral optico-carotid recess, the paraclinoid segment of the internal carotid (ICA), the cavernous sinus and the clival segment of the ICA (Figure 2).

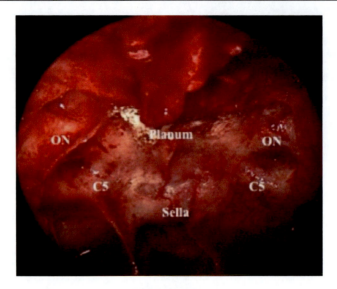

Figure 2. Endoscopic endonasal view inside the sphenoid sinus. Sella turcica on the midline is boarded anteriorly by the planum, and laterally optic nerves (ON) and the parasellar carotid artery (C5).

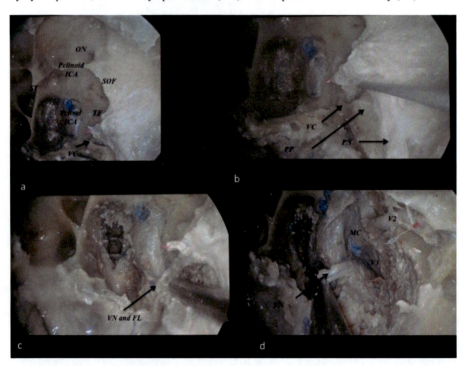

Figure 3. Anatomic study. The sphenoid sinus is largely opened up and the posterior wall of the maxillary sinus removed. A: the vidian canal (VC) is exposed with the paraclival ICA on the back. The usual landmarks can also be identified, the optic nerve: ON, the paraclinoid ICA: Pclinoid. Laterally up is the superior orbital fissure: SOF and below the entry point to the meckel's cave: TF (temporal fossa). The cavernous sinus locates between the Pclinoid and Pclival ICA. B: another view with the pterygoid process: PP with the vidian canal: VC and the palatine nerve and vessels: PN. C: the vidian nerve clearly leads the surgeon to the foramen lacerum and the petrous ICA. D: After removing the bone structure of the lateral sphenoid recess as have to be doing during parasellar surgery. VN: vidian nerve, MC: Meckel's cave, V2 and V3 trigeminal branches.

The parasellar compartment (cavernous sinus – CS -, Meckel's cave – MC -) access requires a large aperture of the sphenoid, the posterior ethmoidal cells and the maxillary sinuses for the MC (Figures 3, 4, 5, 6). The MC entry point is located lateral and up to the paraclival and petrous ICA loops just below the superior orbital fissure.

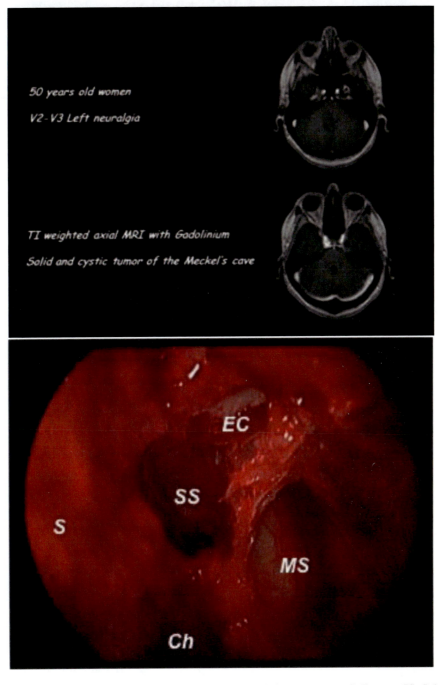

Figure 4. Operative view of a left endonasal approach for a Meckel's cave tumor. S: Septum; SS: Sphenoid Sinus; EC: Posterior Ethmoidal Cells; MS: Maxillary Sinus; Ch: Choanae.

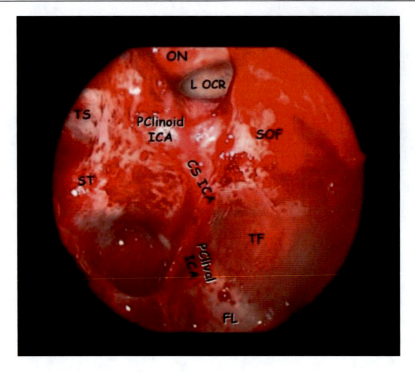

Figure 5. Operative view of the left lateral sphenoid recess in the same patient.
ON: optic nerve; C5: Pclinoid ICA: paraclinoid ICA; Sella: sella turcica; Cav Sinus: cavernous sinus; L OCR: lateral optico-carotid recess; SOF: superior orbital fissure; TF: temporal fossa, entry point to the Meckel's cave; C3: Pclival: paraclival ICA; FL: foramen Lacerum.

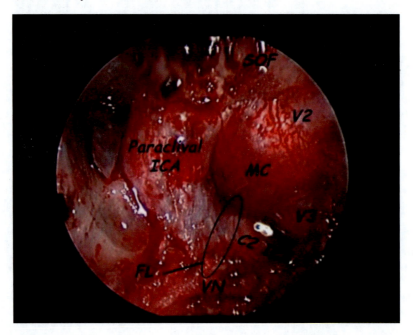

Figure 6. Operative view after removing bones structures. MC: Meckel's cave; VN: vidian nerve; FL: foramen Lacerum; C2 ICA: Horizontal petrous ICA; V2 and V3 trigeminal branches. SOF: superior orbital fissure.

Preoperative Management

Preoperative high-resolution head CT-scan (bone window, slice 1mm in thickness), head MRI, endocrinological assessment for sellar/parasellar lesion and ophthalmological evaluation are systematically performed, as well as an ENT evaluation of the nasal fossas. A CT angiography with occlusion tests [12] is performed when a tumor removal is planned to evaluate the Willis polygone functionality and possibility of vascular occlusion in case of vascular injury. Any suspicion of secondary locations of oncological process must be screened with a whole body CT-Scan and/or PET-Scan.

In all cases of endonasal extended approaches, patients are prepared few weeks before with meningococcal, streptococcus pneumoniae and haemophilus vaccines. Patients undergo polyvidone shower and nasal disinfection with polvidone cream the day before surgery and the morning of the surgery.

Operative Room Settings and Surgical Tools

The authors' endonasal endoscopic procedures derive from Jho [6] and Cappabianca [14], and has been already described in previous publications.

A dedicated instrumentation with straight 0° and angles 30° and 45° endoscopes (diameter 4mm; length: 30 cm or 18 cm) have be especially design for endonasal surgery (STORZ Cappabianca, Kassam, Castelnuevo kits ®). A HD column and endoscope are used from now on. Microprobe Doppler, high-speed bone drills and mucosal shavers are also used.

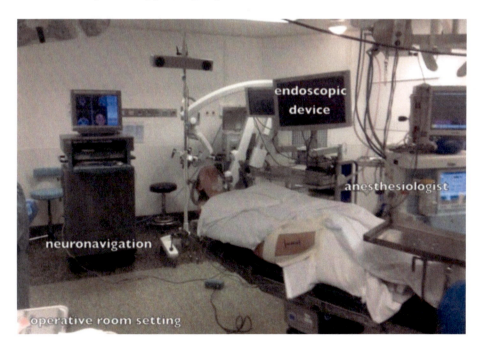

Figure 7. Operative room organization.

Patient is positioned supine with 20° elevation of the trunk and the head fixed with a three-pin Mayfield skull clamp system to be immobile for neuronavigation while right turned to face the neurosurgeon (Figure 7). This positioning is the same than that use for regular pituitary case. We rather prefer this positioning to facilitate the blood falling down during surgery and favor venous drainage. However, some authors use a strict horizontal positioning working either at this head or lateral to the patient.

As seen in the figure 6, the neuronavigation and the endoscopic column face the surgeon the first on the left side, the latter on the right. Neuronavigation is always used for extended approaches with fused CT and MR images.

During the patient positioning, the cottons soaked with an adrenalin and xylocain solution are put on each nasal cavity to favor mucosae retraction and diminish bleeding.

The lateral side of the right thigh is systematically draped in case of need of fascia lata and/or fat harvesting for reconstruction.

Surgical Steps

Two hands working is favored by the authors and a free hand during the nasal phase and with an endoscope holder for the intradural one is also preferred.

Nasal Step

During this step, a rigorous control of the blood pressure (systolic pressure around 10) and high level of sedation and analgesia will be asked to the anesthesiologist to prevent bleeding.

The surgical corridor with large aperture of maxillary, ethmoid and sphenoid sinuses are prepared on the side of the lesion but the two nostrils are used during surgery. When an intradural surgery is planned, a pedicle nasoseptal flap is useful and should have to be done in the opposite nostril.

Indeed, lateral exposition of the maxillary sinus, the pterygopalatin fossa and vidian nerve require cutting the sphenopalatine artery that is the vascular supply of the nasal flap. Moreover if the sphenopalatine artery can be preserved, the flap can be damage or annoying if done on the side of the lesion even placed into the cavum.

Until the opening of the sellar floor, a hand-held short endoscope mainly the 0° or 30° (diameter: 4 mm; length: 18 cm) will be used. The middle and superior turbinates ipsilateral to the lesion are removed by cutting and shaving with high precaution to avoid skull base fracture and CSF leaks during the mobilization of these turbinates. This middle turbinectomy is required to enlarge the nasal corridor thus facilitate access to surgical tools and identification of the sphenopalatine artery just behind the conchal crest. This turbinate will be keep intact and sometimes used for the closure time.

Sinuses Steps (Figures 3-6)

The maxillary, sphenoid and ethmoid sinuses are widely opened up. The posterior thin bone wall of the maxillary sinus is removed to get access to the pterygopalatine fossa. The sphenopalatine artery is coagulated or clipped, drilling or bone punches opens up the great palatine canal at the medial part of the pterygoid process. Pushing away laterally both the sphenopalatine artery and the great palatine nerve lead to the vidian canal that is exactly at the crossing line of the paraclival ICA, pterygoid process and the sphenoid sinus floor (Kassam neurosurgery 2008, Figure 3). Drilling the pterygoid process backward around the vidian nerve will lead to the foramen lacerum. Every effort will have to be made keeping intact the palatine nerve to avoid anesthesia of the ipsilateral superior palate. The conservation of the vidian nerve is trickier and its sacrifice exposes "only" to a dry eye. Exposition of the petrous and paraclival ICA (give the proximal and distal control of the ICA) is necessary as well as bone removal of the lateral wall of the sphenoid recess until the V2 trigeminal branch (Figures 4, 5). All the anatomical landmarks are therefore exposed to begin parasellar surgery.

Is has to be noted that for a tumor biopsy a more simple process can be done with a bone aperture of the lateral wall of the sphenoid recess to get access to the Meckel's Cave. This is relatively easy to be done especially in sellar sinus type coming from the opposite nostril.

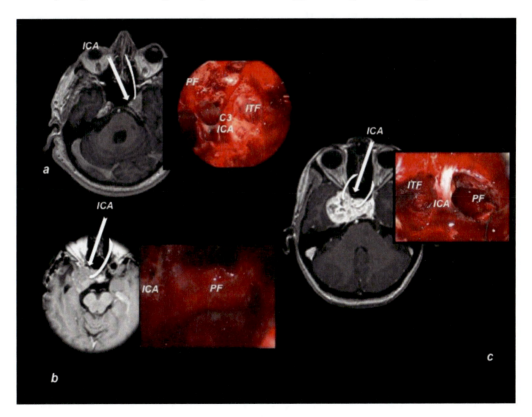

Figure 8. Importance of ICA position.

Cavernous Sinus via the Pituitary Fossa Window: Working Medial to the ICA

To approach the parasellar compartment, position of the ICA is of paramount importance (Figure 8). When pushed laterally, the cavernous sinus can be approach using the pituitary fossa. In such situation when the tumor extended laterally, this window has been already enlarged. Once the bone covering the anterior part of the cavernous sinus removed (often very thin because of the tumor growth) and the pituitary fossa opened up, the dura mater can be pushed away laterally and the medial wall of the cavernous sinus exposed (Figures 9, 10).

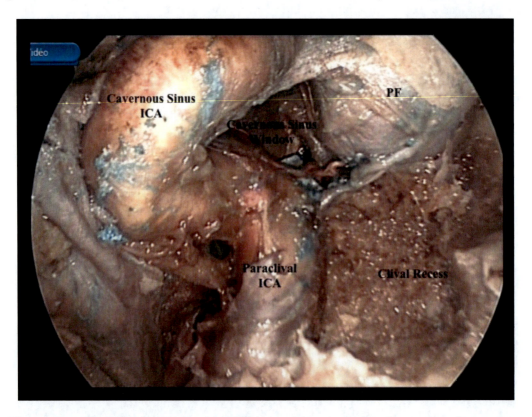

Figure 9. Anatomical study. The pituitary fossa (PF) is one of the entry doors to the cavernous sinus passing behind the anterior ICA loop.

Meckel's Cave Approach: Working Lateral to the ICA

For tumors growing in the Meckel's Cave, the window used is lateral to the paraclival and up to the petrous ICA. The dura mater is opened up behind the V2 trigeminal branch. The tumor removal used regular microsurgical techniques.

In rare case, ICA is on the midline with tumor on each side as in the case of a huge neurinoma presented in the figure 7c and both windows medial and lateral to the ICA will have to be use. In such cases, when tumor are suppose to be highly vascularized and firm, a carotid occlusion or an extra-intra vascular anastomosis depending of the functionality of the Willis polygon is an option to discuss before the tumor removal.

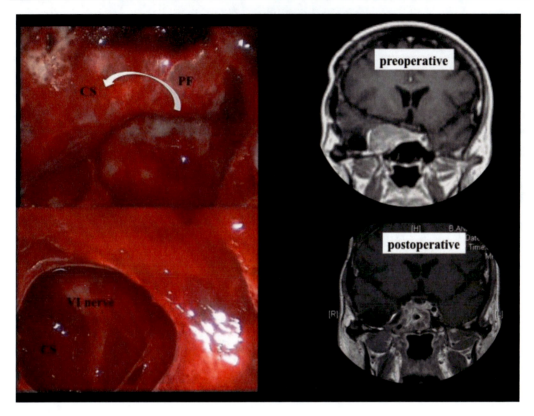

Figure 10. Operative views of a recurrent right cavernous sinus (CS) chondrosarcoma that has been removed using the pituitary fossa window (PF). Ending the removal, an angle 45° endoscope is used to explore the surgical field and the VI nerve visible at the back.

Oculo-motor Nerves Monitoring (Figure 11)

As previously mentioned, the key point to control is the ICA to prevent dramatic vascular damage. The neurological prognosis is mainly to the oculo-motor nerves and surgeon cannot trust the anatomy largely modified when the tumor growth. The V3 can be easily monitored. Mini recording electrodes are from now on available and can be placed inside the lateral and superior rectus muscles to monitor the III and VI nerves. The IV nerve monitoring is not developed yet. A short conjunctival incision is use to accurately place the electrodes without significant morbidity.

Closure Step

For a biopsy, there is not reconstruction to be made and no nasal packing necessary.

For a parasellar surgery, most of the time tumors are extradurally and we have less CSF leakage issue except if tumor extended to the petroux apex.

For an intradural surgery, the closure time is critical and all efforts have to be made to rebuild all anatomical planes. The same multilayer technique is used whatever the extended endonasal approaches.

We rather prefer use fibrin glue injected inside the surgical field than fat. Over plugging is avoided and injection of fibrin glue stopped as soon as CSF exit interrupts. No inlay graft is used but 2 onlay layers of resorbable artificial dura mater placed extradurally. A posterior septal bone piece when available or the middle turbinate can be used for bone reconstruction keep in place by fibrin glue. The nasoseptal flap is placed to cover all bone surfaces. Fat harvested from the tight or belly plugs at this end the sphenoid sinus. Bilateral nasal packing is placed for 2 to 3 days when extensive nasal surgery has been performed to ensure hemostasis and avoid synechias.

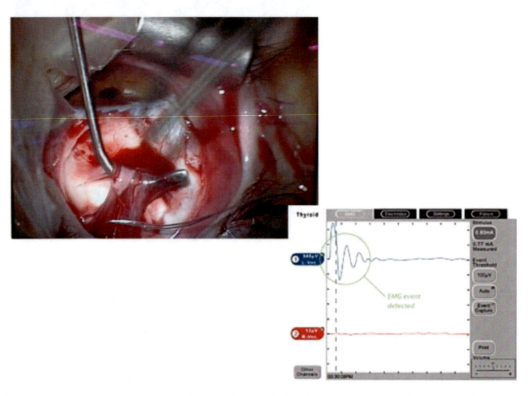

Figure 11. The recording electrode has been put in place inside the lateral rectus muscle after a transconjunctival incision.

Postoperative Management – Complications

After such procedures, patients are monitored in an intensive care unit during at least the 24 first hours and undergo control CT-scan or MRI at day 1.

Endocrinological and ophthalmological tests are performed to assess improvement or deterioration of these functions. Corticosteroid replacement is given each time the pituitary gland was involved during surgery until the endocrine check. Permanent DI is reported in 1% to 5% [15-18, 22, 23] in case of pituitary sella involvement.

A three days lumbar puncture is favored to lumbar drainage in case of intradural surgeries.

Daily nasal instillations are started as soon as the nasal packing is removed to treat crusting rhino-sinusitis and used for weeks or months. If a CSF leakage is checking, nasal instillations are postponed until to be sure of the wondering of the skull base.

Patients must be aware that nasal crusting syndrome is systematic for months in case of large sinuses aperture. When opening up a sinuses, the aperture have to be large for a better post-operative drainage [17]Nasal vascular complications are reported in 0.7% to 7% of cases and are either major bleedings secondary to sphenopalatine artery (endovascular management) or minor epistaxis (nasal packing) [15-18].

Fracture of the sphenoid bone, injury of the optic nerves or lesions of the carotid artery [19] are less rare complications. Injury to the internal carotid artery typically occurs during aggressive surgery of either macroadenomas extending in the cavernous sinus or huge and firm tumors encasing the ICA. These lesions are exceptional (0% to 0.68%) but may be responsible of death [20]. Intracranial hemorrhage is very rare and results from operative ICA lesion, carotid-cavernous fistulas or pseudoaneurysm with secondary rupture [21]. Each surgery with parasellar tumor removal planned have to benefit from a preoperative vascular check up that have to be done also in case of an operative suspicion of vascular damage.

The most annoying complication is obviously the occurrence of CSF leaks and meningitis rates that pedicle flap and the multilayer technique for closing help to decrease below to 10%.

Management Strategy

Further management is based on histopathological patterns of the tumors. Meckel's cave and cavernous sinus surroundings is the location of a wide variety of lesions either benign or malignant. Some inflammatory and benign processes may mimic other aggressive tumors [1, 24]. Each type of lesion requires individual consideration for management that may be medical, oncological or aggressive surgery. Clinical findings and neuroimaging is not always sufficient to make an exact diagnosis and may induce unnecessary aggressive procedure with high morbidity. The percutaneous biopsy through the foramen ovale should be performed in cases of tumor located in the posterior part of the cavernous sinus, upper part of the petroclival region and the Meckel's cave because the needle inserted through the foramen ovale can reach these regions. This diagnostic tool has a high sensitivity and specificity. Messerer et al.[24] reported an excellent kappa coefficient correlation between histological diagnosis at biopsy and open craniotomy. The advantage is a far lower morbidity than the transcranial route. However, in a series of 50 patients, 14% of these procedures were unproductive. In these cases, we recommend the use of endoscopy to make biopsy. The endoscopic procedure is also recommended in cases of lesions located at the anterior part of the cavernous sinus, which cannot be reached by the percutaneous needle. In cases of necessary surgical removal after histopathological results, the endoscopic approach is also recommended.

Figure 12 shows the algorithm proposed by the authors.

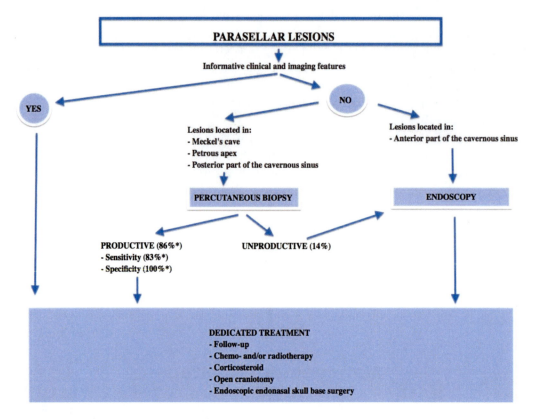

Figure 12. Flow diagram for decision-making process.

Conclusion

The parasellar region is the location of a wide variety of lesions. Clinical and neuroimaging findings are often insufficient to provide an accurate diagnosis. The diagnosis certainty is required as each lesion has its individual management. Percutaneous biopsy of these lesions through the foramen ovale avoids unnecessary aggressive procedure but it is sometimes unproductive. Percutaneous biopsy is also not suitable for lesions located anteriorly to cavernous sinus. Endoscopy allows an additional approach to biopsy and orientates management. Endoscopy may also be used in lesion removal, keeping in mind that conventional aggressive craniotomies are still required in selected cases.

References

[1] Janjua RM, Wong KM, Parekh A, van Loveren HR. Management of the great mimicker: Meckel cave tumors. *Neurosurgery.* 2010;67(2 Suppl Operative):416-21. Epub 2010/11/26.

[2] Ahn JY, Kwon SO, Shin MS, Joo JY, Kim TS. Chronic granulomatous neuritis in idiopathic trigeminal sensory neuropathy. Report of two cases. *Journal of neurosurgery.* 2002;96(3):585-8. Epub 2002/03/09.

[3] Attout H, Rahmeh F, Ziegler F. [Cavernous sinus lymphoma mimicking Tolosa-Hunt syndrome]. *La Revue de medecine interne / fondee par la Societe nationale francaise de medecine interne.* 2000;21(9):795-8. Epub 2000/10/20. Lymphome du sinus caverneux mimant un syndrome de Tolosa-Hunt.

[4] Gottfried ON, Chin S, Davidson HC, Couldwell WT. Trigeminal amyloidoma: case report and review of the literature. *Skull base : official journal of North American Skull Base Society* [et al]. 2007;17(5):317-24. Epub 2008/03/12.

[5] Hartel F. Die leitungsanasthesie und injektionbehandlung des ganglion gasseri und der trigeminumsstamme. *Arch Klin Chir.* 1912;100:193-292.

[6] Jho HD, Carrau RL. Endoscopic endonasal transsphenoidal surgery: experience with 50 patients. *Journal of neurosurgery.* 1997;87(1):44-51. Epub 1997/07/01.

[7] Alfieri A, Jho HD. Endoscopic endonasal approaches to the cavernous sinus: surgical approaches. *Neurosurgery.* 2001;49(2):354-60; discussion 60-2. Epub 2001/08/16.

[8] Alfieri A, Jho HD. Endoscopic endonasal cavernous sinus surgery: an anatomic study. *Neurosurgery.* 2001;48(4):827-36; discussion 36-7. Epub 2001/04/27.

[9] Rivera-Serrano CM, Oliver C, Prevedello D, Gardner P, Snyderman C, Kassam A, et al. Pedicled Facial Buccinator (FAB) flap: a new flap for reconstruction of skull base defects. *The Laryngoscope.* 2010;120 Suppl 4:S234. Epub 2011/01/13.

[10] Rivera-Serrano CM, Terre-Falcon R, Fernandez-Miranda J, Prevedello D, Snyderman CH, Gardner P, et al. Endoscopic endonasal dissection of the pterygopalatine fossa, infratemporal fossa, and post-styloid compartment. Anatomical relationships and importance of eustachian tube in the endoscopic skull base surgery. *The Laryngoscope.* 2010;120 Suppl 4:S244. Epub 2011/01/13.

[11] Prevedello DM, Pinheiro-Neto CD, Fernandez-Miranda JC, Carrau RL, Snyderman CH, Gardner PA, et al. Vidian nerve transposition for endoscopic endonasal middle fossa approaches. *Neurosurgery.* 2010;67(2 Suppl Operative):478-84. Epub 2010/11/26.

[12] Gardner PA, Kassam AB, Rothfus WE, Snyderman CH, Carrau RL. Preoperative and intraoperative imaging for endoscopic endonasal approaches to the skull base. Otolaryngologic clinics of North America. 2008;41(1):215-30, vii. Epub 2008/02/12.

[13] Ramachandran R, Singh PM, Batra M, Pahwa D. Anaesthesia for endoscopic endonasal surgery. *Trends in Anaesthesia and Critical Care.* 2011;1(2):79-83.

[14] Cappabianca P, de Divitiis E. Endoscopy and transsphenoidal surgery. *Neurosurgery.* 2004;54(5):1043-48; discussions 8-50. Epub 2004/04/29.

[15] Frank G, Pasquini E, Farneti G, Mazzatenta D, Sciarretta V, Grasso V, et al. The endoscopic versus the traditional approach in pituitary surgery. *Neuroendocrinology.* 2006;83(3-4):240-8. Epub 2006/10/19.

[16] Jho HD. Endoscopic transsphenoidal surgery. *Journal of neuro-oncology.* 2001;54(2):187-95. Epub 2002/01/05.

[17] Cappabianca P, Cavallo LM, Colao A, de Divitiis E. Surgical complications associated with the endoscopic endonasal transsphenoidal approach for pituitary adenomas. *Journal of neurosurgery.* 2002;97(2):293-8. Epub 2002/08/21.

[18] Dehdashti AR, Ganna A, Karabatsou K, Gentili F. Pure endoscopic endonasal approach for pituitary adenomas: early surgical results in 200 patients and comparison with previous microsurgical series. *Neurosurgery.* 2008;62(5):1006-15; discussion 15-7. Epub 2008/06/27.

[19] Dolenc VV, Lipovsek M, Slokan S. Traumatic aneurysm and carotid-cavernous fistula following transsphenoidal approach to a pituitary adenoma: treatment by transcranial operation. *British journal of neurosurgery.* 1999;13(2):185-8. Epub 2000/01/05.

[20] Raymond J, Hardy J, Czepko R, Roy D. Arterial injuries in transsphenoidal surgery for pituitary adenoma; the role of angiography and endovascular treatment. *AJNR American journal of neuroradiology.* 1997;18(4):655-65. Epub 1997/04/01.

[21] Benoit BG, Wortzman G. Traumatic cerebral aneurysms. Clinical features and natural history. *Journal of neurology, neurosurgery, and psychiatry.* 1973;36(1):127-38. Epub 1973/02/01.

[22] Semple PL, Laws ER, Jr. Complications in a contemporary series of patients who underwent transsphenoidal surgery for Cushing's disease. *Journal of neurosurgery.* 1999;91(2):175-9. Epub 1999/08/05.

[23] Ciric I, Ragin A, Baumgartner C, Pierce D. Complications of transsphenoidal surgery: results of a national survey, review of the literature, and personal experience. *Neurosurgery.* 1997;40(2):225-36; discussion 36-7. Epub 1997/02/01.

[24] Messerer M, Dubourg J, Saint-Pierre G, Jouanneau E, Sindou M. Percutaneous biopsy of lesions in the cavernous sinus region through the foramen ovale: diagnostic accuracy and limits in 50 patients. *Journal of neurosurgery.* 2012;116(2):390-8. Epub 2011/11/22.

In: Minimally Invasive Skull Base Surgery
Editor: Moncef Berhouma
ISBN: 978-1-62808-567-9
© 2013 Nova Science Publishers, Inc.

Chapter XI

Extended Endoscopic Endonasal Approaches to the Pterygopalatine and Infratemporal Fossae

Bashar Abuzayed

Department of Neurosurgery, Al-Bashir Governmental Hospital,
Amman, Jordan

Abstract

Our aim in this study was to recognize the endoscopic anatomy of the pterygopalatine fossa (PPF) and the anatomic variations of the related neurovascular structures, to define the endoscopic endonasal approach to this region. To reach the PPF endonasally, the middle meatus transpalatine approach, the middle meatus transnasal approach and the inferior turbinectomy transnasal approaches can be performed. The PPF region was best exposed by the middle meatus transnasal approach. The superior and posterior walls of the maxillary sinus are defined, the sphenopalatine foramen is widened by drilling the orbital process of the foramen and the sphenopalatine artery is then exposed. The posterior wall of maxillary sinus is opened to expose the pterygopalatine fossa and its neurovascular contents. During the endoscopic transnasal transmaxillary approach to the PPF, it is possible to face wide range of variations in every phase of the approach. Understanding the anatomy of this region and the neurovascular relations from the endoscopic view by cadaver dissections is necessary to perform safer surgery.

Abbreviations

DPA   descending palatine artery.
IOA   infraorbital artery.
PPF   pterygopalatine fossa.
PSAA posterior superior alveolar artery.

| V2 | maxillary nerve. |
| VA | vidian artery. |

Introduction

Approach to the pterygopalatine fossa (PPF) is considered a surgical challenge due to its location deep in the mid-third of the face and its complex neurovascular relations. Because of the open communications between the fossa and surrounding anatomic spaces, tumoral of infectious processes can spread through the PPF. Direct involvement of the PPF by malignancy may occur by contiguous spread, perineural invasion, or vascular seeding of metastatic lesions. Parotid gland tumors, adenocarcinomas, and chondrosarcomas, juvenile nasopharyngeal angiofibromas, as well as benign neoplasms such meningiomas, schwannomas, and neurofibromas, may extend to the PPF [1].

The endoscopic endonasal approach to the PPF has become a popular approach for the surgical management of pathologies occupying this region. This popularity was gained due to the advantages offered by these techniques; such as minimal invasive access, less neurovascular retraction, and thus, lower risk of neurovascular injury, wide view of exposure, no disruption of the facial skeleton – in contrast to traditional transmaxillary – transantral approach, and better postoperative performance of the patients with shorter hospitalization period. However, only a few anatomical and clinical papers describing the endoscopic endonasal approach to the PPF have been published [2-10].

Traditional approaches to the PPF usually involve an anterior transmaxillary technique [2]. The transoral-transantral approach requires the transsection of the anterior and posterior walls of the maxillary sinus. It is associated with complications, such as facial edema and pain, oroantral fistula, chronic maxillary sinusitis, and high risk of vascular, neural and injury [2, 5]. Also, lateral approaches to expose and resect malignant or invasive tumors have been described [11, 12].

Endonasal endoscopic approaches avoid external incisions and flaps, facial skeleton disturbance and cosmetic problems. These approaches also provide wide neurovascular exposure (especially the internal maxillary artery) with easy manipulation, and it can be tolerated better by patients with oncologic pathologies [2, 6-10]. In addition, it can be used to treat lesions in the lateral recess of the sphenoid sinus which represents an extended pneumatization of the sphenoid bone laterally just posterior to the PPF [11, 12].

Anatomic Considerations

The PPF is a small pyramidal area (its height is ~2 cm and base ~1 cm) with incomplete osseous boundaries located among maxillary, sphenoid, and palatine bones [4]. It is located over the virtual line that separates splanchnocranium and neurocranium, and represents a natural intersection among anatomic compartments of the skull base, facial, intracranial, oral and neck regions.

The PPF opens laterally to the infratemporal fossa, and is bordered posteriorly by the pterygoid plates of the sphenoid bone, which fuses at the skull base into the pterygoid process. The perpendicular plate of the palatine bone forms the medial wall of the PPF, which tapers inferiorly in a funnel-shaped space at the level of the attachment between maxillary bone and the pyramidal process of the palatine. The maxillary tuberosity constitutes the anterior boundary of the PPF, which communicates with the orbital apex through the medial portion of the inferior orbital fissure (Figure 1). Through the pterygomaxillary fissure, the PPF merges laterally the infratemporal fossa and the infraorbital nerve in its superior course. The rotundum and sphenopalatine foramina, the vidian and palatovaginal canals, and the greater and lesser palatine foramina open the fossa contents to the surrounding anatomic regions (Figure 2A). Topographically, the PPF may be divided into two compartments. The anterior space is occupied by the third segment of the maxillary artery (and its branches), whereas the posterior compartment is represented by the maxillary nerve and the sphenopalatine ganglion (and its branches).

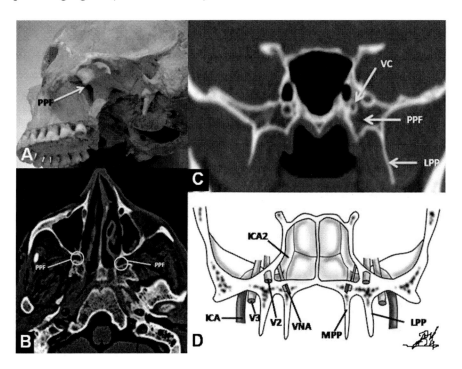

Figure 1. Skull photograph (A), Paranasal sinus computed tomography scan (B, C) and schematic drawing (D) demonstrating the bony anatomy of the pterygopalatine fossa. PPF: pterygopalatine fossa, VC: vidian canal, LPP: lateral pterygoid plate, MPP: medial pterygoid plate, ICA: internal carotid artery, ICA2: second genu of internal carotid artery, VNA: vidian nerve and artery, V2: maxillary nerve, V3: mandibular nerve.

The sphenopalatine ganglion (also known as the pterygopalatine ganglion) is located in the posterior-inferior portion of the PPF just anterior to the pterygoid canal that conveys the vidian nerve (Figure 2B). The maxillary nerve enters the fossa through the foramen rotundum and, remaining superolateral to the ganglion, gives rise to its zygomatic branch while continuing to the infraorbital groove as the infraorbital nerve. During its anterolateral course, the infraorbital nerve constitutes an important surgical landmark that defines the PPF (medial) from the infratemporal fossa (lateral). Neural rami of distribution from the sphenopalatine ganglion are divisible into a few groups: ascending (or orbital), posterior (to the pharynx), internal (to the nose), and descending (to the palate). Communicating rami connect the maxillary nerve to the inferiorly located sphenopalatine ganglion. These nerves, along with the presence of orbital rami, contribute to form the superior apex of the small ganglion conferring to it its characteristic macroscopic triangular-shaped appearance. The thin orbital rami carry parasympathetic neurons and erratically reach the nasal mucosa through the anterior or posterior ethmoidal canals. The sphenopalatine ganglion also receives sympathetic and parasympathetic fibers through the vidian nerve. The vidian nerve is formed by the fusion of greater and deep petrosal nerves at the level of the foramen lacerum, enters the PPF through the pterygoid (vidian) canal along with the vidian branch of the maxillary artery. The greater and lesser palatine nerves (descending branches) arise from the inferior-posterior portion of the ganglion and through the palatine canals reach the oral cavity. The greater and lesser palatine arteries accompany their course (to vascularize mucosal membranes of hard palate, gums, and palatine glands) and may be responsible of intra-operative hemorrhage if inadvertently injured.

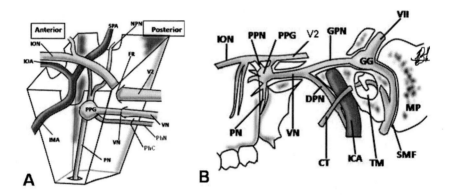

Figure 2. Schematic drawing demonstrating the neurovascular relations in the pterygopalatine fossa. IMA: internal maxillary artery, ION: infraorbital nerve, SPA: sphenopalatine artery, ICA: internal carotid artery, FR: foramen rotundum, VC: vidian canal, PhC: pharyngeal canal, PhN: pharyngeal nerve, VN: vidian nerve, V2: maxillary nerve, NPN: nasopalatine nerve, ION: infraorbital nerve, PPG: pterygopalatine ganglion, PN: palatine nerves, PPN: pterygopalatine nerves, GPN: great petrosal nerve, DPN: deep petrosal nerve, GG: geniculate ganglion, CT: chorda tympany, TM: tympanic membrane, VII: facial nerve, SMF: sternomastoid foramen, MP: mastoid process.

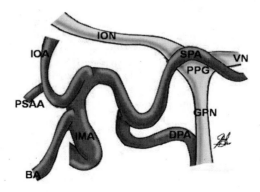

Figure 3. Schematic drawing demonstrating the vascular and neural contents of the pterygopalatine fossa. IMA: internal maxillary artery, IOA: infraorbital artery, SPA: sphenopalatine artery, DPA: descending palatine artery, PSAA: posterior-superior alveolar artery, BA: buccal artery, ION: infraorbital nerve, VN: vidian nerve, GPN: great palatine nerve, PPG: perygopalatine ganglion.

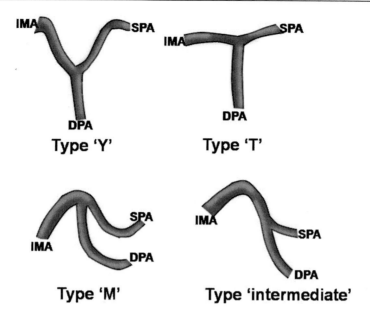

Figure 4. Morton and Khan morphologic classification of the third portion of the maxillary artery. IMA: internal maxillary artery, SPA: sphenopalatine artery, DPA: deep palatine artery.

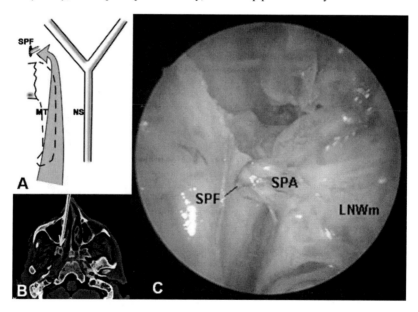

Figure 5. The entry point is between the middle turbinate and lateral nasal wall in the middle meatal transpalatine approach. When the sphenopalatine artery is identified, the sphenopalatine foramen is enlarged to expose the pterygopalatine fossa. NS: nasal septum, MT: middle turbinate, SPF: sphenopalatine foramen, SPA: sphenopalatine artery, LNWm: mucosa of the lateral nasal wall.

The anterior space is occupied by the third segment of the maxillary artery (and its branches) (Figure 3). The pterygopalatine segment of the maxillary artery (third segment) reaches the PPF through the pterygomaxillary fissure. In its variable and tortuous course, the artery provides terminal and anastomotic branches which, because of the diffuse presence of fat tissue, may sometimes result of difficult recognition. The posterior superior alveolar artery

(PSAA) and the infraorbital artery (IOA) rise proximally in the PPF followed distally by the descending palatine (DPA), the vidian (VA), and the sphenopalatine arteries (SPA) in that order. According to Morton and Kahn [15], the third portion of the maxillary artery is classified into four types, 'intermediate' type (33-50%), 'M' type (16-33%), 'T' type

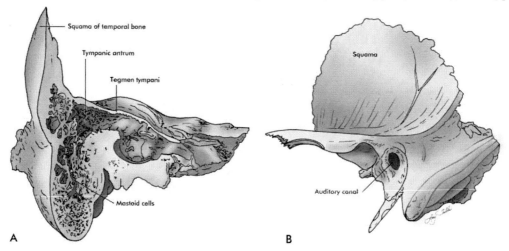

(17-24%) and 'Y' type (17-33%) [4, 15] (Figure 4).

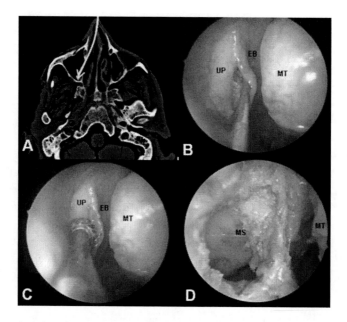

Figure 6. A) Paranasal computed tomography scan demonstrating the entrance path in the Middle metal transnasal approach (arrow). B) The uncinate process is identified lateral to middle turbinate and mucosal incision is made. C) After exposing the bone, the maxillary sinus is entered by drilling. D) infundibulotomy is completed and maxillary sinus is exposed. UP: uncinate process, EB: ethmoid bulla, MT: middle turbinate, MS: maxillary sinus.

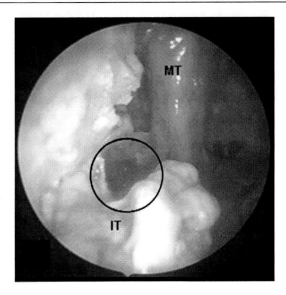

Figure 7. Endoscopic view of inferior turbinectomy, with resection made in the angle between the middle and inferior turbinates *(circle)*. MT: middle turbinate, IT: inferior turbinate.

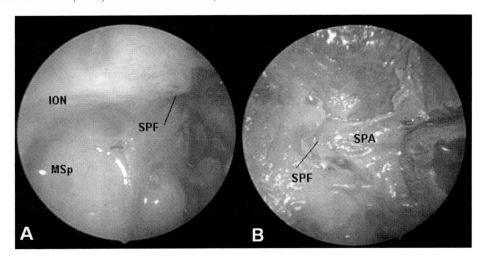

Figure 8. A) Endoscopic view of the posterior wall and the posterior portion of the superior wall of the maxillary sinus. B) The sphenopalatine foramen is enlarged. ION: infraorbital nerve, MSp: posterior wall of the maxillary sinus, SPF: sphenopalatine foramen, SPN: sphenopalatine nerve.

Surgical Technique

The patient is placed in supine position, with the head in neutral position and 10 to 15 degrees adducted toward the left shoulder [4, 16]. The face of the patient is covered keeping the nose and the eyes exposed with transparent drape. Intraoperative C-arm fluoroscopy, neuronavigation and intraoperative imaging techniques can be used for more intraoperative orientation.

A 30-degree endoscope is introduced from the nostril of the side of the pathology. The inferior and middle turbinates are identified. The PPF is entered through the maxillary sinus by three endoscopic endonasal approaches [5]:

1. Middle meatal transpalatine approach, in which the entry point is between the middle turbinate and the lateral nasal wall. When the sphenopalatine artery is identified, the sphenopalatine foramen is enlarged to expose PPF (Figure 5).
2. Middle meatal transnasal approach, in which maxillary ostium is enlarged, with or without infundibulotomy. This is the preferred surgical approach to maxillary sinus (Figure 6).
3. Inferior turbinectomy transnasal approach, in which inferior turbinate is removed and the entry point is lateral nasal wall between the inferior and middle turbinates. This approach provides us the widest exposure of maxillary sinus, but it is not the preferred surgical approach because the great importance of inferior turbinate in areolation (Figure 7).

After entering the maxillary sinus, we could visualize the posterior and superior walls, the infraorbital nerve in the superior wall, the sphenopalatine artery and foramen at the medial side of the posterior wall (Figure 8A). The infraorbital nerve is a consistent landmark that delimits the surgical boundaries between the pterygopalatine fossa and the infratemporal fossa. In fact, the pterygopalatine fossa is located medially to it, whereas the infratemporal fossa is located laterally and inferiorly to it. The posterior wall of the maxillary sinus, which is located medially to the infraorbital nerve, constitutes the anterolateral limit of the pterygopalatine fossa, and the posterior wall of the maxillary sinus located laterally and inferiorly to the infraorbital nerve delimits the infratemporal fossa anteriorly.

The orbital process of the palatine bone is removed and the sphenopalatine foramen is enlarged (Figure 8B). The posterior wall of the maxillary sinus is then removed up to the vertical process of the palatine bone medially, and up to the angle between the lateral and posterior wall of the maxillary sinus laterally to expose the pterygomaxillary fissure. The posterior wall of the maxillary sinus is usually very thin and easily removed by drilling or by a Kerrison rongeur. A wide opening of the posterior wall of the maxillary sinus allows access to the infratemporal fossa structures. However, the presence of a great quantity of fat and of the internal and external pterygoid muscles makes it difficult to identify the mandibular nerve and its branches.

The anterior surface of the pterygoid process now becomes visible, and the pterygoid canal, the foramen rotundum, and the inferior part of the superior orbital fissure, which is external to the pterygopalatine fossa, are identifiable. Located posteriorly and medially to the pterygopalatine fossa is the sphenoid sinus, which is full of landmarks that are useful in orienting the surgeon during live operations on lesions extending to the pterygopalatine fossa.

After the fascia that covers the pterygomaxillary fossa is incised and the fat inside the fossa is removed, the maxillary artery (that is the first vessel to be identified) becomes visible (Figure 9). The maxillary artery runs on the anterior edge of the lateral pterygoid muscle and reaches the pterygopalatine fossa through the pterygomaxillary fissure. This artery has a tortuous and variable route, but is always on an anterior plane with respect to the nerves inside the fossa. The maxillary artery passes through the pterygopalatine fossa and ends with the origin of the sphenopalatine and the descending palatine arteries. As soon as the maxillary artery enters the pterygomaxillary fossa, it branches into two collateral vessels: the posterosuperior alveolar branch, which is small, and the infraorbital branch, which courses with the infraorbital nerve in the homonymous canal. The sphenopalatine artery is the uppermost and medially located vessel in the fossa and is sometimes hidden by the orbital apophysis of the palatine bone. Once the nasal fossa is reached posterior to the tail of the middle turbinate, the sphenopalatine artery divides in two branches: one, called the "nasopalatine artery," is medial and directed to the nasal septum; the other, called the "posterior nasal artery," is directed to the tails of the turbinates. The two most important landmarks inside the pterygopalatine fossa are the vidian and the maxillary nerves. Both reach the pterygopalatine fossa from its upper part, and their identification allows the definition of a surgical corridor between them that enables the exposure of the lateral wall of the sphenoid sinus. This surgical corridor has a quadrangular shape; it is delineated posteriorly by the intrapetrous segment of the intracavernous carotid artery and by the inferior segment of the vertical tract of the same vessel, and anteriorly by the pterygoid bone extending from the foramen rotundum to the pterygoid canal. This area is bordered superiorly by the maxillary nerve and inferiorly by the vidian nerve. The quadrangular area can be involved by lesions arising in the pterygopalatine fossa, extending toward the middle cranial fossa and/or the cavernous sinus, and vice versa.

The posterior compartment of the PPF is occupied by the neural elements (Figure 10). In the PPF, V2 gives a branch to the pterygopalatine ganglion (PPG), before continuing as the infraorbital nerve (ION). ION continues through the inferior orbital fissure, which is continuous with the pterygomaxillary fissure. The latter marks the limit between the PPF and infratemporal fossa; therefore, ION may be used as a proxy landmark to indicate the boundary of the infratemporal fossa. The pterygopalatine ganglion has a triangular shape and lies posterior to the sphenopalatine artery. The three principal branches of the pterygopalatine ganglion can be identified: the descending palatine nerve inferiorly, a branch of V2 superolaterally, and the vidian nerve superomedially. Proximal to the pterygopalatine ganglion, the vidian nerve runs in its canal along the sphenoid sinus floor in a posterolaterally direction. This nerve meets the second genu of the internal carotid artery (between the horizontal and vertical segments). Cavallo et al. and others (Kassam) proposed the vidian canal as a consistent landmark to localize the anterior genu of the horizontal segment of the petrous internal carotid artery safely [3].

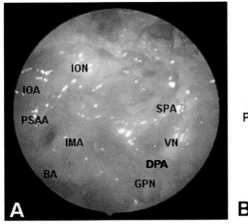
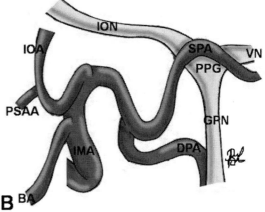

Figure 9. Endoscopic view (A) and schematic drawing (B) demonstrating the contents of the pterygopalatine fossa. IMA: internal maxillary artery, IOA: infraorbital artery, SPA: sphenopalatine artery, DPA: descending palatine artery, PSAA: posterior-superior alveolar artery, BA: buccal artery, ION: infraorbital nerve, VN: vidian nerve, GPN: great palatine nerve, PPG: perygopalatine ganglion.

Surgical Outcome

The three different approaches all showed the structures of the pterygopalatine fossa. The middle meatal transpalatine approach allows a medial exposure of the pterygopalatine fossa. The vidian nerve is easily found superomedially and can be followed to identify the pterygopalatine ganglion. The middle meatal transantral approach allows a wider approach, particularly indicated to approach the lateral pterygopalatine fossa, in which the infraorbital nerve is the first landmark identified. When it is hard to identify the sphenopalatine artery, this approach is a valid alternative. The inferior turbinectomy transantral approach allows the widest view and room for surgical maneuver, so that the infratemporal fossa can also be approached easily.

The wide opening of the maxillary sinus created by endoscopic medial maxillectomy provides easy access to the PPF with a wide working space, thus avoiding the more invasive skull base approaches based upon extensive craniofacial resections. In selected patients, endoscopic approaches to surgical resection have yielded good results with the potential benefits of improved quality of life, decreased short-term morbidity, and fewer long-term side effects [6-10].

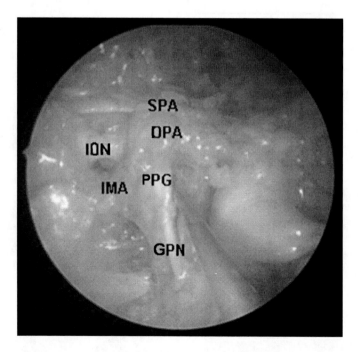

Figure 10. Exposing the pterygopalatine ganglion beneath the internal maxillary artery. IMA: internal maxillary artery, SPA: sphenopalatine artery, DPA: descending palatine artery, ION: infraorbital nerve, GPN: great palatine nerve, PPG: pterygopalatine ganglion.

Complications

The inferior turbinectomy transantral approach allows the widest view and room for surgical maneuver, so that the infratemporal fossa can also be approached easily [4]. However, removal of the inferior turbinate has adverse effect on areolation and phonation [5].

The greater and lesser palatine nerves arise from the inferior-posterior portion of the ganglion and through the palatine canals reach the oral cavity. The greater and lesser palatine arteries accompany their course (to vascularize mucosal membranes of hard palate, gums, and palatine glands) and may be responsible of intraoperative hemorrhage if inadvertently severed [3, 5]. This artery, its branches, and the neural content of the fossa are exposed to the potential risk of damage in the course of both open and endoscopic procedures. In particular, the usual location of the DPA and its anteromedial branching from the maxillary artery should be taken into conscientious account while performing a pterygomaxillary disjunction [3, 5]. Although open approaches may favor an easier management of intraoperative hemorrhages, control of such complications could be a more difficult achievement in the course of minimally invasive procedures. Ligation of the sphenopalatine artery may, for instance, be essential to minimize the intraoperative blood [3, 4].

Visualization and manipulation of normal and pathologic tissues by means of endoscopic instruments may sometimes be problematic as a result of space confinements and distorted anatomy. Diffuse presence of fat tissue surrounding the maxillary artery, anatomic re-arrangement caused by neoplastic or infective diseases, and presence of an engorged venous plexus may also further complicate the correct identification of neurovascular structures during the surgical act [3, 5]. Also, lateral extension is should be avoided due to the risk of destabilization of the pterygoid process and, consequently, problems with mastication due to malfunction of the lateral pterygoid muscle (which opens the jaw) and the medial pterygoid muscle (which governs the slight lateral shift of the mandible during chewing) [3, 5].

References

[1] Jian X. C., Wang C. X., Jiang C. H. Surgical management of primary and secondary tumors in the pterygopalatine fossa. *Otolaryngol Head Neck Surg.* 132(1):90-94, 2005.

[2] del Gaudio J. M. Endoscopic transnasal approach to the pterygopalatine fossa. *Arch Otolaryngol Head Neck Surg.* 129(4):441-446, 2003.

[3] Cavallo L. M., Messina A., Gardner P., Esposito F., Kassam A. B., Cappabianca P., de Divitiis E., Tschabitscher M. Extended endoscopic endonasal approach to the pterygopalatine fossa: Anatomic study and clinical considerations. *Neurosurg Focus* 19 (1):E5, 2005.

[4] Abuzayed B., Tanriover N., Gazioglu N., Cetin G., Akar Z. Extended endoscopic endonasal approach to the pterygopalatine fossa: Anatomic study. *J Neurosurg Sci.* 53 (2):37-44, 2009.

[5] Alfieri A., Jho H. D., Schettino R,. Tschabitscher M. Endoscopic endonasal approach to the pterygopalatine fossa: anatomic study. *Neurosurgery* 52(2):374-378, 2003.

[6] Kodama S., Kawano T., Suzuki M. Endoscopic transnasal resection of ectopic esthesioneuroblastoma in the pterygopalatine fossa: technical case report. *Neurosurgery* 65(6 Suppl):E112-113, 2009.
[7] Kodama S., Mabuchi H., Suzuki M. Endoscopic endonasal transturbinate approach to the pterygopalatine fossa in the management of juvenile nasopharyngeal angiofibromas. *Case Report Otolaryngol.* 2012: 786262, 2012.
[8] Martínez Ferreras A., Rodrigo Tapia J. P., Llorente Pendás J. L., Suárez Nieto C: [Endoscopic nasal surgery for pterigopalatine fossa schwannoma]. *Acta Otorrinolaringol Esp.* 56(1):41-43, 2005. Spanish.
[9] Rosique-López L., Rosique-Arias M., Sánchez-Celemin F. J. [Pterygopalatine fossa schwannoma. Endoscopic approach]. *Neurocirugia* (Astur) 21(5):405-409, 2010. Spanish.
[10] Tosun F., Durmaz A., Deveci M. S., Hidir Y. Endoscopic transnasal transpterygoid approach for parashpenoidal myxoma. *Skull Base* 19(5)349-352, 2009.
[11] Douglas R., Wormald P. J. Endoscopic surgery for juvenile nasopharyngeal angiofibroma: Where are the limits. *Curr Opin Otolaryngol Head Neck Surg.* 14(1):1-5, 2006.
[12] Sennes L. U., Butugan O., Sanchez T. G., Bento R. F., Tsuji D. H. Juvenile nasopharyngeal angiofibroma: the routes of invasion. *Rhinology* 41(4):235-240, 2003.
[13] Al-Nashar I. S., Carrau R. L., Herrera A., Snyderman C. H. Endonasal transnasal transpterygopalatine fossa approach to the lateral recess of the sphenoid sinus. *Laryngoscope* 114(3):528-532, 2004.
[14] Magro F., Solari D., Cavallo L. M., Samii A., Cappabianca P., Paternò V., Lüdemann W. O., de Divitiis E., Samii M. The endoscopic endonasal approach to the lateral recess of the sphenoid sinus via the pterygopalatine fossa: comparison of endoscopic and radiological landmarks. *Neurosurgery* 59(4 Suppl 2):ONS237-242, 2006.
[15] Morton A. L., Khan A. Internalmaxillary artery variability in the pterygopalatine fossa. *Otolaryngol Head Neck Surg.* 104: 204-209, 1991.
[16] Abuzayed B., Tanriover N., Ozlen F., Gazioglu N., Ulu M. O., Kafadar A. M., Erasaln B., Akar Z. Endoscopic endonasal transsphenoidal approach to the sellar region: Results of endoscopic dissection on 30 cadavers. *Turk Neurosurg.* 19(3):237-244, 2009.

In: Minimally Invasive Skull Base Surgery
Editor: Moncef Berhouma

ISBN: 978-1-62808-567-9
© 2013 Nova Science Publishers, Inc.

Chapter XII

Management of Middle Fossa CSF Leak

*Chiazo Amene, Sudheer Ambekar, Cedric Shorter,
Osama Ahmed, Justin Haydel, Anil Nanda and Bharat Guthikonda*
Department of Neurosurgery, LSU HSC Shreveport, Shreveport, LA, US

Abstract

Defects in the middle cranial fossa can present a daunting challenge to skull base surgeons, both in diagnosis and treatment. Anatomically, this area includes the temporal bone and its extensions and defects are classified as congenital, spontaneous or acquired. Acquired middle fossa defects are more commonly seen and are often iatrogenic or traumatic in origin. Defects may present clinically as middle ear effusions, hearing loss, cerebrospinal fluid (CSF) otorrhea or meningitis. Various imaging modalities can aid in the diagnosis of a middle fossa skull base defect and these include thin-cut spiral CT scans of the temporal bone, 3D-ciss MR imaging and intrathecal CT or MRI cisternography. Traumatic middle fossa defects are usually small and uncomplicated with a minuscule amount of CSF leakage and may be managed conservatively. However, spontaneous defects and defects acquired after surgical manipulations often do not respond to conservative management and require further surgical intervention to avoid possibly fatal complications like meningitis and brain herniation.

In the literature, there are multiple reports of repair of middle fossa skull base defects using various approaches which include the transmastoid approach, the middle fossa approach, combined transmastoid/middle fossa approaches and more recently, 'minimally-invasive' techniques that often involve the use of an endoscope. There have also been different modalities utilized in these repairs including the use of vascularized local flaps, microvascular free tissue transfers and synthetic agents like fibrin glue, collagen-based dural substitute and bone cement.

While there does not appear to be any consensus in the treatment of these defects, both with the approach and agent used, there is general agreement that a multi-layered closure is tantamount and the approach used should not create further defects. The following chapter is an overview of current approaches to the repair of middle cranial fossa skull base defects, with emphasis on minimally-invasive techniques, their complications and outcomes and a review of the authors' current series.

Introduction

Cerebrospinal fluid leaks pose a significant concern to skull base surgeons secondary to their often-difficult accessibility and morbid, possibly fatal, sequelae. CSF leaks from the skull base arise when there is a defect in the bony calvarium as well as a breach in the overlying dura. These defects may be in the anterior cranial fossa, middle fossa or posterior fossa. Defects from the middle fossa prove especially difficult to manage as the primary bone that makes up the base of the middle cranial fossa, the temporal bone, accounts for a complex anatomy, especially when discussing surgical approaches. Diagnosing a CSF leak from a middle fossa cranial defect may be straightforward when it occurs after a traumatic head injury or after surgical resection of a lesion in the area. In the absence of any mechanism of injury, the index of suspicion must be high to diagnose middle fossa defects, as they can occur spontaneously. A detailed neurologic exam is crucial in making the diagnosis. The evaluation process can also be facilitated by one or more imaging modalities, like thin-cut computed tomography (CT), CT cisternography or magnetic resonance imaging (MRI) with or without cisternography. While some middle fossa CSF leaks can be managed conservatively, others do require surgical repair and this is often dependent on the etiology of the defect.

The history of middle fossa CSF leak repairs dates back to the early 1900s where Canfield was the first to describe the use of a xenograft (canine dura) to repair a post-operative leak. Dandy later established the use of autologous tissue to repair CSF otorrhea [1]. Approaches and mechanisms of repair vary widely in the current literature and are dependent on a multiple factors including size and location of defect and also the surgeon's personal preference and experience level. Both the bony defect and the dural breach need to be addressed to efficiently repair a CSF leak and the end goal is to prevent further CSF leakage and therefore avoid complication like hearing loss, recurrent meningitis/encephalitis, or death.

In this chapter, we will discuss the complex anatomy of the middle fossa, types of defects, their clinical presentation and steps in the diagnosis of CSF leaks. We will also detail both clinical and surgical management and discuss the advantages and disadvantages of the different surgical approaches and materials available for repair.

Brief Anatomy

Diagnosing and treating middle cranial fossa defects require an in-depth knowledge of the anatomy of the area. The temporal bone is the most complex bone in the human body and consists of 5 major parts, the squama, the petrous portion, the mastoid and tympanic portions and the styloid process (Figure 1). The squama is the only portion that is not an integral part of the skull base. The mastoid process contains air cells which are highly variable from individual to individual in terms of size, number and extent of pneumatization. These air cells connect to a large pocket known as the tympanic antrum which in turn connects the tympanic cavity. Above the tympanic antrum is a very thin plate of bone which separates the temporal lobe of the brain from the middle ear. This plate of bone is called the tegmen [2]. The tegmen is covered on its superior surface by the middle fossa floor dura and by mucosa on its inferior (or mastoid) surface.

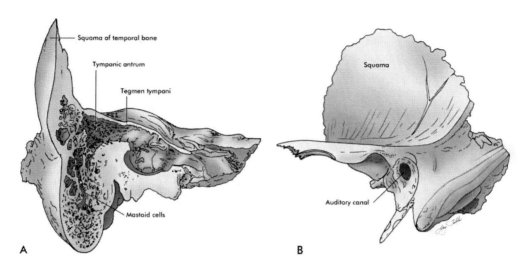

Figure 1. Coronal (A) and lateral (B) views of the temporal bone. Note the tegmen, the area of most defects.

Most defects in the middle fossa are found in this area and, with an associated dural breach, lead to a fistula between the intradural cavity and the tympanic cavity, resulting in CSF otorrhea. A defect also creates the potential for passage in the other direction of microbes into the intracranial cavity leading to infections like meningitis or intracranial abscesses. Autopsy studies have found that single tegmen defects occurred in about 15-34% of specimens [3] while there have been many cases where multiple tegmen defects were noted [4-6].

The other part of the temporal bone that often features in CSF leaks is the petrous portion, which is a common site of temporal bone fracture after head injury. These are typically grouped into longitudinal or transverse fractures. Longitudinal fractures parallel the long axis of the petrous pyramid, starting at the mastoid or external auditory canal (EAC), extending through the superior bony external canal and across the roof of the middle ear and ending anteromedially in the middle cranial fossa. With this connection, CSF otorrhea may occur, although it is often transient, as discussed below. Transverse fractures across the petrous pyramid may also cause otorrhea but are less likely to do so.

Types of Middle Fossa Defects

There are a wide variety of etiologies that can lead to a middle fossa CSF leak. These can be grouped into three broad categories: congenital, spontaneous or acquired defects.

Congenital conditions, which tend to occur in the younger population, can be related to developmental aberrations like a Hyrtl fissure, Mondini dysplasia, a petromastoid canal fistula or through an enlarged facial nerve canal. A Hyrtl fissure is a transient landmark in the fetal petrous temporal bone that connects the fossula of the round window to the posterior cranial fossa [7]. It usually ossifies by the 24th week of gestation. Persistence of this fissure can lead to a CSF fistula [7-8].

Mondini dysplasia is a congenital anomaly of the cochlea, where portions of the cochlea are underdeveloped, leading to vestibular dysfunctions and hearing loss. Patients with Mondini dysplasia are predisposed to developing middle fossa CSF leaks, although some controversies about this exist in the literature [9]. The petromastoid canal carries subarcuate vessels beneath the superior semicircular canal to the periantral mastoid air cells. A fistula at this location will lead to a CSF accumulation in the mastoid air cells. An enlarged proximal facial nerve canal has also been reported as the etiology for middle fossa CSF leak [9-10]. Congenital etiologies usually present early in life, often with sensorineural hearing loss or meningitis.

Spontaneous middle fossa CSF leaks are the most difficult to diagnose and are considered when there is no obvious etiology responsible for the defect (Figure 2). One theory as to the cause of spontaneous middle fossa CSF leak is the arachnoid granulations theory postulated by Gacek [11]. This essentially states that arachnoid granulations that do not find a venous termination during embryonic development come to lie in a blind-end against the inner bony surface of the skull. With time, the pulsations of CSF from the subarachnoid space causes erosion through the fibrous covering of the arachnoid granulation and eventually through the thin bone on the floor of the cranial fossa, thus resulting in a CSF leak [11]. This particular condition usually presents later in life. Ommaya theorized that bony areas with focal atrophy could be subject to changes in intracranial pressure [12]. Along these lines, intracranial hypertension as the cause of CSF leaks has been investigated [13]. Spontaneous CSF leaks do not respond to conservative management and usually require surgery [5, 14].

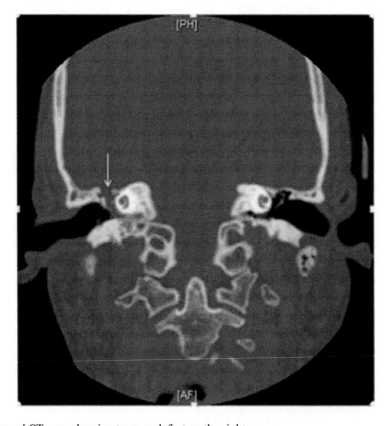

Figure 2. Coronal CT scan showing tegmen defect on the right.

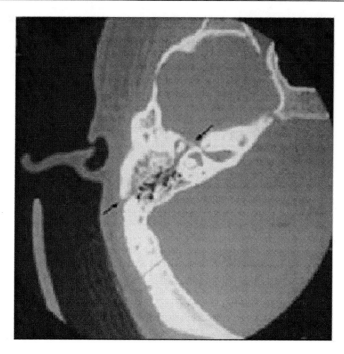

Figure 3. Axial CT scan showing a longitudinal fracture through the right temporal bone.

Acquired middle fossa defects can be related to a local, destructive pathology, like a benign or malignant tumor, infection, prior surgery, or trauma of the temporal bone. The most common causes of acquired middle fossa CSF leaks, and all middle fossa leaks in general, are trauma and post-surgical defects [15]. In traumatic cases, it is usually associated with blunt force trauma to the cranium which can result in fracture of the temporal bone (often longitudinal) which in turn may tear the dura, involve the otic capsule, or may lacerate the tympanic membrane (Figure 3). The reported incidence of CSF leak after closed head injury varies in the literature from 2-5% [16] and is higher when specifically evaluating temporal bone fractures [17] and as discussed below, most of these resolve with conservative management. CSF leaks after surgery may occur after many types of surgery including the translabyrinthine approach, middle fossa approach or retrosigmoid approach [18]. Conservative management may be attempted in these cases but often, the leaks are refractory and require surgical intervention to effectively treat them [14].

Presentation and Diagnosis

Middle fossa CSF leaks may present in a variety of ways. The patient may present with headache, frank or intermittent clear fluid leakage from the ear (otorrhea) either before or after myringotomy, ear fullness, hearing loss, serous or purulent otitis media, meningitis or even seizures [14, 19-20]. As mentioned earlier, the clinical suspicion for middle fossa CSF leak should be high if a patient presents with any of these symptoms, even in the absence of trauma or history of surgery. When otorrhea is present, the fluid can be collected and analyzed for β-2 transferrin, a protein that is present only in CSF. This test is therefore highly sensitive and specific in diagnosing CSF in the fluidleaks [21].

However, in some cases, the test may take days to return. A faster but less specific test is the glucose test using test strips. This may, however, produce false-positives from blood and mucous contamination, or false-negatives, possibly from low CSF glucose in the setting of an infection [22].

Appropriate imaging is essential in diagnosing middle fossa defects causing CSF leaks, although patients with vague symptoms and a small defect may go undiagnosed because neuroimaging may fail to show a middle fossa defect. If a defect in the temporal bone is suspected, high-resolution, thin-cut CT scans obtained through that area can confirm the presence and location of the defect with a high level of sensitivity and also aid in surgical planning [19, 23]. If a CT scan does not reveal a defect, CT cisternography with or without radionuclide isotopes is another option. CT cisternography may be more useful in patients with active CSF leaks as opposed to intermittent or low-volume leaks and is also position-dependent [6, 19-20]. Patients with intermittent or low-volume leaks tend to have a higher false-negative rate with CT cisternography and in these cases, MRI cisternography may be an appropriate next step as these appear to be independent of an active leak, can show soft tissue masses like encephaloceles and also has the added advantage of not requiring a lumbar puncture [24-25].

Clinical Management

Conservative management has shown to be an effective method in treatment middle fossa CSF leaks, especially secondary to traumatic injury. Greater than 80% of CSF leaks secondary to traumatic injury resolve spontaneously [15-16], often within the first week. Avoiding Valsalva maneuvers, like nose blowing or straining, and elevating the head of bed has been shown to be helpful. The next step in clinical management is CSF diversion, usually with the placement of a lumbar drain. CSF is diverted via this approach usually for 5-7 days and this has been shown to be successful in terminating leaks [16, 26]. The success of these conservative treatments is directly related to the etiology of the CSF leak and is important to note that many studies have shown a higher success rate of conservative management in CSF leaks related to trauma when compared to spontaneous leaks or those acquired after surgery [15]. If conservative therapy fails, surgical intervention is indicated.

Secondary meningitis is well documented in the literature, with an incidence of 10-27% of post-traumatic cases and 30-60% of post-operative CSF leaks [19, 27]. Evidence for the use of prophylactic antibiotics to prevent meningitis in cases of CSF leak has been conflicted [28-29] and should be assessed on a case-to-case basis. The fact remains that in the presence of a persistent leak, early surgical intervention is recommended.

Surgical Management

When a middle fossa CSF leak cannot be resolved via conservative measures, surgical intervention is necessary to prevent complications. Three main surgical approaches for treating these defects have been reported in the literature; transmastoid, middle fossa approach and a combined transmastoid/middle fossa approach.

Historically, the transmastoid approach has been used for smaller, well-defined defects while the middle fossa approach used for larger (>1cm) or multiple defects [1, 15, 20, 30]. Also, there has been a tendency for surgeons to first attempt the less invasive transmastoid approach, and then go on to the middle fossa approach if the leak recurs [6, 30]. Appropriate patient selection remains a crucial part of the process of choosing a surgical approach. With the advent and increased popularity of the neuro-endoscope, there have also been a few attempts at studying endoscopic-assisted repairs [31-32]. Variations exist on all the approaches and especially in materials used for repair of the defects.

Transmastoid Approach [19, 27, 30]

The indications for a transmastoid approach include a small well-delineated defect that is located on the posterior aspect of the tegmen and can be fully appreciated on pre-operative imaging. It can also be used in cases where the patient is unable to undergo a more extensive surgery. This approach is essentially a completely extradural and extracranial procedure although there are a few reports where it was combined with a mini-craniotomy [33-34] (Different from a full combined approach).

A C-shaped incision is made at the post-auricular area and carried down through the subcutaneous tissue to the periosteum. A standard mastoidectomy is performed by drilling in a triangular fashion, with one side of the triangle parallel to the external auditory canal and another parallel to the slope of the tegmen. The sigmoid sinus, sino-dural angle and middle fossa are skeletonized. Once the antrum is exposed, an anthrostomy is performed, in addition to a posterior tympanotomy, disarticulation of the incus and removal of the ossicular heads. This enables further anterior dissection and may not be a requisite if the defect is more posterior, that is, in the tegmen mastoideum. At the tegmen defect, the dura is separated from the bone using blunt dissection, with utmost care to avoid further tears in the dura. This is important as the dura is usually thin and may be adherent to the bone. The technique for repair is variable as is discussed below. After repair, the mastoid is often obliterated, or the incus may be replaced, to decrease the incidence of conductive hearing loss and this is accomplished by interposing it between the remnant neck of the malleus and the stapes (Figure 4).

Advantages for this approach include its less invasive nature as most cases are performed as same-day surgeries or with 24-hr observation. It also provides the surgeon with the ability to access both middle and posterior fossa defects and while avoiding complication that may occur secondary to a craniotomy or temporal lobe retraction. Disadvantages include post-operative conductive hearing loss, and increased incidence of CSF leak recurrence (Table 1).

Middle Fossa Approach [1, 8, 20]

Indications for the middle fossa approach include multiple defects, a large anterior defect or inability to fully analyze the defect on pre-operative imaging.

The patient is placed in a lateral decubitus position and a lumbar drain is inserted. Facial nerve monitoring is also utilized. Then turning to a supine position, the head is fixated with a 3-point Mayfield head holder.

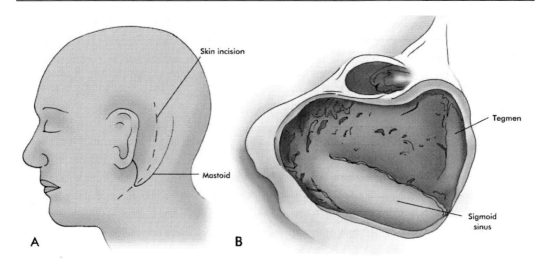

Figure 4. A: Site of incision for the transmastoid approach. B: Drawing of the exposure after mastoidectomy.

Table 1. Advantages and disadvantages of the transmastoid and the middle fossa approaches

Approach	Advantages	Disadvantages
Transmastoid	1. Technically simpler 2. No brain retraction 3. Able to access posterior fossa defects	1. Limited anterior exposure 2. Cannot address multiple defects 3. Higher rate of graft failure 4. Higher risk of hearing loss
Middle Fossa	1. Wider exposure 2. Can address multiple defects 3. Allows for more secure placement of grafts	1. Technically demanding 2. Requires brain retraction 3. Cannot access posterior fossa defects

The head is rotated towards the opposite shoulder and extended so that the squamous portion of the temporal bone is parallel to the floor. A standard reverse-U or reverse question mark incision is made and a subgaleal dissection is performed. The temporalis muscle is sharply incised and dissected away from the calvarium. Two bur holes are placed, one just superior to the zygomatic root and the other posterior to the EAC. A rectangular craniotomy, approximately 3x4 cm is then carried out, with the superior border just above the squamosal suture and the EAC at the line demarcating the posterior one-third of the rectangle from the anterior two-thirds. The inferior edge of the craniotomy is drilled down further until it is flush to the floor of the middle fossa. Lumbar drainage may commence after the bone flap is elevated. Extradural dissection is then begun, dissecting the dura away from the floor moving in an anterior to posterior fashion. The middle meningeal artery is identified as it enters through the foramen spinosum and it is coagulated and divided. Lateral to medial extradural dissection is dissuaded to avoid elevating the greater superior petrosal nerve (GSPN) from its groove. The extent of the dissection is anteriorly to the mandibular division of the trigeminal nerve and medially to petrous ridge (Figure 5).

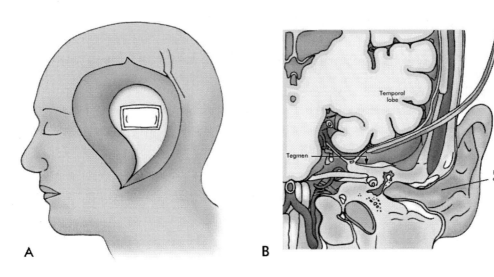

Figure 5. A: Skin incision and area of craniotomy for the middle fossa approach. B: Schematic view of the approach.

This allows for full examination of the tegmen and temporal dura. Dural defects are primarily repaired, when possible. Materials and technique for the actual repair are again detailed below. After repair, the craniotomy flap is replaced and fixated with titanium mini-plates and screws. The temporalis muscle is re-approximated and the galeal layer is closed tightly. Skin is closed with staples.

Advantages for the middle fossa approach include providing a wide exposure of the entire tegmen plate, therefore enabling the repair of large or multiple defects.

Also, hearing is preserved as there is no risk to the ossicular chain and there is easier and more secure placement of materials for sealing the defect, thereby decreasing risk of recurrence.

Disadvantages include that this is a more technically challenging approach. Mobilization and retraction of the temporal lobe and associated vascular structures, like the vein of Labbé, can lead to seizures or venous infarction. Extradural dissection may also put the geniculate ganglion and GSPN at risk for avulsion (Table 1).

Combined Middle Fossa/Transmastoid Approach [5, 35]

Indications for a combined approach are the same as for a middle fossa approach, and it is especially beneficial when the defect or defects cannot be localized on pre-operative imaging. Technique involves combining the two detailed approaches above.

Advantages are also the same as for the middle fossa approach with the added benefit of enabling the visualization of intradural and extracranial aspects of the defect as well as facilitating secure placement of graft materials, possibly with sutures. It carries the morbidity of a middle fossa approach, in addition to post-operative conductive hearing loss.

Minimally Invasive Approaches

With the advent of endoscopic techniques, great leaps have been made in the repair of anterior cranial fossa CSF leaks and defects through an endonasal approach [36]. In the literature, data on endoscopic approaches to repair middle fossa defects is essentially non-existent. There are 2 cadaveric studies evaluating an endoscope-only approach to the middle fossa [31-32] with one of the studies evaluating ability for pedicled deep temporal fascial flap for repair of a defect [32]. Clinical studies are needed to further evaluate the efficacy of these approaches. Possible disadvantages could include difficulty in manipulating graft materials through an endoscope.

Materials and Technique for Repair [1]

A multilayered closure is the gold-standard in middle fossa CSF leak repair. As mentioned earlier, both the bony defect and the dural breach need to be addressed to effectively treat a CSF leak. One study found 100% vs. 78% rate of success of repair with multilayered closure as against single layered closurewhen [15]. There are many studies that have advocated the use of autologous tissue, fibrin glue, hydroxyapatite (HAC) cement or a combination of these to effectively repair these defects [1, 15, 20, 33, 37].

Autologous tissue is often considered superior to allogenic materials in closure of skull base defects. These can be free grafts like temporalis fascia, fascia lata, abdominal fat grafts and calvarial bone grafts, some of which may be harvested from the same incision, avoiding potential donor site complications. Another type of autologous tissue is a vascularized pedicled flap which, when available, is even more ideal for closing defects and preventing CSF leaks. In the middle fossa, it can be a pedicled pericranial flap or pedicled temporalis fascial flaps. With pedicled flaps, a vascular graft is provided that speeds healing and there are no donor site complications, like infection or hematoma. Sometimes, however, these are difficult to harvest or are unavailable.

Dural substitutes are essential for repair, especially when the dural defect cannot be primarily approximated with stitches. Multiple synthetic collagen-based (bovine) dural substitutes are commercially available (Table 1). Their safety and efficacy are well documented in the literature. They essentially work by providing a low-pressure surface for CSF absorption, which enhances graft-dura interface, and chemical signaling for native fibroblasts.

Polyethylene glycol (PEG) hydrogel sealant has also been well described as a successful adjunct in the prevention and repair of CSF leaks. It is biocompatible, provides a strong adherence to tissue and is absorbable.

Hydoxyapatite cement (HAC) is composed of tetracalcium and dicalcium phosphate salts that react to form hydroxyapatite, the major component in skeletal bone. It has osteoconductive and osteointegrative properties that allow the formation of new bone with minimal inflammatory changes. It is malleable for approximately 20 minutes and has been used successfully to repair temporal bone defects. When using HAC in the transmastoid approach, care must be taken to avoid placing the material too close to the ossicles which can result in ossification and conductive hearing loss.

The technique for sealing defects varies depending on approach and surgeon preferences. For the transmastoid approach, we recommend a multi-layered closure with autologous temporalis muscle or fascia and/or dural substitute placed over the temporal dura defect and augmented with PEG hydrogel sealant. As this approach may result in exposure of the ossicular chain, we dissuade the use of HAC for the tegmen defect. Instead, bone dust can be collected during the mastoidectomy and used to seal the bony defect. If obliteration of the mastoid is necessary, this can be accomplished with fascia, abdominal fat graft and again augmented with PEG hydrogel sealant. For the middle fossa or combined approaches, we recommend a multi-layered closure with autologous temporalis muscle or fascia and/or dural substitute placed on both sides of the dural defect, an inlay/onlay technique. This is then augmented with PEG hydrogel sealant. The tegmen defect is sealed with HAC except for cases where the ossicular chain is exposed; for these, a split-thickness calvarial bone graft large enough to cover the defect is fashioned (Figure 6) [1].

Review of Series/Outcomes

In the authors' personal 2 year retrospective review, 7 patients were indentified that required surgical intervention for repair of middle fossa CSF leak. The most common initial presenting symptom was CSF otorrhea (85%).

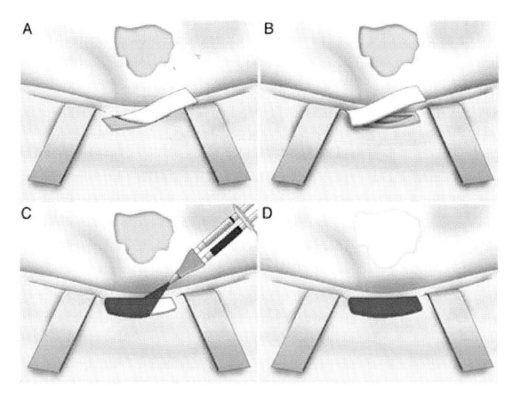

Figure 6. Schematic representation of middle fossa leak repair technique. A: Inlay placement of dural substitute into the temporal dural defect. B: Onlay placement of dural substitute over the defect. C: Sealing of the inlay/onlay construct with PEG hydrogel. D: Tegmental defect covered with HAC.

Table 2. Materials used in repair (HAC: hydroxyapatite cement, PEG: polyethylene glycol)

Autologous	Non-autologous
Calvarial bone	HAC
Bone dust	
Fascia (temporals, fascia lata)	Collagen-based dural substitutes (DuraGen™, DURAFORM™, Dura-Guard™, Durepair™)
Muscle (temporalis)	PEG hydrogel sealant (DuraSeal™)
Fat (abdominal)	Titanium mesh

Less common presenting symptoms were hearing loss (42%) and rhinorrhea (14%). Five (71%) of the patients had a history of previous otological procedures and two (29%) had a history of chronic otitis media. The defects were diagnosed with a combination of CT, cisternography, and MRI. The middle fossa approach with intraoperative placement of lumbar drain was utilized in all patients. A dural substitute, PEG hydrogel sealant and HAC were used in all cases. 4 cases were supplemented with autologous material, which included split thickness bone graft, temporalis muscle and fascia. Mean follow-up was 12 months (range 5-33 months) and there was a 100% repair rate [1]. Common complications seen after middle fossa CSF leak repair include recurrence of the leak, graft failure hearing loss and wound infections. Due to the variability in techniques and repair materials, it is difficult to accurately assess the overall complication rate of middle fossa CSF leak repairs. In a retrospective review of 92 patients, Savva et al. noted a 25% recurrence rate at 3-year follow-up among patients undergoing surgical repair. Factors related to successful surgical management included primary closure enforced by both autologous and artificial grafts in a multi-layered fashion [15]. The implantation of any graft carries the risk for potential migration. The use of HAC can be used to fix bone graft in place. Similarly, PEG hydrogel can be used to prevent dural graft migration. In our small series, one patient developed an epidural hematoma at the operative site on postoperative Day 2 requiring surgical evacuation, and another patient had a superficial wound dehiscence seen on postoperative Day 48. During the follow-up period, there was no evidence of wound infections, neurovascular damage, or CSF leakage requiring re-operation [1].

Conclusion

Middle fossa CSF leaks pose a challenge in their evaluation and treatment. Basic recommendations in their management include:

- The diagnosis of middle fossa defect should be considered in an adult patient presenting with unilateral middle ear fullness and/or hearing loss even in the absence of trauma or meningitis. A history of trauma, meningitis or frank CSF otorrhea raises the index of suspicion.
- When a CSF leak is suspected, obtain appropriate imaging including CT and MRI.

- If an active leak is seen and a sample can be collected, the specimen should be sent for β-2 transferrin analysis. However, results must be timely and a negative result should not preclude further evaluation.
- Traumatic CSF leaks can be managed conservatively. Spontaneous or iatrogenic defects almost always need surgical repair.
- Appropriate patient selection is tantamount in selecting the surgical approach. The transmastoid approach is appropriate for small, well-delineated defects. If any uncertainty exists about the location or multiplicity of the defect, a middle fossa approach or a combined approach should be used.
- More studies are needed to evaluate endoscopic and other minimally invasive approaches.
- A multi-layered closure is very important and should include a vascularized pedicled flap when available. Otherwise, a combination of a collagen-based dural substitute matrix, a polyethylene glycol hydrogel and hydroxyapatite cement in an inlay/overlay method should be used.

References

[1] Shorter, C. D., Connor, D. E. Jr, Thakur, J. D., Gardner, G., Nanda, A., Guthikonda, B. Repair of middle fossa cerebrospinal fluid leaks using a novel combination of materials: technical note. *Neurosurg. Focus*. 2012 Jun.; 32(6):E8.

[2] Glasscock, M. E., Dickins, J. R. E., Jackson, C. G., Wiet, R. J., Feenstra, L. Surgical management of brain tissue herniation into the middle ear and mastoid. *Laryngoscope*. 1979 Nov.; 89(11):1743–54.

[3] Ahren, C., Thulin, C-A. Lethal intracranial complications following inflammation in the external ear canal in treatment of serous otitis media and due to defects in the petrous bone. *Acta Otolayrngol.* 1965; 60:407–21.

[4] Merchant, S. N., McKenna, M. J. Neurotologic manifestations and treatment of multiple spontaneous tegmental defects. *Am. J. Otol*. 2000 Mar.; 21(2):234-9.

[5] Brown, N. E., Grundfast, K. M., Jabre, A., Megerian, C. A., O'Malley, B. W., Jr, Rosenberg, S. I. Diagnosis and management of spontaneous cerebrospinal fluid-middle ear effusion and otorrhea. *Laryngoscope*. 2004; 114(5):800–5.

[6] Pelosi, S., Bederson, J. B., Smouha, E. E. Cerebrospinal fluid leaks of temporal bone origin: selection of surgical approach. *Skull Base*. 2010 Jul.; 20(4):253-9.

[7] Jégoux, F., Malard, O., Gayet-Delacroix, M., Bordure, P., Legent, F., Beauvillain de Montreuil, C. Hyrtl's fissure: a case of spontaneous cerebrospinal fluid otorrhea. *AJNR Am. J. Neuroradiol.* 2005 Apr.; 26(4):963-6.

[8] Gacek, R. R., Gacek, M. R., Tart, R. Adult spontaneous cerebrospinal fluid otorrhea: diagnosis and management. *Am. J. Otol*. 1999 Nov.; 20(6):770-6.

[9] Petrus, L. V., Lo, W. W. Spontaneous CSF otorrhea caused by abnormal development of the facial nerve canal. *AJNR Am. J. Neuroradiol*. 1999 Feb.; 20(2):275-7.

[10] Gacek, R. R., Leipzig, B. Congenital cerebrospinal otorrhea. *Ann. Otol*. 1979; 88:358–65.

[11] Gacek, R. R. Arachnoid granulation cerebrospinal fluid otorrhea. *Ann. Otol. Rhinol. Laryngol.* 1990 Nov.; 99(11):854-62.
[12] Ommaya, A. K.: Cerebrospinal fluid rhinorrhea. *Neurology* 1964; 14:106–113.
[13] Kenning, T. J., Willcox, T. O., Artz, G. J., Schiffmacher, P., Farrell, C. J., Evans, J. J. Surgical management of temporal meningoencephaloceles, cerebrospinal fluid leaks, and intracranial hypertension: treatment paradigm and outcomes. *Neurosurg. Focus.* 2012 Jun.; 32(6):E6.
[14] Pappas, D. G., Hoffman, R. A., Holliday, R. A., Hammerschlag, P. E., Pappas, D. G., Swaid, S. N.: Evaluation and management of spontaneous temporal bone cerebrospinal fluid leaks. *Skull Base Surg.* 1995; 5:1–7.
[15] Savva, A., Taylor, M. J., Beatty, C. W. Management of cerebrospinal fluid leaks involving the temporal bone: report on 92 patients. *Laryngoscope.* 2003; 113:50–6.
[16] Bell, R. B., Dierks, E. J., Homer, L., Potter, B. E. Management of cerebrospinal fluid leak associated with craniomaxillofacial trauma. *J. Oral Maxillofac. Surg.* 2004 Jun.; 62(6):676-84.
[17] Brodie, H. A., Thompson, T. C. Management of complications from 820 temporal bone fractures. *Am. J. Otol.* 1997 Mar.; 18(2):188-97.
[18] Becker, S. S., Jackler, R. K., Pitts, L. H. Cerebrospinal fluid leak after acoustic neuroma surgery: a comparison of the translabyrinthine, middle fossa, and retrosigmoid approaches. *Otol. Neurotol.* 2003 Jan.; 24(1):107-12.
[19] Bento, R. F., Pádua, F. G. de M: Tegmen tympani cerebrospinal fluid leak repair. *Acta Otolaryngol.* 2004; 124:443–8.
[20] Dutt, S. N., Mirza, S., Irving, R. M. Middle cranial fossa approach for the repair of spontaneous cerebrospinal fluid otorrhoea using autologous bone pate. *Clin. Otolaryngol. Allied Sci.* 2001 Apr.; 26(2):117-23. Review.
[21] Skedros, D. G., Cass, S. P., Hirsch, B. E., Kelly, R. H. Beta-2 transferrin assay in clinical management of cerebral spinal fluid and perilymphatic fluid leaks. *J. Otolaryngol.* 1993; **22**:341–344.
[22] Chan, D. T., Poon, W. S., IP CP, Chiu, P. W., Goh, K. Y. How useful is glucose detection in diagnosing cerebrospinal fluid leak? The rational use of CT and Beta-2 transferrin assay in detection of cerebrospinal fluid fistula. *Asian J. Surg.* 2004 Jan.; 27(1):39-42.
[23] Stone, J. A., Castillo, M., Neelon, B., Mukherji, S. K. Evaluation of CSF leaks: high-resolution CT compared with contrast-enhanced CT and radionuclide cisternography. *Am. J. Neuroradiol.* 1999; **20**:706–12.
[24] El Gammal, T., Sobol, W., Wadlington, V. R., Sillers, M. J., Crews, C., Fisher, W. S. 3rd, Lee, J. Y. Cerebrospinal fluid fistula: detection with MR cisternography. *AJNR Am. J. Neuroradiol.* 1998 Apr.; 19(4):627-31.
[25] Selcuk, H., Albayram, S., Ozer, H., Ulus, S., Sanus, G. Z., Kaynar, M. Y., Kocer, N., Islak, C. Intrathecal gadolinium-enhanced MR cisternography in the evaluation of CSF leakage. *AJNR Am. J. Neuroradiol.* 2010 Jan.; 31(1):71-5.
[26] Allen, K. P., Isaacson, B., Purcell, P., Kutz, J. W. Jr, Roland, P. S. Lumbar subarachnoid drainage in cerebrospinal fluid leaks after lateral skull base surgery. *Otol. Neurotol.* 2011 Dec.; 32(9):1522-4.
[27] Kveton, J. F., Goravalingappa, R. Elimination of temporal bone cerebrospinal fluid otorrhea using hydroxyapatite cement. *Laryngoscope.* 2000 Oct.; 110(10 Pt 1):1655-9.

[28] Klastersky, J., Sadeghi, M., Brihaye, J. Antimicrobial prophylaxis in patients with rhinorrhea or otorrhea: a double-blind study. *Surg. Neurol.* 1976 Aug.; 6(2):111-4.

[29] Brodie, H. A. Prophylactic antibiotics for posttraumatic cerebrospinal fluid fistulae. A meta-analysis. *Arch. Otolaryngol. Head Neck Surg.* 1997 Jul.; 123(7):749-52.

[30] Oliaei, S., Mahboubi, H., Djalilian, H. R. Transmastoid approach to temporal bone cerebrospinal fluid leaks. *Am. J. Otolaryngol.* 2012 Feb. 29. [Epub. ahead of print]

[31] Mourgela, S., Sakellaropoulos, A., Anagnostopoulou, S. Middle cranial fossa endoscopy using a rigid endoscope. *Minim. Invasive Ther. Allied Technol.* 2007; 16(6): 355-9.

[32] Komatsu, M., Komatsu, F., Di Ieva, A., Inoue, T., Tschabitscher, M. Endoscopic reconstruction of the middle cranial fossa through a subtemporal keyhole using a pedicled deep temporal fascial flap: a cadaveric study. *Neurosurgery.* 2012 Mar.; 70(1 Suppl. Operative):157-61; discussion 162.

[33] Adkins, W. Y., Osguthorpe, J. D. Minicraniotomy for management of CSF otorrhoea from tegmen defects. *Laryngoscope.* 1983; 93:1038–40.

[34] Kuhweide, R., Casselman, J. W. Spontaneous cerebrospinal fluid otorrhea from a tegmen defect: transmastoid repair with minicraniotomy. *Ann. Otol. Rhinol. Laryngol.* 1999 Jul.; 108(7 Pt 1):653-8.

[35] Agrawal, A., Baisakhiya, N., Deshmukh, P. T. Combined Middle Cranial Fossa and Trans-Mastoid Approach for the Management of Post-Mastoidectomy CSF Otorrhoea. *Indian J. Otolaryngol. Head Neck Surg.* 2011 Jul.; 63(Suppl. 1):142-6. Epub. 2011 Apr. 10.

[36] Briggs, R. J., Wormald, P. J. Endoscopic transnasal intradural repair of anterior skull base cerebrospinal fluid fistulae. *J. Clin. Neurosci.* 2004 Aug.; 11(6):597-9.

[37] Ota, T., Kamada, K., Saito, N.: Repair of cerebrospinal fluid leak via petrous bone using multilayer technique with hydroxyapatite paste. *World Neurosurg.* 2010 Dec.; 74 (6):650–3.

In: Minimally Invasive Skull Base Surgery
Editor: Moncef Berhouma

ISBN: 978-1-62808-567-9
© 2013 Nova Science Publishers, Inc.

Chapter XIII

The Epidural Anterior Petrosectomy: A Minimally Invasive Skull Base Approach to the Posterior Fossa

P. H. Roche[*]*, L. Troude, A. Melot and R. Noudel*

Department of Neurosurgery, CHU Nord, Assistance Publique, Hôpitaux de Marseille,
Aix-Marseille University, France

Abstract

The petroclival region harbors multiple diseases of various origins and extensions. The depth of this area and its highly functional neurovascular content explain how challenging the surgery of this target is. At present, endoscopic endonasal approach cannot manage routinely most of these diseases, neither the regular intradural approaches. The epidural anterior petrosectomy (EAP) is a lateral epidural route that preserves the intrapetrosal neuro-otological structures. The EAP can be reproduced safely with minimal knowledge of the anatomy and expertise in the field of skull base surgery. The goal of this study is to describe the surgical technique, the indications and the limitations of this approach.

Keywords: Anterior petrosectomy, meningioma, minimal invasive surgery, petroclival region, skull base

Abbreviations

EAP   Epidural anterior petrosectomy
GG   Geniculate ganglion
GSPN   Great superficial petrosal nerve

---

[*] Corresponding author: Pierre-Hugues Roche, M. D. Email: proche@ap-hm.fr.

IAC	Internal auditory canal
ICA	Internal carotid artery
IPS	Inferior petrosal sinus
MC	Meckel's cave
SOF	Superior orbital fissure
SPS	Superior petrosal sinus

Introduction

In the last decades, the field of skull base surgery has been enriched by the development of a large variety of transpetrosal approaches. The epidural anterior petrosectomy (EAP) that was originally dedicated for the control of vertebrobasilar aneurysms [5] rapidly appeared useful for the resection of petroclival tumors. When compared to regular intradural approaches or endoscopic endonasal techniques, the EAP offers an optimized exposure of the petroclival and related areas that are frequently involved in those tumors. The safety and simplicity of the EAP is correlated to the surgeon's ability to identify reliable and permanent osteodural landmarks while proceeding. Considering that the EAP offers a direct access to the target, that it avoid any cerebellar or brainstem retraction and that critical cranial nerves and vessels are usually not interposed in the operative field, we claim that EAP belongs to the category of minimally invasive procedures. The aim of this chapter is to describe this surgical technique and discuss its fields of applications along with its limitations.

Surgical Anatomy

The Bony Anatomy

The critical intrapetrous structures that are located in the close environment of the approach and that must be preserved are the horizontal segment of the intrapetrous internal carotid artery (ICA), the great superficial petrosal nerve (GSPN), the geniculate ganglion (GG) and the cochlea.

The petrous apex will be reached and resected once the floor of the middle fossa will be exposed in an epidural way and particularly once the superior surface of the petrous pyramid will be identified. The key point is to focus on the anteromedial half of the superior surface of the petrous bone. The concept of "rhomboid fossa" [1] is helpful to provide the landmarks and the boundaries of the area that will be drilled as the initial step of the petrosectomy.

The limits of the rhomboid fossa are represented by the petrous ridge which is the groove of the superior petrosal sinus (SPS) medially, the posterior border of the mandibullary branch of the trigeminal nerve (V3) anteriorly, the GSPN laterally and arcuateeminencia posteriorly, (Figure 1a). The rhomboid fossa can be divided into two areas by a vertical straight line drawn from the GG toward the petrous ridge. This line materializes the superficial landmark of the roof of the internal auditory canal (IAC). In front of this line is the pre-meatal area, which corresponds to the posteromedial middle fossa triangle, also known as Kawasetriangle

[5, 6]. Posterior to the line is the retro-meatal area, which does not need to be drilled in the regular approach.

It is of utmost importance to understand that the true petrous apex is under the Meckel'scave (MC). This area is an ovaleshapeoval shape concave bony depression posteriorly marked by a tubercle, so called Princeteau tubercle that delineates the posterior aspect of the porustrigeminus.

The Dura Anatomy

The dura that covers the middle fossa and the superior surface of the petrous apex can be divided into 2 parts. The one that covers the petrous bone backward: the two layers will be elevated simultaneously as far as the petrous ridge and the wall of the SPS

The other one that covers the true petrous apex, the MC and the parasellar compartment frontward: Here the two dural layers are elevated as far as the limit of rotundum and ovale foramen. Then, at the level of the lateral wall of the MC, one layer will be peeled and elevated. The medial layer will be left on the MC to protect the pars triangularis and GG except if there is a need to enter inside the MC [2, 7].

The Intradural Anatomy

The EAP gives access to the petroclival regionthat can be described by its limits and its content.

The limits: The top of the field consists in the lower surface of the tentorium from the SPS to the free edge – Laterally is the dura that covers the posterior surface of the petrous bone in front of the IAC from the SPS to the IPS. Frontward is the dura that covers the mid and upper clivus [2], excluding the posterior clinoid process. The medial limit is given by the ventrolateral pons. The petroclival region is opened backwards to the cerebellopontine angle, upward toward the cerebral isthmus and downward toward the premedullary space and the cerebellomedullary cistern.

The content is represented by the elements from the prepontine cistern and from the anterior expension of the cerebellopontine cistern.

Surgical Technique

Prerequisite

Before starting the approach, the surgeon should have fulfil the following rules and criterions:

- Good knowledge of the surgical anatomy
- Sufficient training in a skull base laboratory

- Need for an extensive preoperative work-up of the case including a bone window CT scan of the petrous bone (Figure 2)
- Assessment of the natural history of the disease (Figure 3)
- Be prepared to suspend the surgery in case of inextirpable tumor because of tumor texture or adhesion to critical neurovascular structures
- Consider the other toolsin the management of the disease (radiosurgery, wait and see, ...)
- Be able to communicate clearly with the patient and relatives before surgery and during the early and delayed postoperative course

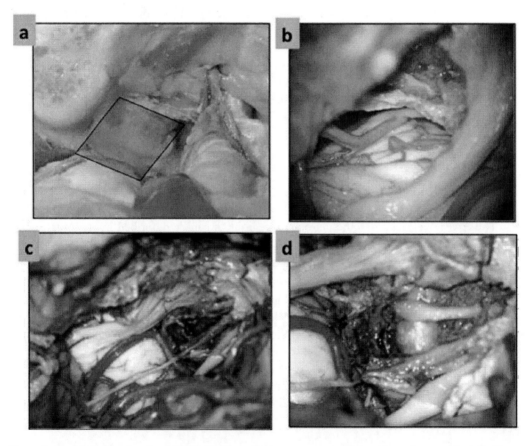

Figure 1. Anatomical landmarks and steps for the anterior petrosectomy. Injected specimen on the left side. a. Limits of the rhomboid fossa delineated by the lozenge. b. View obtained after bone resection and dura opening. Trigeminal and abducens nerve are shown, as the AICA, MCA and ventrolateral pons. c. Extension of the approach toward the MC and the CS: The MC is opened showing the pars triangularis and Gasserian ganglion. The trochlear and the third nerve are exposed while they penetrate inside the CS. d. For anatomical purpose, the posterior wall of the cavernous sinus is opened, showing the carotid artery and abducens nerve overcrossing it.

Positioning and General Setting

The patient is positioned supine with the head rotated approximately 80° toward the opposite side and maintained in a 3-pin Mayfield head clamp. Care should be taken to avoid contralateral jugular vein compression. While rotating the head at 90°, the external auditory canal projects at the plumb line of the IAC. The surgeon is positioned in front of the vertex. Another field is draped to harvest abdominal fat for closure and this point is mentioned to the patient beforehand. Antibiotics are administered 30 minutes before skin incision. Neurophysiological monitoring includes monitoring of the facial (VII) and abducens (VI) nerves. Auditory-evoked brain stem responses are monitored when hearing preservation is being attempted. Perioperative cerebrospinal fluid lumbar drainage and neuronavigation are optional.

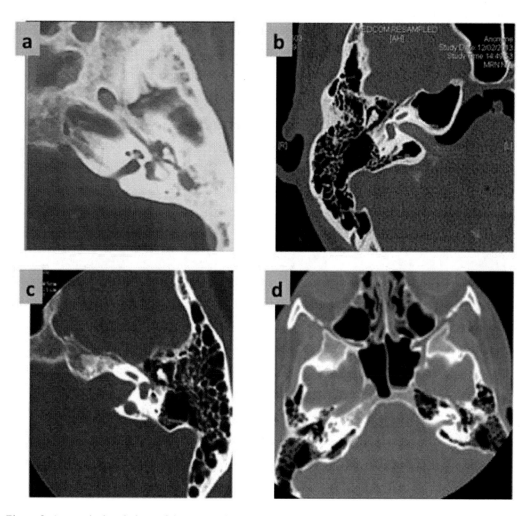

Figure 2. Anatomical variations of the petrous bone anatomy. a. Compact petrous bone with rare pneumatisation. b. Highly pneumatized petrous bone and petrous apex. c. Variations of the IAC which is significantly widened in this case. d. Absence of carotid canal and of carotid artery on the left side with asymmetrical pneumatisation.

Soft Tissue Dissection

The skin incision is carried out following a reverse question mark. This starts 1 cm anterior to the tragus at thelevel of the zygomatic arch and runs above the external ear, ending in the frontal region. This incision can be modified following the extent of the drilling that is required. The frontotemporal pericranial flap is dissected separately in two layers, and the temporal muscle is detached with its deep aponeurosis and reflected anteroinferiorly.

The Bone Flap

One burr hole is made above the root of zygoma and is usually sufficient to allow a safe detachment of the dura from the inner table of the bone flap (Figure 4a). A temporo- pterional craniotomy is undertaken reaching the floor of the middle fossa downward. The lesser sphenoid wing is drilled with a 5-mm diamond drill. The contours of the bone flap stay under the landmark of the superior linea.

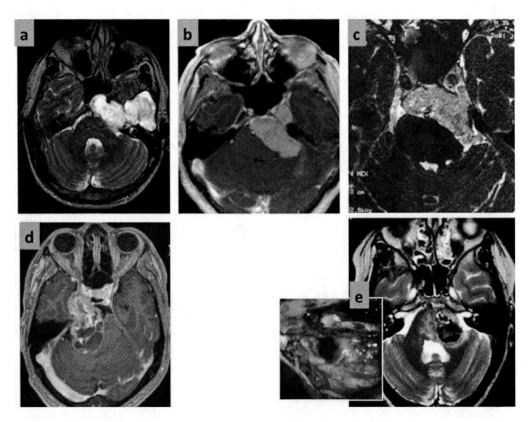

Figure 3. Examples of diseases with various origin involving the petroclival area and suitable for EAP. a. Large epidural petrous apex epidermoid cyst on the left side. b. Petroclival meningioma involving the MC on the left side. c. Cisternalepidermoid cyst between both trigeminal nerves. d. Dumbble shape trigeminal schwannoma involving the MC and the PC area on the right. d. Pontine cavernoma at the early phase of the bleeding on the left. T2 weight MRI and operative view showing the lateral surface of the pons and the subpial hematoma.

Superficial Drilling

The drilling of the pterion is conducted as far as the external margin of the superior orbital fissure (SOF). Elevation of the dura from the anterior part of the floor of the middle fossa is conducted laterally, medially and posteriorly. It is now possible to identify the external margins of the foramen rotundum, foramen ovale and foramen spinosum successively. The 180° external borders of these foramina are skeletonized using a 5 mm diamond drill and under copious irrigation of saline. Coagulating and dividing the middle meningeal artery a few millimeters above the spinosum is necessary. The SOF, the maxillary (V2) and the mandibular (V3) branches of the trigeminal nerve are now identified into their foraminal course. The next step is to peel and elevate the durapropria that covers the MC under high magnification. This step is required to translocate V3 anteriorly and then to expose the true petrous apex that lies under the MC [4]. It is not unusual to experience some venous bleeding from the posterior margin of the foramen ovale. Such bleeding is controlled by gently packing with small pieces of oxidized cellulose.

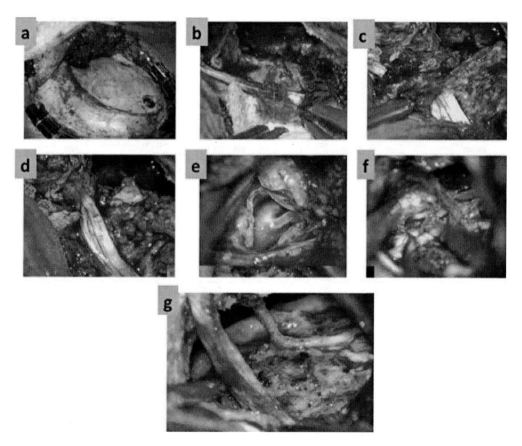

Figure 4. Surgical steps of the resection of apetroclivalmeningioma (right side) using the EAP. a. Regular pteriono-temporal bone flap; b. The petrous apex has been drilled. The tentorium is exposed and division is started after coagulation of the SPS; c. The meningioma encases the trigeminal sensory root; d. The meningioma is dissected form the trigeminal nerve as far as the MC; e. The tumor is dissected toward the basilar artery and AICA origin; f. Both abducens nerves are exposed in their whole cisternal course; g. Final view after resection showing the trigeminal and abducens nerves on the right side. The basilar artery is clearly shown. Note that a piece of tumor has been left over the pial surface of the pons.

Upon elevating the dura behind V3, the great superficial petrosal nerve (GSPN) is identified. Being detached from the dura using a sharp dissector preserves the nerve. The GSPN running over the horizontal intra-petrous carotid artery leads to the geniculate ganglion (GG). The exact location of the GSPN is easily checked using a facial nerve stimulator (0.3 to 0.5 mA). It is noteworthy that the bony roof of the horizontal intrapetrous carotid is frequently dehiscent, thereby requiring the use of a diamond drill while flattening the middle fossa until the carotid landmarks have been clearly identified.

The Petrous Bone Resection

While proceeding to elevate the dura, the surgeon reaches the petrous ridge medially and behind V3. The limits of the rhomboid fossa are now delineated as aforementioned: the petrous ridge medially, posterior border of V3 anteriorly, GSPN laterally and arcuate eminence posteriorly. Once the landmarks of the pre-meatal triangle have been identified, drilling of the petrous apex is undertaken. The drilling is facilitated by a high level of pneumatization of the petrous pyramid in some cases (Figure 2); in others, the fatty aspect of the bone is usually helpful. Approaching the cortical bone of the inner surface, close to the dura of the posterior petrous bone, more cautious drilling is needed to avoid opening the dura prematurely. Indeed, an early and inappropriate opening jeopardizes the structures the dura protects. Inferiorly, the drilling is conducted until the bluish tinge of the inferior petrosal sinus (IPS) is seen. If a tear occurs in the IPS, there is always abundant bleeding, and gentle packing of the sinus lumen should be carried out with small pieces of oxidized cellulose. Such an event can be prevented by preoperative embolization of the sinus, though the proximity of the abducens nerve may lead to the risk of a subsequent palsy following this technique. It is sometimes quite difficult to reach the extreme tip of the petrous apex; in that situation, upward retraction of MC is useful. At this point, it is necessary to at least expose the dura of the anterior border of the porus, because this point corresponds to the posterior limit of the EAP approach. However, attempting to skeletonize the whole length of the canal, and particularly the fundus, may damage the GG and the first turn of the cochlea more in the depth. Once the drilling is completed, meticulous hemostasis of the dura and the epidural space is achieved before opening the dura.

The Meningeal Step: Opening the Dura

Proper opening of the dura has to fit with the bony exposure in order to optimize the operative corridor and avoid undue retraction of the temporal lobe. Care should be taken to proceed in a stepwise manner to avoid damage of the underlying cranial nerves, excessive bleeding of the venous sinuses and closure problems. Horizontal sectioning of the dura is carried out above and alongside the SPS, from MC to the retro-meatal field backward. An excessive posterior opening puts the vein of Labbé at jeopardy. Then, horizontal sectioning of the posterior fossa dura is carried out just inferior and alongside the SPS, from the porustrigeminus to the anterior margin of the auditory meatus. At that time, attention should be paid to the trigeminal sensory root, which can be laterally displaced by the tumor mass against the dura. Coagulation and division of the SPS is followed by the transverse section of the tentorium toward its notch (Figure 4b). When the tentorium cut is too oblique backward, the working distance increases, resulting in temporal lobe retraction and damage to bridging veins. Conversely, when the cut is too close to the cavernous sinus, the trochlear nerve that lies under the petroclinoid ligament may be injured. Suture stitches at both edges (silk 4/0) of the cut section and traction elevation of the threads widen the operative field. At this point a tailored tentorial resection can be performed if invaded by a tumor. The approach is now achieved and the operative field is focused on the cisternal part of the trigeminal nerve as well the ventral and lateral pons.

Closure

At the end of surgery, a watertight stitched closure of the dura is usually hampered by its resection and retraction. Therefore, the use of an abdominal fat graft is the best option to cover the osteodural defect. The fat is divided into several pieces of various sizes. The first one plugs the bony defect created by the petrous apex resection. The other pieces cover the dural defect and are maintained with fibrin glue. It is advised to overlay the superior surface of the petrous bone at the level of the tegmen tympani with strips of fat or fascia because this area is usually thin and can be sometimes dehiscent. The dura is stitched at the margins of the bone flap as well as at its central part using tack-up sutures to prevent postoperative epidural hematoma. The bone flap is repositioned and maintained with titanium plates. Drainage of the epidural space should be avoided to avoid creating a CSF channel and predisposing the patient to postoperative leaks. Re-application and suturing of the temporal muscle are done with absorbable sutures. Skin suturing is carried out in two layers.

Postoperative Course and Instructions from the Surgeon

During the 24 h following surgery, the patient is observed in an intensive care unit, and a postoperative CT scan (without contrast) is done. Low-molecular-weight heparin is started the day after surgery to prevent thromboembolic complications. It is recommended that all Valsalva-related maneuvers should be avoided to prevent CSF leaks. Postoperative checks of the cranial nerves should be carried out with particular attention to the trigeminal and facial nerves in order to avoid ocular complications.

Table 1. Indications for epidural anterior petrosectomy. R means routinely, O means occasionally

Origin of the Disease	Tumor	Cyst	Vascular disease
Bone	Chondrosarcoma–Chordoma (R)	Epidermoidcyst – Cholesterin granuloma Aneurysmal bone cyst (R)	
Dura mater	Meningioma (R)		
Cisterna	Trigeminal schwannoma (R)	Epidermoid–dermoid (R)	Posterior circulation aneurysms (O)
Brain stem/ventral pons	Glioma – Metastasis (O)		Cavernoma (O)

Indications

Originally dedicated to the management of posterior circulation aneurysms, the indications have changed while other techniques have been developed: endovascular techniques for vascular diseases and endoscopic surgery in skull base. Today the main indications are mainly focused on several types of tumors from various origin that are summarized in Table 1.

Comments

The EAP is a straightforward procedure that offers optimal corridor with minimal petrous bone drilling [3, 6, 8]. On average, from skin incision to full exposure, it takes an hour and half to be completed. If needed by the extension of the disease, this approach can include the opening of the MC and cavernous sinus. In special circumstances, the EAP can be part of a largest approach named combined petrosectomy [4, 9] for giant tumors extending to the retromeatal region or to the lower cranial nerves and as far as the foramen magnum. The concept is to offer a conservative petrosectomy while preserving the cochlea and avoiding any mobilization of the intrapetrous facial nerve. In spite of these advantages, the EAP has several limitations.

We outlined that a fully achieved EAP exposes a large area through a small window. However, it does not control safely the cerebellomedullary cistern at the bottom of the field. Backward it is impossible to expose what is behind the acoustic of acialacousticofacial bundle. Upward, what is above the posterior clinoid process cannot be reached unless a critical retraction of the temporal lobe is to be done. In special circumstances, the tumor mass which is in front of the clivus may push backward the basilary artery and the ventral brain stem, giving enough space to reach the opposite side (Figure 3c). However, this cannot be predicted and manipulation of the contralateral trigeminal and abducens nerves is usually at risk.

Protection of the temporal lobe is a major concern. In up to 15% of cases, we personally experienced at least transient dysphasia, seizure, memory changes or behavioural changes due to the temporal lobe dysfunction. Venous mechanisms are the most suspected ones. These events usually come at 2 to 3 days after surgery and have to be mentioned while counselling the patient before surgery. Moreover, in up to 40% of cases postoperative MR shows that the temporal lobe display mild or moderate permanent changes of signal on T2 sequences. So far, I we did not observed any permanent clinical sequella of this injury.

The surgical alternatives to the EAP may be either intradural or extradural. The endoscopic endonasal approaches are extremely valuable for midline and epidural diseases. They are also logical options for the petrous apex cysts. Whether they carry the same ratio of safety/efficacy than EAP for intradural petroclival diseases remains to be shown by additional publications. From the intradural side, the regular or modified retrosigmoid route is simple, routinely performed by experienced neurosurgical hands. The target is the cerebellopontine angle for vestibular schwannomas or for meningiomas that are inserted around the porus of the IAC. Drilling of the suprameatal bone will expose the content of the MC but looks a narrow corridor with permanent interposition of the acousticfacial bundle in the operative field.

Conclusion

The epidural anterior petrosectomy is one of the safest skull base approaches to appropriately expose the petroclival area and ventral pons. This route belongs to the group of conservative petrosectomies which spare the neurotological structures. In addition, EAP offers an excellent modularity to expose diseases that affect simultaneously the parasellar compartment, the floor of the middle fossa and the petroclival are area that cannot be achieved routinely by endonasal routes. The EAP is carried out in a relative short time and closed in a simple manner with a very low rate of CSF leaks. Such advantages bring the EAP in the group of minimal invasive procedures. Nevertheless, tumor resection carries its own difficulties depending on its texture and adhesion to the vascular and neural structures.

References

[1] Day J. D., Fukushima T., Giannotta S. L. (1994) Microanatomical study of the extradural middle fossa approach to the petroclival and posterior cavernous sinus region: description of the rhomboidconstruct. *Neurosurgery* 34:1009–1016.

[2] Destrieux C., Velut S., Kakou M. K., Lefrancq T., Arbeille B., Santini J. J. (1997) A new concept of Dorello's canal microanatomy: the petroclival venous confluence. *J Neurosurg* 87:67–72.

[3] Fukushima T., Day J. D., Hiraha K. (1996) Extradural total petrous apex resection with trigeminal translocation for improved exposure of the posterior cavernous sinus and petroclival region. *Skull Base Surg* 6:95–103.

[4] Hakuba A., Nishimura S., Inoue Y. (1985) Transpetrosal- transtentorial approach and its application in the therapy of retrochiasmaticcraniopharyngiomas. *SurgNeurol* 24: 405–415.

[5] Kawase T., Toya S., Shiobara R., Mine T. (1985) Transpetrosal approach for aneurysms of the lower basilar artery. *J Neurosurg* 63:857–861.

[6] Kawase T., Shiobara R., Toya S. (1991) Anteriortranspetrosaltranstentorial approach for sphenopetroclivalmeningiomas: surgical method and results in 10 patients. *Neurosurgery* 28:869–876.

[7] Ozveren M. F., Uchida K., Aiso S., Kawase T. (2002) Meningovenous structures of the petroclival region: Clinical importance for surgery and intravascular surgery. *Neurosurgery* 50:829–837.

[8] Roche P. H., Lubrano V., Noudel R. Epidural Anterior petrosectomy. *Acta Neurochirurgica,* 2011, 153 (6): 1161-7.

[9] Spetzler R. F., Daspit C. P., Pappas C. T. E. (1992) The combined supra- and infratentorial approach for lesions of the petrous and clival regions: experience with 46 cases. *J Neurosurg* 76:588–599.

Chapter XIV

Pediatric Endoscopic Skull Base Surgery

Smriti Nayan,[1,*] ***Kesava Reddy,***[2]
Adam M. Zanation,[3] ***and Doron D. Sommer***[1]

[1]Division of Otolaryngology and Head and Neck Surgery,
McMaster University, Hamilton, ON, Canada
[2]Division of Neurosurgery, McMaster University, Hamilton, ON, Canada
[3]Department of Otolaryngology–Head and Neck Surgery, University of
North Carolina, Chapel Hill, North Carolina, US

Abstract

Continued advances in instrumentation and experience of skull base teams has led to the expanded endonasal approach (EEA) becoming a useful tool in the management of selected pediatric skull base pathologies [1-3]. Endoscopic approaches can now be applied to benign and malignant processes including infections, neoplasms, and CSF leak repair. Effective and consistent results are based on a sound knowledge of the variable pediatric anatomy and its changes through maturation [2-4]. Insights into the limitations of reconstruction will also be discussed [1]. A variety of approaches with clinical examples will be discussed in the anterior to posterior direction including access to the frontal region [5], the upper c-spine [6] and those in between.

Outcomes have improved with a reduction in traditional external approaches in selected cases. EEA, however, is not without risk and should be generally undertaken by experienced teams. Pre-operative planning and evaluation using computed tomography (CT) or magnetic resonance imaging (MRI) is vital to assess appropriateness of this approach. Depending on the age and size of the patient, limitations may include nostril size, pneumatisation of the sinuses and skull base, reduced inter-carotid distance, and variations in the nasoseptal flap size [1, 4]. Studies based on CT anatomical measurements relevant to endoscopic approaches will also be highlighted [4]. This

---

[*] Corresponding author: Smriti Nayan. E-mail: smriti.nayan@gmail.com.

chapter aims to review the anatomical and surgical considerations of EEA in pediatric patients as well as special limitations for this population.

Introduction

The combined effort of neurosurgeons and otolaryngologist-head and neck surgeons specializing in skull base surgery has allowed for increasing applicability of the expanded endonasal endoscopic approaches (EEAs) to access the midline skull base in pediatric patients. [4] Endoscopic techniques have allowed for instrumentation of the anterior skull base via a variety of approaches, resulting in decreased disruption of normal anatomy. Pediatric patients present a unique challenge due to the anatomical and developmental differences compared to adults. Furthermore, the options of skull base reconstruction may be more limited and must be planned carefully.

The traditional craniofacial approach to the anterior skull base offers good exposure and access to pathology of the skull base. It also allows for true three-dimensional visualization of the pathology and for wider control of potential bleeding. However, there are significant potential side effects and complications, which may occur from this approach. These morbidities may arise from the incision and disruption of the growth centers in the craniofacial skeleton, which may result in facial asymmetry or disfigurement. The open approach may require retraction of the brain or other vital structures depending on the location of the lesion [7]. There is also an increased risk of CSF leak, meningitis, injury to the cranial nerves and brain, palatal dehiscence, velopharyngeal insufficiency, wound infection, dental malocclusion, and orbital injury.

EEAs for certain pathologies of the anterior skull base may have several advantages in the pediatric population compared to the traditional craniofacial approaches. The endoscope offers a wider, more magnified and dynamic view of the operative field [2, 7]. Thus, for example, the pituitary gland and stalk are better visualized, reducing the risk of endocrinopathy, incomplete resection as well as reduced risk of neurovascular injury [2, 7]. However, the view with the endoscope is two-dimensional which may result in certain limitations for the inexperienced surgeon. Furthermore, EEAs do not require a skin incision, external craniotomy, translocation of the maxillofacial skeleton and retraction of the brain or other neurovascular structures [7, 8].

For selected skull base pathologies, there are multiple advantages of EEA. These advantages may include the potential for less overall morbidity, allowing for faster patient recovery, and reduced postoperative pain. There may also be a reduction in the overall cost with the use of EEA [4], although this requires further evaluation. The surgical goals when using the endoscopic approach should generally be the same as those during open surgery.

Embryology and Development

Paranasal sinus development has a multitude of influences including congenital anomalies, environmental conditions and past infections. Anterior skull base lesions are often accessed through the sphenoid sinus, thus, knowledge of the size, capacity and shape of the sphenoid sinus is vital to the management of these complex lesions. At birth, the sphenoid bone contains only red bone marrow at its center.

In their radiological study, Adibelli et al., noted that the sphenoid sinus may begin to pneumatize between 6 and 9 months of age [8].

Pneumatization of the sinus generally occurs only after the marrow converts to fatty marrow in the third year of life [2, 8]. The sphenoid sinus rapidly develops to extend posteriorly towards the sella turcica around the age of 7 years, reaching adult size at around age 12. Reports in the literature state that the sphenoid sinus volume ranges between 7.5-8.65 cm^3 in pediatric patients [8-10].

Computed tomography scans have demonstrated growth trends through craniofacial measurements. Cranial growth appears to be most rapid in the first few years of life, followed by a leveling off in later childhood [1]. The skull base continues to develop for at least the first ten years of life, whereas, the upper midface does not show a significant increase in size early in life. The upper midface region continues to grow later in life at a more rapid rate than the cranium [11]. At 10 years of age, Scott's craniofacial analysis demonstrated that cranial measurements reached or exceeded 95% of adult size, whereas facial measurements had only reached 85% of adult size [1, 11].

Anatomical Considerations

The endonasal corridor can be divided into surgical modules based on their orientation in sagittal and coronal planes [3]. The sphenoid sinus is a central landmark between the sagittal and coronal planes, making it an important starting point for many of the surgical modules. The sagittal plane extends from the frontal sinus to the body of C2 in the midline corridor. The coronal plane is subdivided into anterior, middle and posterior coronal planes corresponding to the cranial fossae [2, 3].

There are six different surgical modules in the saggital plane, which are classified into the transfrontal, transcribriform, transplanum, transsellar, transclival and the transodontoid approach [2, 12] (Table 1). Essential neurovascular structures, namely, the optic nerves and internal carotid arteries, can be followed to other areas of the skull base from the sphenoid sinus [3]. Three major bony anatomical landmarks are important in expanded EEAs in pediatric patients. The most superficial bony structure to be encountered is the anterior nasal aperture [1]. Care must be taken to not damage the developing facial skeleton as the endoscope and instruments pass through the nasal aperture. Once the nasal cavity is entered, the incompletely pneumatized sinuses must be dissected followed by exit through the skull base into the anterior, middle or posterior cranial fossae [4].

Table 1. Surgical Modules in the Sagittal Plane [2, 13]

Surgical Module	Anatomical Boundaries
Transfrontal	Floor and posterior wall of the frontal sinus
Transcribriform	Cristal galli to the planum sphenoidale, across the fovea ethmoidalis to the roof of the orbit
Transplanum	Crista galli to the sella
Transsellar	Sella through the posterior clinoid process, lateral limit is the cavernous internal carotid artery
Transclival	Posterior clinoid process through the foramen magnum
Transodontoid	Foramen magnum to C2

Piriform Nasal Aperture

The piriform nasal aperture represents a potential anatomical limitation to endonasal approaches in pediatric patients as multiple instruments are required to pass through this area [4]. Piriform aperture width does not change significantly after the age of 9 or 10 [4]. Tatreau et al. demonstrated that the piriform aperture diameter in patients less than 24 months of age was 17.2 ± 0.5 mm compared to adults where it was 22.2 ± 1.3 mm ($p < 0.0003$) (Figure 1, reproduced with permission) [4]. The small piriform aperture size is a relative contraindication to EEA only when patients are less than two or three years old or in those that have midline developmental anomalies [4]. In these patients, an endoscopic approach is still feasible but may have to be modified with the use of such techniques such as sublabial incision, midfacial degloving, [13] maxillary sinus windows, or external rhinoplasty incisions.

Sphenoid Sinus Pneumatization

The pneumatization of the sphenoid sinus varies in pediatric patients [4]. Developmental patterns of the sphenoid sinus have been well studied in the literature (Figure 2, reproduced with permission) [4, 7, 8, 14].

The radioanatomical study by Tatreau et al. demonstrated that by 6 or 7 years of age, the anterior wall of the sphenoid sinus is fully pneumatized [4]. Pneumatization progresses from an anterior to posterior direction and an inferior to superior direction [4]. Pneumatization continues to progress in the posterior direction, first reaching the floor of the sphenoid bone, then the planum sphenoidale and the anterior sellar wall [4].

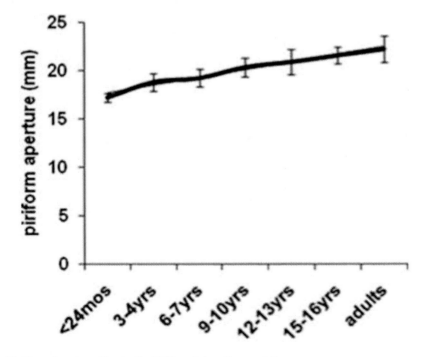

Figure 1. Piriform Aperture Size in the Different Age Groups [4].

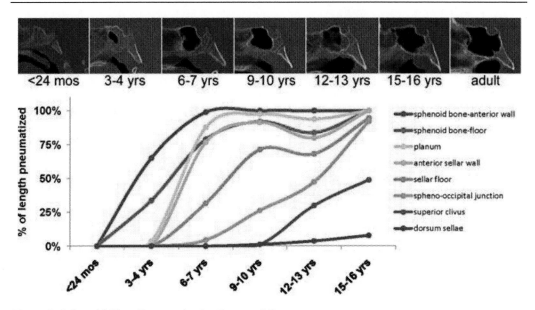

Figure 2. Sphenoid Sinus Pneumatization Patterns [4].

Pneumatization patterns have been categorized on the basis of aeration along the anterior-posterior axis with classification into specific categories: conchal/presellar, sellar, or postsellar [4, 7]. This criteria, however, does not predict the feasibility of certain approaches or the depth of drilling required to penetrate the sphenoid bone [4].

Complete pneumatization of the planum facilitates entry into the anterior cranial fossae. Due to the incomplete pneumatization in pediatric patients, it was felt in the past that patients under the age of 3 would not be candidates for trans-spehnoidal surgery [4]. However, precise drilling in these areas may result in surgical pneumatization, allowing access to the skull base [7].

Intercarotid Distance

The internal carotid arteries (ICAs) are closely associated with the sphenoid sinuses. The bony impressions of the ICAs within the sphenoid sinus are absent in younger patients. In 73% of adults, the ICA prominences can be seen in the posterolateral wall of the sphenoid sinus [4]. Early experience with EEA reported that approximately 1-9% of EEAs in adults have ICA injury, despite visualization of the arteries within the sphenoid sinus wall [4, 15]. Although this complication is now less frequently encountered, the clival intercarotid distance remains an important relationship within the developing sphenoid sinus and ICAs. Trans-clival approaches may be limited by a narrow intercarotid distance [4], however, the intercarotid distances do not vary significantly with age. Furthermore, although more technically difficult and associated with increased risk, the internal carotid artery may be mobilized to allow for better access.

Thus, intercarotid distances may not be an anatomic limitation to EEA in children (Figure 3, reproduced with permission) [4].

In summary, these anatomical factors may be considered relative contraindications in the pediatric population.

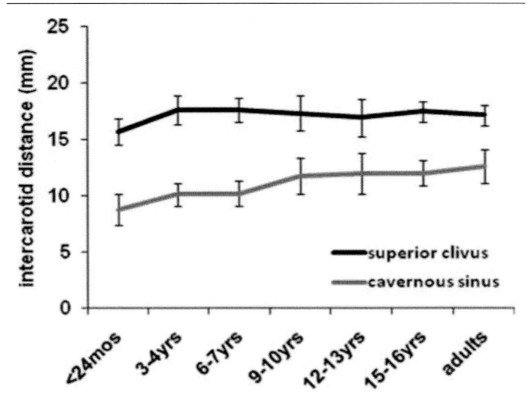

Figure 3. Intercarotid Distance [4].

Evaluation of these individual anatomical factors should be carefully considered preoperatively in pediatric patients undergoing endoscopic endonasal skull base surgery.

Surgical Technique

Coordination between the otolaryngologist-head and neck surgeon and the neurosurgeon is essential and this team requires skill and experience in this area [3]. The EEA has more utility for certain pathologies and locations, specifically, midline lesions along the anterior skull base. In pediatric patients, the EEA can be utilized in providing a biopsy for diagnosis and definitive treatment for benign or malignant neoplasms [13].

The endoscopic surgical technique for removal of skull base lesions does not differ significantly in pediatric patients compared to adult patients [2]. The goals of endoscopic resection should be identical to that of open craniofacial approaches i.e. complete resection of disease with minimal morbidity [3].

Occasionally, residual disease may have to be left behind to preserve critical neurovascular structures.

Pre-operative surgical planning involves a thorough history and physical examination, as well as review of the preoperative radiographic scans. Coordination with the anesthesiologist, endocrinologist, ophthalmologist and pediatrician is necessary in certain cases. Image guidance may be used with either computed tomography (CT) +/- contrast images, CT angiography (CTA) images and/or MR imaging.

CTA imaging aids in identifying the relationship of the lesion to various vascular structures including the ICAs [3], basilar artery, the anterior cerebral arteries and other blood supply. Image guidance has been determined to be safe and effective in the pediatric population [3, 16, 17]. The indications for image guidance are similar to adults and include chronic rhinosinusitis, complications of acute sinusitis, allergic fungal sinusitis and tumor [3, 18].

The use of a sublabial incision or an external septorhinoplasty approach through a columellar incision may be used with a trans-septal, trans-sphenoidal approach to help aid in access and help overcome anatomical limitations [18]. However, this combination of techniques may only be necessary in the very young subgroup of patients. Avoidance of a transmaxillary approach helps preserve dentition and facial growth plates [2]. Post-operatively, pediatric patients better tolerate their recovery without any nasal packing [2].

Dehdashti et al. [19] state that the EEA may not be suitable for lesions greater than four cm or those that have significant lateral extension beyond the optic canals, encasement of neurovascular structures or brain invasion by malignant lesions [3].

There are several descriptions of the endoscopic approach to the clivus in the literature [20, 21]. Burkart et al. [20] described the approach to the clivus and noted that the lateral limitations were the medial pterygoid plate and the eustachian tubes [3]. Coronal plane approaches help to overcome this limitation e.g. through the transpterygoid approach. The extent of caudal exposure to the upper cervical spine is limited by the nasal bones superiorly and the hard palate posteriorly which delineates the nasopalatine line (NPL) or K-line (Figure 13A) [1-3, 22]. The angle between the NPL and the plane of the hard palate is termed the nasopalatine angle.

Aside from anatomical limitations, other potential disadvantages of the EEA have been discussed. The technique does not allow for true 3D vision and requires other spatial and proprioceptive cues to aid with the surgery, though the development of 3-D endoscopes may obviate this issue. Furthermore, due to the limitation in access with EEA, major bleeding was earlier thought to be a challenge to control. However, the development of management strategies [15], pre-operative evaluation with embolization of selective cases, and newer instrumentation including various bipolar options and topical hemostatic agents have been major developments in our ability to prevent and manage major intra-operative bleeding.

Surgical Reconstructive Options

The goals of effective and safe reconstruction following EEA comprise complete separation of the cranial cavity from the sinonasal tract, obliteration of dead space, preservation of neurovascular and ocular function, and reconstruction of tissue barriers [1].

The pedicled nasoseptal flap (NSF) has become the work horse for vascularized endoscopic skull base reconstruction [1, 9, 18, 23]. The nasoseptal flap is based on the nasoseptal artery, which is a branch of the posterior septal branch of the sphenoplatine artery. This neurovascular pedicled flap includes the nasal septal mucroperiosteum and mucoperiochondrium [1]. The NSF is potentially able to provide coverage for maximal dural defects, extending from the frontal sinus to the sella and from orbit to orbit in adults [3].

Preoperative planning of the NSF is essential for success. CT scans may aid in determining the dimensions of the flap [1].

The flap is harvested at the start of the case and is then shifted posteriorly into the nasopharynx until it is required for closure of the surgical defect. Flap elevation aims to preserve the olfactory epithelium and the posterolateral neurovascular pedicle. The NSF must be planned prior to surgical resection to preserve the pedicle [1, 18]. The NSF may be placed as an overlay with possible use of an underlay dural graft. The repair may be further reinforced and bolstered with absorbable dressings, tissue sealant, stenting and removable packing [18].

Logically, the NSF is of smaller size in pediatric patients thus limiting its use for skull base reconstruction. It has been demonstrated that the average potential flap length is less than the average anterior skull base length until about age 9-10 years and less than the average trans-sellar defect until age 6-7 years (Figure 4, reproduced with permission) [1, 3]. Shah et al. [1] concluded that the NSF would provide adequate coverage in children greater than fourteen years of age as they would have adult-sized septal area (Figure 5, reproduced with permission).

The NSF has been demonstrated to reduce the incidence of postoperative cerebrospinal fluid (CSF) leaks following EEA in adults [1]. This type of reconstruction may be used to repair high-flow intraoperative CSF leaks [24]. In the study by Zanation et al. [24], a multivariate analysis demonstrated that the pediatric age group was the only statistically significant factor found to correlate with increased incidence of postoperative CSF leaks [18, 24]. Due to the relatively late septal growth in the sagittal plane, the NSF may only be reliable for CSF leak repair after the age of 10-13 years [1, 2].

Alternatively, in certain cases, a CSF leak is not anticipated and thus the NSF is not elevated at the beginning of the operation. In these cases, a nasoseptal rescue flap may be used instead if the pedicle is preserved [25]. The rescue flap requires partial harvest of the posterior and superior aspect of the flap to preserve the pedicle and provide surgical access to the face of the sphenoid.

The rescue flap can then be harvested in its entirety if required in the setting of an unanticipated CSF leak. This technique thus reduces septum donor morbidity in those patients who do not have a CSF leak. Rawal et al. demonstrate the use of the nasoseptal rescue flap in pediatric patients aged fifteen or greater [26].

The anteriorly based pericranial flap is a viable option when the NSF cannot be used [27, 28]. It may be harvested in an external coronal approach and transposed to the skull base via a nasionectomy into the nasal cavity [3, 18, 27]. Another alternative for vascular reconstruction may include the use of the temporoparietal fascial flap [3, 18]. However, the effectiveness and safety of these flaps have not yet been evaluated in pediatric patients [18, 19].

Finally, if locoregional flaps are not available for reconstruction, free flaps may be used such as the rectus abdominis myocutaneous or the anterolateral thigh free flaps [12]. Important considerations with the use of free flaps include increased total operative time and intraoperative blood loss [12].

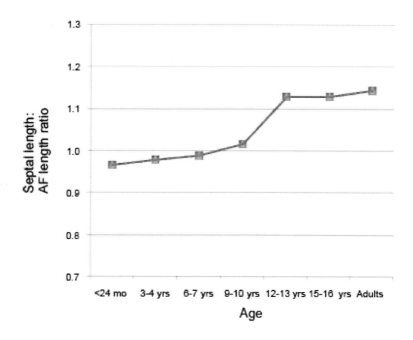

Figure 4. Ratio of Flap Length to Anterior Skull Base Approach Defect According to Age [1].

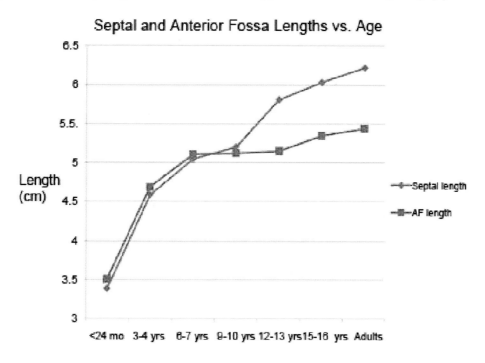

Figure 5. Ratio of Septal Length Compared to Anterior Cranial Fossae Length According to Age [1].

Complications

The most common complication encountered in this population is CSF leak. The largest reported series in the literature by Kassam et al. [2], demonstrated an 8% risk, though this appears to be declining as more experience with vascularized reconstruction develops (Kassam, personal communication). None of these cases developed bacterial meningitis or required shunt for persistent CSF leak.

In this series, all the leaks were controlled with revision surgery involving an onlay graft and repacking the sphenoid sinus with fat [2]. Additionally, if a NSF has been used for reconstruction, repositioning and rebolstering of the flap is often possible. Other rare potential complications are similar to those in the adult literature and include bleeding, orbital/optic nerve injury, endocrinopathy and cranial nerve or other intracranial injury.

Pathology

Skull base lesions are relatively uncommon in children (29). The largest series in the literature by Kassam et al. demonstrated only 25 cases out of 430 consecutive case [2]. In the authors' experiences, at McMaster University there have been 18 consecutive pediatric cases excluding primarily intranasal pathologies. This amounts to approximately 3.6% of the total of over 500 cases (unpublished data). Pediatric skull base lesions include a wide variety of pathologies. These pathologies may be congenital (eg encephalocele), infectious, traumatic or neoplastic in origin. Neural, mesenchymal, notochordal, vascular and epithelial tissues may be the origins of neoplastic masses within the anterior skull base. Pituitary adenomas and lateral skull base lesions may be encountered, but are relatively uncommon in the pediatric population [18]. Encephaloceles and other skull base defects may be amenable to endoscopic repair depending on their size, location and the age/size of the patient. These lesions are often noted in infancy, which often precludes this approach. Juvenile nasopharyngeal angiofibroma (JNA) is one of the most common pathological entities treated with EEA [9]. Smaller JNAs have low morbidity and decreased rates of persistent or recurrent disease [18]. For the purposes of this section, JNA represents a primarily intranasal pathology and thus falls beyond the scope of this chapter. Craniopharyngiomas are benign but locally aggressive lesions of the skull base that often adhere to the surrounding neurovascular structures [18]. A large review by Jane et al. [30] with 22 patients demonstrated that recurrence is higher in revision surgery (30%) compared to primary surgery (9%) [9, 30]. As in adults, the majority of patients developed a new endocrinopathy [30]. Rathke cleft cysts (RCC) may be in the sellar or suprasellar space and thus are also amenable to endoscopic treatment [18]. Fibro-osseous lesions (e.g. fibrous dysplasia) are occasionally seen in the pediatric population and may necessitate surgical intervention for biopsy purposes or if impinging on vital structures (e.g. optic nerve).

Cases

The following section will review case examples, which are anatomically described in an anterior to posterior direction.

Frontal Epidural Abscess

A fourteen year old boy presented with a two day history of severe right sided headaches associated with nausea and vomiting.

He also noted a four week history of symptoms consistent with sinusitis including purulent nasal discharge, obstruction and midfacial pressure. Upon presentation, he was febrile (38.7°C) and in significant discomfort. Appropriate intravenous antibiotic therapy was initiated, however there was symptomatic worsening during the following 2 days. A preoperative CT scan (Figure 6) demonstrated a right frontal epidural abscess. The intra-operative and post-operative images can also be seen (Figures 7-8).

As with other pathologies, the endoscopic approach to frontal epidural abscesses offers significant potential advantages over traditional approaches. This trans-frontal approach, involves a standard endoscopic uncinectomy and anterior ethmoidectomy. Next, a Draf II a or b frontal approach is performed, depending on access and location of the abscess.

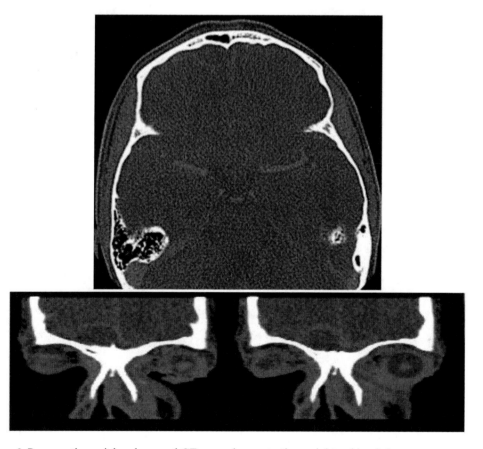

Figure 6. Preoperative axial and coronal CT scans demonstrating a right epidural abscess.

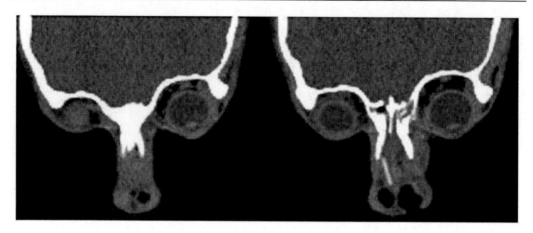

Figure 7. Postoperative CT demonstrating resolution of the abscess, the skull base defect in the fronto-ethmoid region and a stent in the right nasal cavity.

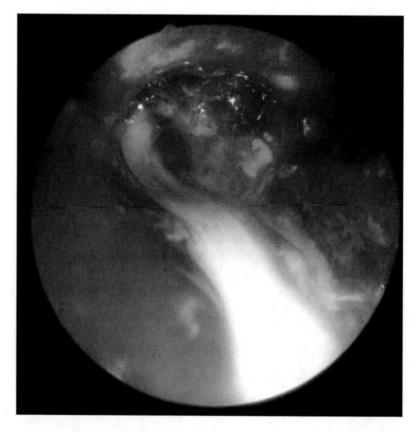

Figure 8. Intraoperative photo demonstration a wide frontal Draf IIa sinusotomy with drainage of the abscess.

If the abscess is more superiorly or laterally based, a Draf II b or III may be necessary for additional access. The skull base may be partially thinned out due to the infectious process and may require only a curette or frontal angled forceps to traverse the bone and enter the abscess/extradural space. However, in some of these cases a 2 mm diamond drill or angled Kerrisson's forceps may be required to access the abscess.

The cavity is then gently irrigated and an optional drain may be left in for 1-2 days (Figure 9). There is no breach of the dura in this technique and thus there is no resultant CSF leak or need for intranasal repair. Our experience involves an age range of 8 years old up to adulthood for this approach (5).

In our experience, this has generally led to rapid clinical resolution or improvement within 1-2 days and has become the favored method for approach to this pathology at our institution.

This approach, however, depends on the pneumatization of the frontal sinuses, which may be variable. The frontal sinuses are generally the last to develop, and may occasionally under-develop. Typically, the frontal sinuses extend above the orbital roof after age 8-10. However, sinogenic epidural abscesses are typically found in older children, usually over the age of 12 [31].

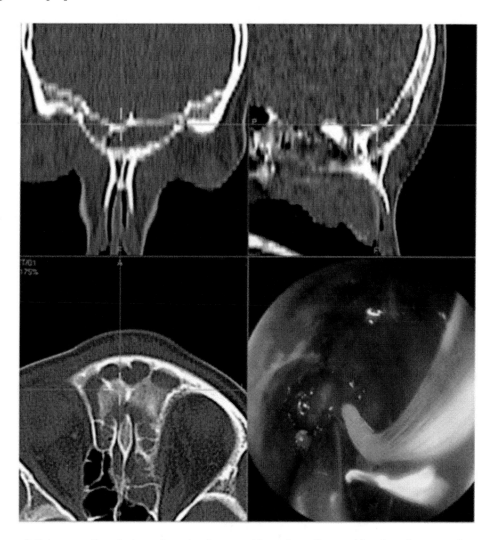

Figure 9. Intra-operative photographs using image guidance in an 8 year old patient demonstrating location of skull base fenestration and insertion of a soft silicone drain into the frontal epidural abscess cavity via a wide frontal sinusotomy.

Approach to the Sella, Parasellar, Planum Sphenoidale Regions

A previously well 16 year old boy presented with delayed onset of puberty. He had a five year history of occasional headaches. Endocrine testing revealed low gonadotropic and testosterone levels as well as a slightly low TSH. Visual field testing revealed bilateral temporal field deficits. MRI revealed a large partially cystic lesion in the retrochiasmatic region consistent with a craniopharyngioma (Figure 10)

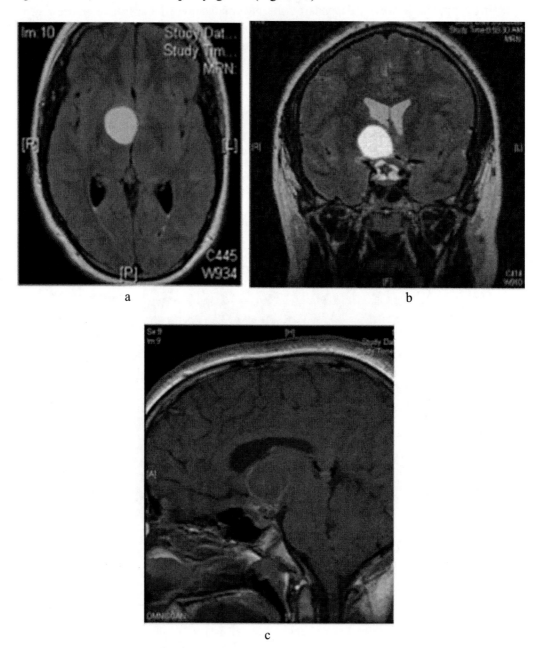

Figure 10. Preoperative MRI showing a large complex cystic lesion consistent with a craniopharyngioma extending to the lateral ventricles. A: axial cut, B: coronal cut, C: sagittal cut.

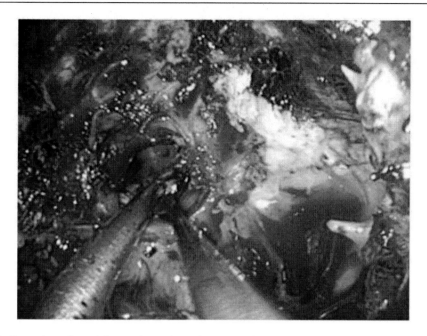

Figure 11. Resection of cranipharyngioma through a trans-planum approach.

The craniopharyngioma was resected via an endonasal endoscopic trans-planum, transtuberculum approach (Figure 11). Standard techniques were utilized and a wide exposure was provided. As is the norm for this pathology, a high volume CSF leak was evident during resection and was repaired with a fascia lata underlay graft and a vascularized naso-septal mucosal graft. Post-operatively, the patient's headaches and visual fields improved, however he required supplemental hormonal therapy to treat his pituitary dysfunction.

His post-operative imaging showed a persistent 7 mm lesion in the area of resection (Figure 11), for which he underwent radiation therapy. This area has shown no change after over 5 years of follow-up.

Patients in this age group typically present with pituitary tumors (32), Rathke's cysts and craniopharyngiomas in this region. Other less common pathologies are occasionally seen. In the teenage group, lesions in this area are treated very similarly to those in the adult population. Skull base repair with the use of a vascularized septal flap is generally achievable in those older than thirteen with careful pre-operative planning and evaluation of imaging [1].

Approach to the Craniocervical Junction

A seven year old boy presented with significant neck pain increasing over 2 months. His past medical history was relevant for a left cheek non-metastatic alveolar rhabdomyosarcoma, which was treated 2 years previously with radiation. Imaging revealed a lytic lesion in the C2 vertebral body (Figure 12).

Pre-operative evaluation of the sagittal CT showed that the inferior extent of access at the nasopalatine line (Figure 13A) [21, 33] was below the lesion, allowing for an endonasal approach to the lesion.

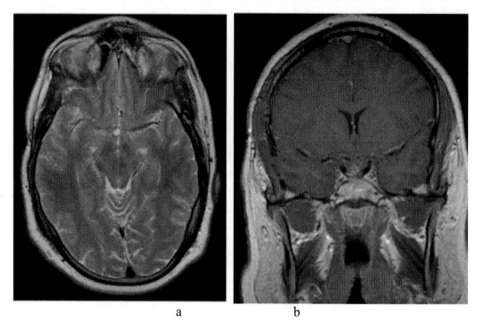

Figure 12. Post-operative MRI demonstrating 7mm Residual Craniopharyngioma. A: axial cut B: coronal cut.

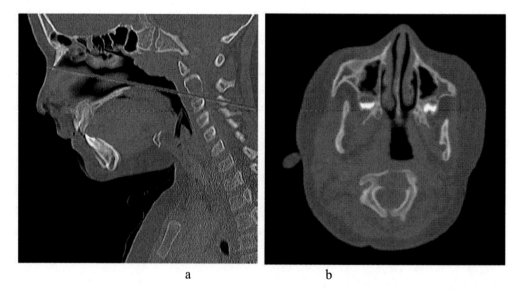

Figure 13. Preoperative imaging demonstrating a lytic lesion in the C2 vertebral body. A: Sagittal cut, the red line delineates the nasopalatine line. B: axial cut.

The patient underwent an EEA approach to the odontoid for biopsy of his lesion. Intraoperatively, an adenoidectomy was performed to improve access to the lesion (Figure 14). The remainder of the approach was similar to that in adults [6]. Pathological evaluation of the lesion resulted in diagnosis of an aneurysmal bone cyst. The smaller working corridor does make this approach more challenging in young children, however the use of standard adult endoscopes and instruments are usually feasible. As is common in this approach [6], occipitocervical instrumented fusion was required for cervical spine stability.

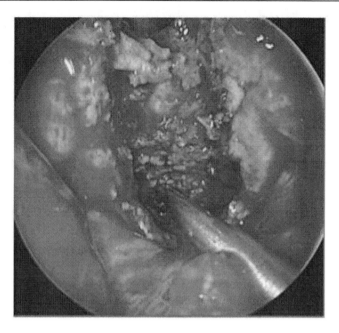

Figure 14. Intraoperative photo demonstrating a sphenoidotomy superiorly and drilling inferiorly to the level of C2.

Conclusion

Advances in EEA through newer instrumentation and the experience of skull base teams has allowed for optimized management of certain pediatric skull base pathologies. Outcomes have improved with decreased need for traditional external approaches in selected cases. Successful EEA requires experienced teams, careful pre-operative planning and evaluation of appropriate imaging to minimize risk and enhance success.

References

[1] Shah, R. N., Surowitz, J. B., Patel, M. R., et al. Endoscopic Pedicled Nasoseptal Flap Reconstruction for Pediatric Skull Base Defects. *Laryngoscope* 2009. P: 1067-1075.

[2] Kassam, A., Thomas, A. J., Snyderman, C., et al. Fully endoscopic expanded endonasal approach treating skull base lesions in pediatric patients. *J. Neurosurg.* 2007;106(2 suppl.):75–86.

[3] Solares, C. A., Ong, Y. K., Snyderman, C. H. Transnasal endoscopic skull base surgery: what are the limits? *Curr. Opin. Otolaryngol. Head Neck Surg.* 18:1–7

[4] Tatreau, J. R., Patel, M. R., Shah, R. N., Kinwei, A. N., Zanation, A. M. Anatomical Considerations for Endoscopic Endonasal Skull Base Surgery in Pediatric Patients. *Laryngoscope* 2010 p:1730–1737.

[5] Sommer, D. D., Minet, W., Singh, S. Endoscopic Transnasal Drainage of Frontal Epidural Abscesses. *Journal of Otolaryngol Head Neck Surg.* 2011; 40(5):401-406.

[6] Lee, A., Sommer, D. D., Reddy, K., Gunnarsson, T. Endoscopic Transnasal Approach to the Cranio-Cervical Junction. *Skull Base.* May 2010; 20(3): 199-205.

[7] Zanation, A. M., Snyderman, C. H., Carrau, R. L., et al. Minimally invasive endoscopic pericranial flap: a new method for endoscopic skull base reconstruction. *Laryngoscope.* January 2009. 119(1): 13-18.

[8] Adibelli, Z. H., Songu, M., Adibelli, H. Paranasal sinus development in children: A magnetic resonance imaging analysis. *Am. J. Rhinol. Allergy* 2011. 25: 30-35.

[9] Amedee, R. Sinus anatomy and function. In: *Head and Neck Surgery—Otolaryngology.* Bailey, B. J. (Ed). Philadelphia: J.B. Lippincott Company, 342–349, 1993.

[10] Yonetsu, K., Watanabe, M. and Nakamura, T. Age-related expansion and reduction in aeration of the sphenoid sinus: Volume assessment by helical CT scanning. *AJNR Am. J. Neuroradiol.* 21:179–182, 2000.

[11] Scott, J. H. The growth of the human face. *Proc. R Soc. Med.* 1954;47:91–100.

[12] Yano, T., Tanaka, K., Kishimoto, S., et al. Review of Skull Base Reconstruction Using Locoregional Flaps and Free Flaps in Children and Adolescents. *Skull Base* 2011. 21(6): 359-364.

[13] Snyderman, C. H., Pant, H., Carrau, R. L., et al. What Are the Limits of Endoscopic Sinus Surgery? The Expanded Endonasal Approach to the Skull Base. *Keio J. Med. Sept.* 2009. 58 (3) .152−160.

[14] Kassam, A., Thomas, A. J., Snyderman, C., et al. Fully Endoscopic Expanded Endonasal Approach Treating Skull Base Lesions in Pediatric Patients. *J. Neurosurg.: Pediatrics.* Feb. 2007. 106: 75-86.

[15] Wormald, P. J., Valentine, R. Carotid artery injury after endonasal surgery. *Otolaryngoogic Clinics of North America.* October 2011. 44(5): 1059-1079.

[16] Parikh, S. R., Cuellar, H., Sadoughi, B., et al. Indications for image-guidance in pediatric sinonasal surgery. *Int. J. Pediatr. Otorhinolaryngol.* 2009; 73:351– 356.

[17] Benoit, M. M., Silvera, V. M., Nichollas, R., et al. Image guidance systems for minimally invasive sinus and skull base surgery in children. *Int. J. Pediatr. Otorhinolaryngol.* 2009; 73:1452 – 1457.

[18] Munson, P. D., Moore, E. J. Pediatric Endoscopic Skull Base Surgery. *Current Opinion in Otolaryngology and Head and Neck Surgery* 2010. 18:571–576.

[19] Dehdashti, A. R., Karabatsou, K., Ganna, A., et al. Expanded endoscopic endonasal approach for treatment of clival chordomas: early results in 12 patients. *Neurosurgery* 2008; 63:299-307.discussion 299-307.

[20] Burkart, C. M., Theodosopoulos, P. V., Keller, J. T., Zimmer, L. A. Endoscopic transnasal approach to the clivus: a radiographic anatomical study. *Laryngoscope* 2009; 119:1672–1678.

[21] Kelley TF, Stankiewics JA, Chow JM et al. Endoscopic Transsphenoidal Biopsy of the Sphenoid and Clival Mass. Am J Rhinol 1999. Jan-Feb 13(1):17-21.

[22] Pornethep, K., Carrau, R. L., Prevedello, D. M., et al. *Indications and Limitations of Endoscopic Skull Base Surgery.* http://www.medscape.org/viewarticle/763293. Accessed: September 12, 2012.

[23] Hadad, G., Bassagasteguy, L., Carrau, R. L., et al. A novel reconstructive technique after endoscopic expanded endonasal approaches: vascular pedicle nasoseptal flap. *Laryngoscope* 2006; 116:1882–1886.

[24] Zanation, A., Carrau, R., Snyderman, C., et al. Nasoseptal flap reconstruction of high flow intraoperative cerebral spinal fluid leaks during endoscopic skull base surgery. *Am. J. Rhinol. Allergy* 2009; 23:518–521.

[25] Rivera-Serrano, C. M., Snyderman, C. H., Gardner, P., Prevedello, D., Wheless, S., Kassam, A. B., Carrau, R. L., Germanwala, A., Zanation, A. Nasoseptal "recue" flap: a novel modification of the nasospetal flap technique for pituitary surgery. *Laryngoscope.* 2011. May;121(5):990-3.

[26] Rawal, R. B., Kimple, A. J., Dugar, D. R., Zanation, A. M. Minimizing Morbidity in Endoscopic Pituitary Surgery: Outcomes of the Novel Nasoseptal Rescue Flap Technique. *Otolaryngology–Head and Neck Surgery September* 2012. 147(3): 434-437.

[27] Zanation, A. M., Snyderman, C. H., Carrau, R. L., et al. Minimally invasive endoscopic pericranial flap: a new method for endoscopic skull base reconstruction. *January* 2009. 119(1): 13-18.

[28] Patel, M. R., Shah, R. N., Snyderman, C. H., et al. Pericranial flap for endoscopic anterior skull-base reconstruction: clinical outcomes and radioanatomic analysis of preoperative planning. *Neurosurgery* 2010; 66:506 – 512.

[29] Bruce, D. A.: Skull base tumors in children, In: Albright, A. L., Pol- lack, I. F., Adelson, P. D. (eds): *Principles and Practice of Pediatric Neurosurgery.* New York: Thieme, 1999, pp 663–684.

[30] Jane, J.A. Jr, Prevedello, D, Alden, T. D., et al. The trans-sphenoidal resection of pediatric craniopharyngiomas: a case series. *J. Neurosurg. Pediatrics* 2010; 5:49 – 60.

[31] Smith, H. P., Hendrick, E. B. Subdural empyema and epidural abscess in children. *Journal of Neurosurgery.* March 1983. 58(3): 392-397.

[32] Kassam, A., Snyderman, C. H., Mintz, A., et al. Expanded endonasal approach: the rostrocaudal axis. Part 1. Crista galli to the sella turcica. *Neurosurg. Focus.* 2005. 19: E3.

[33] DeAlmeida, J. R., Zanation, A. M., Snyderman, C., et al. Defining the nasopalatine line: the limit for endonasal surgery of the spine. *Laryngoscope* Feb. 2009. 119(2): 239-244.

In: Minimally Invasive Skull Base Surgery
Editor: Moncef Berhouma
ISBN: 978-1-62808-567-9
© 2013 Nova Science Publishers, Inc.

Chapter XV

Endoscopic Transnasal Removal of Midline Skull Base Tumors under the Side-Viewing Scopes

Masaaki Taniguchi, Nobuyuki Akutsu,
Kohkichi Hosoda and Eiji Kohmura

Department of Neurosurgery, Kobe University Graduate
School of Medicine, Chuo-ku, Kobe, Hyogo, Japan

Abstract

In this chapter, we demonstrate our technique to deal with laterally extended tumor compartments under the side-viewing endoscope, focusing on the management of laterally extended chordomas and pituitary adenomas with the Knosp grading of 3 and 4. The hybrid integrated endoscope-holder system was used in all cases. The system enabled instant fixation and release of the endoscope allowing bimanual surgery performed by a single surgeon. Basically, the binostril transseptal approach was employed. The tumor compartments extending laterally behind the vital structures were mostly not amenable to the removal under the straight view endoscope. Application of the side-viewing endoscope enhanced the removal rate of those compartments and was especially efficient for tumors extending into the intracavernous and paraclival retro-carotid space. With the use of specially designed malleable / steerable instruments, not only surgical maneuvers such as suctioning and curettage of the tumor but also semi-sharp dissection became feasible. The range of the surgical access at the upper and lower half of the clivus reached to the trigeminal impression and the jugular tubercle, respectively, without the need to displace the internal carotid artery or to remove the pterygoid process. Application of the side-viewing endoscope enabled removal of the tumor compartment, the exposure of which has conventionally required an extensive skull base resection. The armamentarium of the surgical instruments, however, is still not satisfactory and its ongoing development would further advance the less invasive concept in the treatment of the skull base tumors.

Introduction

Recent evolution of the endoscopic surgical technique facilitated to treat a wide range of midline skull base tumors through the transnasal route [1-3]. Though most of the lesions can be approached with the straight view endoscope, exposure of the laterally extended tumor compartment still presents a challenge for the surgeon [4, 5]. To visualize those compartments directly under the straight view endoscope necessitates resection of the anatomical structures obstructing the straightforward route, which may increase the chance of postoperative morbidity.

Moreover, vital structures such as the internal carotid artery are not resectable and their mobilization is possible only in a limited range [6]. One of the prominent properties of the endoscope is the capability to look around the corner with the side-viewing scopes. Taking advantage of this feature in the endoscopic transnasal surgery, the anatomical hindrance can be circumvented without causing injury to them.

In this way the laterally extended tumor compartment can be visualized and removed without additional surgical invasiveness [7]. In this chapter, the authors present their technique of endoscopic transnasal removal of clival chordoma and pituitary adenomas with the Knosp grading 3 and 4, especially focusing on the management of laterally extended tumor compartments under the side-viewing endoscope.

Surgical Technique

Bayonet-style rigid lens-scopes with 4.0-mm diameter and 0-, 30- and 70-degree viewing directions were used in all cases. Scopes were attached to the specially designed pneumatically driven endoscope-holder system (EndoArm, Olympus Co, Tokyo, Japan), which enabled instant fixation and release of the endoscope allowing bimanual surgery performed by a single surgeon [8].

Basically, binostril trans-septal approach was employed. Posterior half of the middle turbinate was resected and wide sphenoidotomy was performed to expose the vidian canal and paraclival carotid artery on the side, to which the tumor demonstrated lateral extension. The micro-doppler probe was used to precisely identify the course of the carotid artery.

Major compartment of the tumor in the midline was removed with the straight surgical instruments. Even during this procedure, 30-degree side-viewing scope was often used to avoid conflict with the surgical instruments ensuring their dynamic range of movement. For visualization of the lateral tumor compartment, 30-degree side-viewing endoscope was placed at the contralateral edge of the surgical field with the lens directed to the lesion side (Figure 1).

Subsequent surgical maneuver under the side-viewing endoscope was performed using various modified instruments with malleable / steerable tips. Essentially, bimanual maneuver was the principle. The curved suction tube held in the left hand supported the dissection and grasping maneuver performed with the right hand. At the final stage of the surgery, the 70-degree scope was introduced to visualize and remove the most lateral tumor compartment.

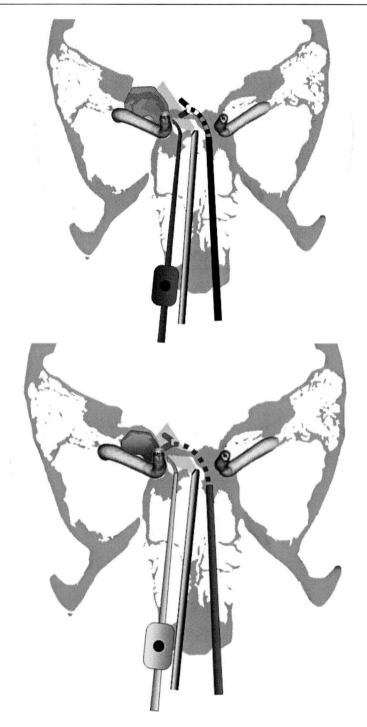

Figure 1. An illustration demonstrating the operative setting for surgical maneuver under the side-viewing endoscope. The endoscope (grey) is inserted in the patient's right nostril keeping some distance from the target. The curved sucker and dissector / forceps were inserted in the rightleft and leftright nostril, respectively, approaching a lesion (dotted green area) behind the paraclival carotid artery from medial to lateral direction.

Case Presentation

Clival Chordoma

Clival chordoma extending into the space behind the upper paraclival carotid artery is an ideal target for this type of procedure (Figure 2 and 3). After removal of the major compartment in the midline, the residual tumor obscured by the paraclival carotid artery can be visualized clearly with the 30-degree side-viewing endoscope. As the lateral compartment of this lesion is mostly extradural and surrounded by bone and dura, which tolerate relatively rough surgical maneuver, the tumor can be dissected from the surrounding tissue with malleable dissector in a semi-sharp fashion. The posterior clinoid process is often eroded by the tumor and its removal is feasible as well. For visualization of the tumor remnant attached to the cavernous sinus wall in the deepest part of the space, introduction of highly angled, 70-degree endoscope is needed. The tumor can be grasped and removed gently using the steerable forceps (Flexible Curette Forceps, Mizuho. Co, Tokyo, Japan), of which the tip can be bended and rotated at an arbitrary angle (Figure 4) [9].

In cases where the tumor extends laterally at the lower clivus, resection of medial subpetrous bone opens the corridor behind the lateral loop of the paraclival carotid artery in the direction to the jugular tubercle (Figure 5). The tumor can be followed from the midline under the side-viewing endoscope and removed together with the invaded bone with the curved drill. In this way the pterygoid plate and structures attached to can be preserved.

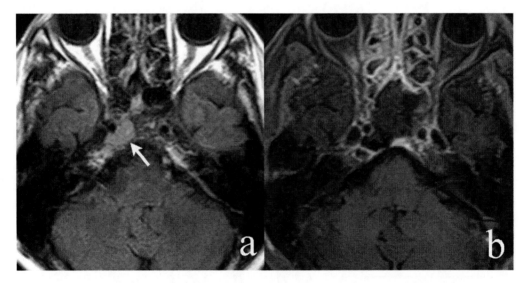

Figure 2. The pre- and postoperative MRIs of a patient with recurrent clival chordoma. The tumor is demonstrated as a high intensity lesion on FLAIR image at the right upper clivus behind the paraclival carotid artery (a, white arrow). Gross total removal of the tumor through the endoscopic transnasal approach is confirmed on postoperative fat suppression T1-weighted image (b).

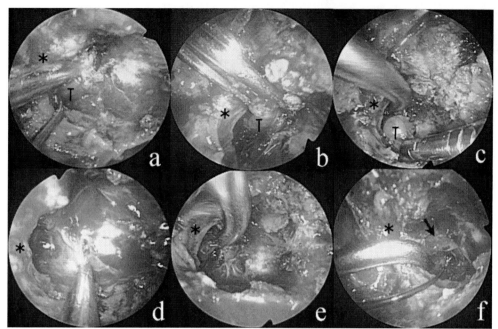

*paraclival internal carotid artery, T: tumor, black arrow: Dorello's canal.

Figure 3. Intraoperative photographs of tumor removal in the patient with recurrent clival chordoma presented in the Figure 2. The tumor dissection is started with the 30-degree side-viewing endoscope in the space behind the paraclival carotid artery at the petrous apex (a). By introducing the 70-degree endoscope, the dissection plane between the carotid artery and the tumor is demonstrated more clearly and becomes visible down to the bottom of the tumor bed (b). This enhances the certainty and safety of the dissection maneuver with bended malleable dissector. Finally the tumor is dissected from the clival dura and removed with the steerable forceps (c). The tumor bed is inspected with 0- (d), 30- (e) and 70-degree endoscopes (f). With the higher angled endoscope, the tumor bed can be visualized in straight direction, thus enabling to detect a residual even in the deepest part of the tumor bed if present.

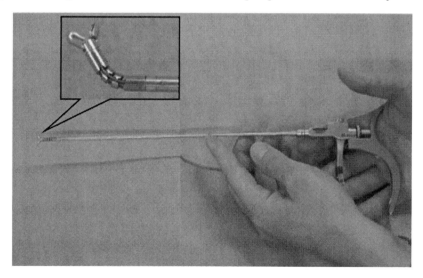

Figure 4. A photograph of the steerable grasping forceps. The whole instrument is 335mm long with the diameter of 3mm and the tip can be bended and rotated. The inset demonstrates the bending and grasping capability of the tip.

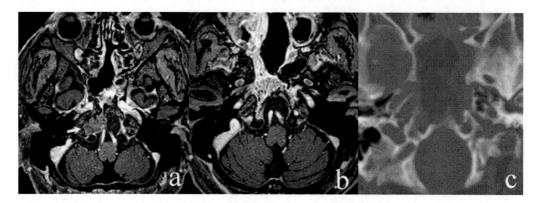

Figure 5. The pre- and postoperative MRIs of a patient with recurrent clival chordoma. The tumor indicated by the arrow is located at the lower clivus and extended to the jugular tubercle (a). Gross total removal was confirmed in the postoperative MRI (b). The bone-window CT demonstrates partial removal of the jugular tubercle (c).

In this way the pterygoid plate and structures attached to can be preserved. The range of the surgical access under the side viewing endoscope reaches trigeminal impression of the petrous apex in the upper and middle clivus and jugular tubercle in the lower clivus without the need to displace the internal carotid artery.

Pituitary Adenoma

In cases of Knosp grade 3 pituitary adenomas, extra-pseudocapsular dissection is not always feasible if the lateral extension to the cavernous sinus is prominent. The tumor mostly resides within the pocket-like space dorsal to the C4 segment of the intracavernous carotid artery, which mandates additional surgical maneuver with curved instruments under the side-viewing endoscope. Even in those cases, the medial wall of the cavernous sinus is often preserved, and semi-sharp dissection with the malleable dissector and grasping maneuver with the steerable forceps along the medial wall are feasible. In certain cases of functioning adenomas, however, medial cavernous sinus wall should be opened and the space within it explored, as residual tumor within the sinus can be sometimes found although the sinus wall looks intact (Figure 6 and 7). In those cases and in all Knosp 4 adenomas, the true intracavernous tumors can be gently curetted, aspirated and grasped with the steerable forceps, but semi-sharp dissection should be withheld to avoid injury to the cranial nerves or to the arterial branches from the carotid artery. Entry into the cavernous sinus through the medial wall can be roughly divided into 4 compartments in relation to the internal carotid artery (Figure 8). The corridor dorsal to the horizontal portion of the cavernous segment carotid artery, the largest one in most of the cases, is relatively safe to enter, as the abducens nerve, which is at most risk during the cavernous sinus exploration, is hidden by the carotid artery, and as other cranial nerves are coursing within the lateral cavernous sinus wall. The dissection is started with the 30-degree endoscope. To visualize the deepest area or the space just behind the anterior loop, introduction of the 70-degree endoscope becomes mandatory. In contrast, care must be taken when entering the corridor ventral to the horizontal segment or dorsal to the most upper portion of the vertical segment, as the abducens nerve traverse the area behind the carotid artery. Moreover, the meningo-hypophyseal trunk may offspring from the carotid artery at those segments, and inadvertent maneuver within this corridor may lead to avulsion of those vessels with subsequent disaster. This area is almost in straight line to the approach direction and can be fully visualized with 30-degree endoscope, and the use of the 70-degree endoscope is seldom needed.

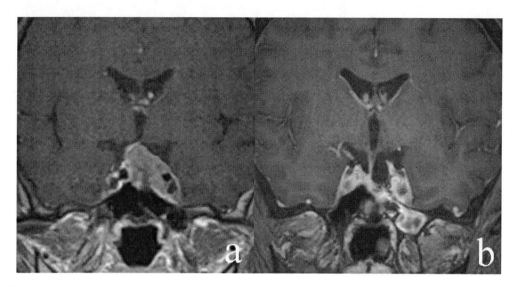

Figure 6. The pre- and postoperative MRIs of a patient with GH secreting adenoma. Note the lateral tumor extension to the left cavernous sinus which is considered as Gr. 3 according to Knosp (a). The tumor is totally removed including the lateral compartment within the cavernous sinus (b) and endocrinological remission is also achieved.

The corridor ventral to the anterior loop can be entered with relative ease in case of Knosp 4 adenoma without the use of side-viewing endoscope. This is also a corridor for direct straight approach to the lateral cavernous sinus. In Knosp 3 adenomas, though exposure of the whole anterior loop is seldom necessary, tiny tumor extension into the cavernous sinus through this corridor may exist. In such a case, the area can be approached with the side-viewing scope and the tiny tumor compartment can be grasped and removed with the steerable forceps.

In exploring the cavernous sinus through any of the corridors, the risk of carotid artery injury cannot be overemphasized. The key for the safe maneuver is the maintenance of a clear surgical view. As the profuse venous bleeding cannot completely be avoided during the cavernous sinus exploration, the surgical maneuver within the sinus should never be continued until the adequate hemostasis is achieved. Bleeding from the cavernous sinus can be often controlled with elevation of the head and insertion of the absorbable hemostatic agents. Inadvertent blind insertion of the hemostatic agents, however, leads to obscuring of the surgical field and incomplete removal of the tumor. The semi-sharp dissection to the cavernous sinus is feasible as far as the medial wall is preserved. But once the lesion extends beyond the medial wall, the maneuver has to be restricted mainly to curettage, suctioning and occasionally grasping of the tumor.

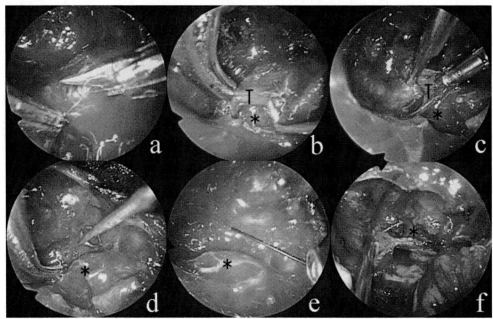

*intracavernous carotid artery, T: tumor.

Figure 7. Intraoperative photographs of left cavernous sinus exploration under the side-viewing endoscope in the patient with GH secreting adenoma presented in the Figure 6. The cavernous sinus medial wall is sharply cut (a) and the tumor within the cavernous sinus is dissected from the underlying cavernous portion carotid artery (b). The tumor is removed in an extracapsular fashion from the corridor dorsal to the upper vertical segment of the intracavernous internal carotid artery (c). Venous bleeding from the cavernous sinus is controlled with plugging of the fibrinogen soaked oxycellulose. The tumor is further curetted out through the corridor dorsal to the horizontal portion of the intracavernous carotid artery (d). With the 70-degree side-viewing endoscope this area is entirely visualized in front (e). Absence of residual tumor is confirmed and the oculomotor nerve coursing within the lateral wall of the cavernous sinus is identified with electrical stimulation. The final view under the 30-degree endoscope demonstrates the whole course of the internal carotid artery within the cavernous sinus (f).

Discussion

The recent evolution of the endoscopic transnasal approach for midline skull base tumors provided practical alternatives to conventional microscopic approach. Owing to the clear demonstration of the depth in the narrow surgical field, the lesions can be now approached without resection of large amount of skull base tissue. But still there are certain tumors exhibiting prominent lateral extension, the removal of which through the midline approach is a challenge [5]. Even in those situations, surgical procedure under the straight view endoscope using straight instruments would be certainly the most reliable form of tumor removal. However, to get a straightforward view of the target necessitates resection of certain amount of anatomical structures which obscure the direct route. Moreover, there are certain vital structures which are even not resectable. The carotid artery is one of such structures becoming most often a hindrance during the transnasal approach and its mobilization is also only feasible for a limited range [4, 6]. Using the side-viewing lenses with its capability to look around the corner, the tumor behind those vital structures can be visualized without injuring them [10]. Adding adequate surgical maneuver under such curved view of the endoscope, the removal of laterally extended skull base tumor becomes feasible. The localized lesion without displacement of surrounding structures can be also removed in a pinpoint way from the space behind the vital structures as demonstrated in Figure 2.However, to get a straightforward view of the target necessitates resection of certain amount of anatomical structures which obscure the direct route. Moreover, there are certain vital structures which are even not resectable. The carotid artery is one of such structures becoming most often a hindrance during the transnasal approach and its mobilization is also only feasible for a limited range [4, 6]. Using the side-viewing lenses with its capability to look around the corner, the tumor behind those vital structures can be visualized without injuring them [10]. Adding adequate surgical maneuver under such curved view of the endoscope, the removal of laterally extended skull base tumor becomes feasible. The localized lesion without displacement of surrounding structures can be also removed in a pinpoint way from the space behind the vital structures as demonstrated in Figure 2.

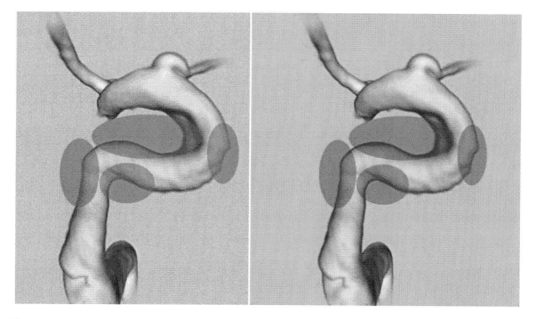

Figure 8. A schema of the left internal carotid artery viewed with the 30-degree endoscope. The 4 corridors, through which the cavernous sinus can be entered, are indicated as shaded red oval areas.

Use of the side-viewing scope is not only restricted for circumventing vital structures, but is also useful for straightforward target to avoid the conflict between the endoscope and the surgical instruments, thus ensuring wide trajectory for surgical maneuver. For this purpose the less degree endoscope i.e. the 30-degree endoscope is mostly used.

By gradually increasing the degree of viewing direction, the lesion can be approached from the medial to the lateral compartment, thus following the tumor along the direction of its growth. The use of the 70-degree endoscope is often essential to visualize the most lateral tumor compartment in the late stage of the surgery.

The development of curved and malleable forms of the conventional instruments increased the variety of surgical maneuver under the side-viewing endoscope [11]. With currently developed malleable dissector and steerable instruments, not only curettage and suctioning of the tumor but semi-sharp dissection and grasping became feasible in the curved surgical view. The resectability of laterally extended tumors depends not only to the degree of lateral extension but also to in which area this lateral compartment exists. An extradural laterally extended lesion is the most amenable to surgical removal while intradural one is the most challenging. Intrasellar laterally extended lesion must be also discriminated on whether it extends beyond the medial cavernous sinus wall. Suctioning of the tumor adjacent to the vulnerable structures needs to be performed with caution, as the suction power can not be adjusted precisely and the adhesion of the tumor to the tiny structures becomes not obvious until the structures are pulled together and sometimes torn off. Instead, grasping of a small piece of tumor with the flexible forceps, while adding countertraction to the structures behind, would be more reliable. At present we don't rely on the intradural surgical maneuver under the side-viewing endoscope. For this purpose, more precise sharp dissection and coagulation technique has to become feasible. An instrument with a steerable tip equipped with cutting and bipolar function would be mandatory.

Lateral tumor extension into the cavernous sinus is one of the main factors which hinder total removal of pituitary adenomas. The previous anatomical studies of the endoscopic transnasal approach demonstrated straight entrance into the cavernous sinus by performing a wide sphenoidotomy and opening the parasellar bone covering the cavernous sinus [12, 13]. This approach provides a wide corridor entrance into the cavernous sinus lateral to the carotid artery, but the exposure of the medial wall and cavernous sinus compartment dorsal to the horizontal portion of the internal carotid artery is rather restricted. For Knosp Gr.3 pituitary adenomas, an approach through the medial wall of the cavernous sinus is more appropriate as the cavernous sinus compartment lateral to the carotid artery is mostly narrow to enter. Moreover, the advantage of this approach is that the tumor can be pursued from the medial to lateral along the direction of its growth using the space already made by the tumor itself. Though the side-viewing endoscope was known to be capable to increase the view on the medial wall, its use during the actual surgery was restricted so far [14]. As in the case illustrated in Figure 7, a variety of surgical maneuvers within the cavernous sinus are now becoming feasible under the side-viewing endoscope. The cavernous sinus exploration is not always necessary for non-functioning adenomas but is effective for certain functioning adenomas to achieve endocrinological remission.

Though this maneuver requires training and may not be feasible in every case, it would further improve the result of the endoscopic surgery, especially for functioning pituitary adenomas.

References

[1] Kassam, A., Snyderman, C. H., Mintz, A., Gardner, P., Carrau, R. L. Expanded endonasal approach: the rostrocaudal axis. Part I. Crista galli to the sella turcica. *Neurosurg. Focus*. 2005;19:E3.

[2] Kassam, A., Snyderman, C. H., Mintz, A., Gardner, P., Carrau, R. L. Expanded endonasal approach: the rostrocaudal axis. Part II. Posterior clinoids to the foramen magnum. *Neurosurg. Focus*. 2005;19:E4.

[3] Schwartz, T. H., Fraser, J. F., Brown, S., Tabaee, A., Kacker, A., Anand, V. K. Endoscopic cranial base surgery: classification of operative approaches. *Neurosurgery.* 2008;62:991-1002.

[4] De Notaris, M., Cavallo, L. M., Prats-Galino, A., Esposito, I., Benet, A., Poblete, J., Valente, V., Gonzalez, J. B., Ferrer, E., Cappabianca, P. Endoscopic endonasal transclival approach and retrosigmoid approach to the clival and petroclival regions. *Neurosurgery.* 2009;65(6 Suppl.):42-50.

[5] Kassam, A. B., Gardner, P., Snyderman, C., Mintz, A., Carrau, R. Expanded endonasal approach: fully endoscopic, completely transnasal approach to the middle third of the clivus, petrous bone, middle cranial fossa, and infratemporal fossa. *Neurosurg. Focus.* 2005;19:E6.

[6] Zanation, A. M., Snyderman, C. H., Carrau, R. L., Gardner, P. A., Prevedello, D. M., Kassam, A. B. Endoscopic endonasal surgery for petrous apex lesions. *Laryngoscope.* 2009;119:19-25.

[7] Taniguchi, M., Kato, A., Taki, T., Tsuzuki, T., Yoshimine, T., Kohmura, E. Microsurgical Maneuvers under Side-Viewing Endoscope in the Treatment of Skull Base Lesions. *Skull Base.* 2011;21:115-22.

[8] Morita, A., Okada, Y., Kitano, M., Hori, T., Taneda, M., Kirino, T. Development of hybrid integrated endoscope-holder system for endoscopic microneurosurgery. *Neurosurgery.* 2004;55:926-31.

[9] Kawamata, T., Amano, K., Hori, T. Novel flexible forceps for endoscopic transsphenoidal resection of pituitary tumors: technical report. *Neurosurg. Rev.* 2008; 31:65-8.

[10] Fraser, J. F., Nyquist, G. G., Moore, N., Anand, V. K., Schwartz, T. H. Endoscopic endonasal transclival resection of chordomas: operative technique, clinical outcome, and review of the literature. *J. Neurosurg.* 2010;112:1061-9.

[11] Taniguchi, M., Kohmura, E. Endoscopic endonasal removal of laterally extended clival chordoma using side-viewing scopes. *Acta Neurochir.* (Wien). 2012;154:627-32.

[12] Alfieri, A., Jho, H. D. Endoscopic endonasal approaches to the cavernous sinus: surgical approaches. *Neurosurgery.* 2001;49:354-60.

[13] Cavallo, L. M., Cappabianca, P., Galzio, R., Iaconetta, G., de Divitiis, E., Tschabitscher, M. Endoscopic transnasal approach to the cavernous sinus versus transcranial route: anatomic study. *Neurosurgery.* 2005;56(2 Suppl.):379-89.

[14] Doglietto, F., Lauretti, L., Frank, G., Pasquini, E., Fernandez, E., Tschabitscher, M., Maira, G. Microscopic and endoscopic extracranial approaches to the cavernous sinus: anatomic study. *Neurosurgery.* 2009;64(5 Suppl. 2):413-21.

Chapter XVI

Percutaneous Biopsy of Parasellar Lesions Through the Foramen Ovale

Mahmoud Messerer[1,2] *and Marc Sindou*[2]

[1]Department of Clinical Neurosciences, Department of Neurosurgery, Centre Hospitalier Universitaire Vaudois, Lausanne, Switzerland
[2]Department of neurosurgery A, Pierre Wertheimer hospital, Hospices civils de Lyon, Lyon, France

Abstract

Parasellar lesions comprise a wide variety of inflammatory and benign or malignant tumorous processes. Each of these lesions requires individual considerations regarding the best management. Percutaneous biopsy through the foramen ovale should be performed in cases of insufficiency of imaging findings to avoid unnecessary open procedures or inappropriate treatment and to guide therapeutic decision. Pathological diagnostic may be sometimes difficult to ascertain with imaging findings. The parasellar region can be reached for biopsy through the foramen ovale percutaneously. After a complete overview of surgical anatomy, the authors describe the used methods. Based on the authors' experience with 50 such procedures, the results are described in terms of diagnostic accuracy and morbidity. Percutaneous biopsy revealed a good diagnostic accuracy with a sentitivity of 83% (CI: 52 – 98) and a specificity of 100% (CI: 79 – 100). No mortality was deplored and morbidity was mostly transient. Indications are lesions located in Meckel's cave, posterior part of the cavernous sinus and upper part of petroclival region.

Introduction

Various types of lesion may appear in the parasellar region, so called "cavernous sinus" and its surroundings. The most frequent encountered tumors are meningiomas but neuromas, chordomas, metastasis and lymphomas are also often seen. Non-tumorous processes such as inflammatory process can be a differential diagnosis of lesions in the cavernous sinus area [1, 2]). Many lesions in this area have been considered unresectable because of the existence and density of important neurovascular structures [3], and total removal of these various lesions are associated with a relatively high rate of morbidity and mortality [4]. The different surgical open approaches are: the fronto-pteriono-temporal craniotomy to access the Parkinson triangle, the subtemporal approach to reach the Mullan triangle, the pterional approach to reach the roof of the cavernous sinus along the third nerve (Dolenc or Hakuba) or through the carotid ring (Perneczky), the inferomedial wall by transsphenoidal approach (Laws), the contralateral subfrontal approach to access the superomedial wall (Sano), the periauricular infratemporal approach to access the Meckel's cave (Sekhar). Although the microsurgical anatomy of this region has been extensively described in the past [5-11], surgical exposure is still challenging for neurosurgeons. However, several treatment strategies are available. A choice has to be made between radical surgical removal, conservative surgery combined with radiotherapy, medical treatment or simple survey. Furthermore, the choice between surgical or non-surgical treatment is still controversial and the treatment of each lesion requires individual consideration [12-14]. Therefore it is necessary to obtain the exact pathological diagnosis of the lesion in order to establish the best strategy. Current neuroimaging can indirectly provide a diagnosis but lesions in this area are sometimes difficult to characterize radiologically so that a histopathological diagnosis can be necessary with this goal. Percutaneous biopsy through the foramen ovale has been developed. The first description was reported by Stechison and Bernstein [15] in 1989. This technique uses the same approach of the technique for percutaneous treatment of trigeminal neuralgia described by Härtel [16] and Kirschner [17]. This percutaneous biopsy allows performing cytological and/or histopathological examination avoiding unnecessary open craniotomy.

The aim of this chapter is to familiarize readers to the decision-making process for percutaneous biopsy through the foramen ovale. For this goal, authors will recap surgical anatomy and highlight method, indications and results. This chapter is based on the experience of the senior neurosurgeon (M.S.) who operated on 306 central skull base tumors between 1991 and 2010, among which 50 patients benefited from percutaneous biopsy through the foramen ovale.

Surgical Anatomy

An exact knowledge of the anatomical region crossed by the biopsy needle is of paramount importance in order to avoid any complication and to reach correctly the cavernous sinus area. There are three main landmarks, which delineate the triangular bases of the 3D pyramid area with its apex at a cutaneous point 3 cm from the labial commissure:

1. The foramen ovale
2. The cutaneous point 3 cm anterior to the tragus, on the orbito-meatal line along the inferior border of the zygoma.
3. The pupilla

The entry point of the percutaneous needle (Figure 1) is the apex of the pyramidal area i.e. the cutaneous cheek point located 3 cm from the labial commissure.

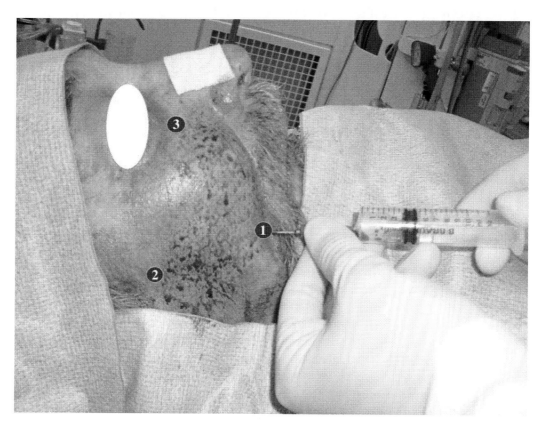

Figure 1. Landmarks for the transjugal-transovale route for percutaneous biopsy. The entry point (1) is located 3 cm from the labial commissure. The trajectory to the foramen ovale is defined by targeting at 35 mm anterior to the tragus (2) in lateral view and the medial border of the pupilla (3) in frontal view. Depth is guided by fluoroscopy.

From this point to the parotid duct, there is the inferior segment that is approximately 13 mm in length. From the parotid duct to the lateral pterygoid muscle, there is the middle segment which lengths approximatively 29 mm. It is constituted of fatty tissue and contains the lingual, the chorda tympani, and, the buccal and the inferior alveolar nerves. This segment should be passed very carefully because it is possible to encounter branched of the maxillary artery. From the lateral pterygoid muscle to the foramen ovale, there is the superior segment. This passage should also be made with caution because there is still the possibility to encounter the maxillary artery located posteriorly to the lateral pterygoid muscle. There is also a risk to encounter the pterygoid venous plexuses.

The passage through the foramen ovale leads the needle to the trigeminal cave where it is necessary to be vigilant to the trigeminal system [18]. When reaching the skull base, many structures may be encountered [19]:

- The internal jugular vein: located approximatively 27 mm posterolaterally to the right needle trajectory

- The internal carotid artery: located either approximatively 25 mm posteriorly to the right needle trajectory into the petrous carotid canal or at the foramen lacerum if cases of deviation of the needle from 10° medially of the correct trajectory
- The inferior and superior orbital fissures, and, the optic nerve, in cases of deviation of the needle from 17° anteriorly of its correct trajectory
- The membranous portion of the auditory tube in cases of deviation of the needle from 9° angle in the anteromedial direction of the correct trajectory

Surgical Procedure

Preoperative Protocol

When preparing a percutaneous biopsy though the foramen ovale, patients should be evaluated preoperatively using Magnetic Resonance Imaging (MRI), Computed Tomography Scanning (CT-scan) and cerebral angiography. MRI evaluation used T1 weighted images with and without enhancement and T2 weighted images in sagittal, coronal and axial planes. This MRI provides information about tumor extent and its relationship with the internal carotid artery. CT-scan with 3D reconstruction provides information regarding bone structures (erosions, calcifications) and the diameter of the foramen ovale. Selective angiographic details tumor's vascularization and allows eliminating an arteriovenous malformation or arterial aneurysm. All these examinations are realized in order to optimize the surgical procedure of the biopsy.

The preoperative evaluation also includes neurootological and ophthalmological examinations.

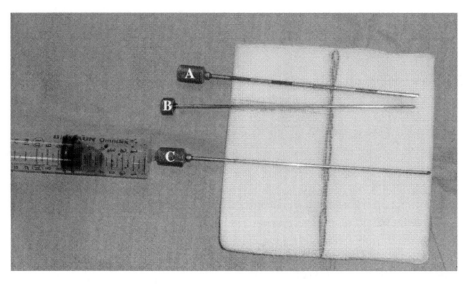

Figure 2. Biopsy needle (ref Sindou Biopsy Needle N°ACS-976 DIXI microtechniques SAS, 4 chemin de Palente - BP 889, 25025 Besançon, Cedex, France). A: Outer needle introduced first with needle B for puncturing; B: inner solid needle withdrawn when the targeted region is reached; C: inner needle placed into Outer needle to aspirate tissue samples, this needle is connected to a 20 ml syringe allowing strong negative pressure to aspire.

Surgical Steps

Patients are operated using light anesthesia with propofol (Diprivan; Zeneca Pharma, Cergy, France) and local anesthesia with xylocaine at the site of the cheek up to the pterygomaxillary fossa. Patients are in supine position and the procedure is performed under fluoroscopy. A biopsy needle, a Tuohy's No. 14 needle [20] (Cordis S.A., Sofia Antipolis, France) or a new designed needle (Sindou 2010) (Figure 2) is introduced through the foramen ovale according to the technique used for trigeminal thermorhizotomy [21, 22].

The entry point as described above is located 3 cm lateral to the labial commissure and the right trajectory is to reach the foramen ovale 3 cm anterior to the tragus on a horizontal line corresponding to the inferior border of the zygoma and the pupilla. The location of the needle tip, which reaches to Meckel's cave is confirmed by a lateral view X-Ray (Figure 3).

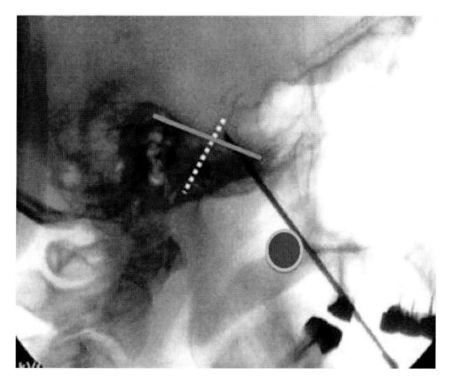

Figure 3. Lateral control X Ray of needle insertion. To be correctly place, the tip of the needle must be approximatively at the intersection between the upper petrous ridge line (continuous line) and the clivus line (dotted line). When entering the pterygo maxillary fossa, note that the projection of the needle passed at the maxillary sinus angle (ovale)

Then, the needle is connected to a 20 ml syringe through which strong negative pressure is applied, and at least two specimens of soft consistency without any trouble are obtained.

In all the cases, the cytological specimens are studied using the May-Grunwald staining technique, either directly or after centrifugation. Preliminary staining and microscopic examination of each specimen are performed extemporaneously at the department of histology. When the samples are big enough, additional conventional histological techniques and study of the immunological markers are added.

Indications for Percutaneous Biospy – Management

A recent published article [23] on a series of 50 patients by Messerer et al. the senior author (MS) discussed the percutaneous biopsy of cavernous sinus lesions, i.e. cavernous sinus, Meckel's cave and petroclival apex region, through the foramen ovale.

There are a great variety of symptoms that can reveal a mass lesion of the cavernous sinus area, including benign or malign tumors. Some symptoms have been described in the literature to be more frequent in certain lesions. For instance, patients with malignant lesions of the Meckel's cave may be more likely to suffer paresthesiae and trigeminal motor deficits than those with meningiomas [24]. However, it is really important to notice that any symptom or association of symptoms cannot be considered as a specific sign of a disease. Every symptom may reveal every disease.

By comparison with this great variety of symptoms, there are also a great variety of lesions that can match with many types of symptoms and each lesion may require its own treatment strategy. The cavernous sinus and its surroundings can be invading by intrinsic tumors such as trigeminal neuromas, meningiomas, congenital tumors (epidermoid cysts, lipomas and dermoids), or, by secondary tumors which consist in retrograde perineural extension along the trigeminal nerve or in subarachnoid dissemination and metastasis from extracranial malignancy [25]. The incidence of cancer first presenting as one or more metastatic intracranial lesions is between 5% and 12% [26]. Furthermore, it is difficult to distinguish metastatic tumors from benign tumors solely based on clinical or radiographic characteristics, especially in those patients without an identified primary malignancy at the time of presentation. In a historic series of 137 cavernous sinus, there were 37% of malignant lesions described by Sekhar et al. [6] in 1991.

In the beginning of the nineties, the treatment of cavernous sinus lesions was practically exclusively surgical with the will of neurosurgeons to obtain gross total removals. This surgical treatment was associated with a relatively high rate of morbidity [4, 9, 27]. However from the end of this decade, a bend was marked with the advent of oncological treatment such as radiosurgery and, other approaches to treatment were advocated with the aim of achieving an optimal life quality [28, 29]. Coppa et al. [30] reported that surgical resection of skull base malignancies may no longer be the "gold standard" or optimal first line treatment thanks to the advent of radiosurgery. For instance, for a meningioma that is not resectable, many studies demonstrated a high degree of local tumor control after radiosurgery [28]. Pollock et al. [31], in a series of 198 meningiomas, demonstrated that radiosurgery provided superior tumor control for patients with either a Grade 2 or Grade 3 – 4 resection. Therefore when mass lesions are revealed during diagnostic evaluation, tissue diagnoses should be obtained because it is absolutely necessary to establish an exact pathological diagnosis in order to provide the best management and to avoid unnecessary surgical procedures and its associated morbidity.

Our series published in the literature [23] represents the largest series of percutaneous biopsy through the foramen ovale for the diagnosis of cavernous sinus area lesions. Since its first description in 1989 by Stechison et al. [15], studies have been involved but they only reported few cases [32-34] preventing the analysis of their diagnostic accuracy. Only one report included more than five patients [20]. Our series allowed us to evaluate the diagnostic accuracy of the percutaneous biopsy in comparison with histopathological examination which came from a second surgery. The sensitivity to distinguish benign and malign tumors was

83% (CI: 52 – 98) and the specificity was 100% (CI: 79 – 100). In addition, the kappa coefficient showed a good correlation in terms of histopatholgical diagnosis between biopsy results and histopathological results from another surgery.

Although this diagnostic test is not perfect, it can be used as a useful add-on test in order to identify the pathological diagnosis of cavernous sinus area lesions. Figures 4, 5, 6 and 7 show sample examples of imaging, biopsy and/or histopathological (from open surgery) findings for some patients.

In addition, percutaneous biopsy via the foramen ovale for cavernous sinus area lesions and percutaneous trigeminal rhizotomies has been reported previously and is considered to be a minimal invasive procedure without major complications [20, 33, 35]. In our published study [23], we encountered two complications: a face cellulitis and a cheek hematoma likely due to a vascular injury of the maxillary artery that is located on the very trajectory of the needle [36].

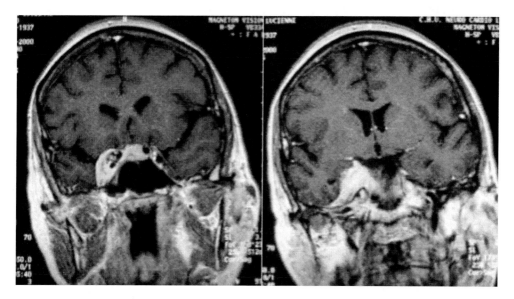

Figure 4. Coronal T1-weighted images with gadolinium enhancement showing a lesion into the right cavernous sinus with a floor middle fossa extension. Percutaneous biopsy stated for a meningioma, which was confirmed at open surgery.

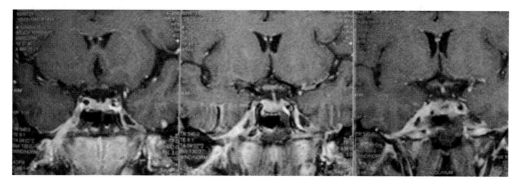

Figure 5. Coronal T1-weighted images with gadolinium enhancement showing a lesion into the right cavernous sinus extending to the tentorial incisura, suggesting a meningioma. Percutaneous biopsy stated for an inflammatory pseudotumor. Lesion regressed after corticosteroid treatment.

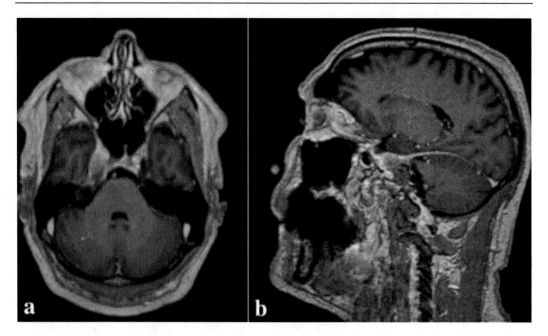

Figure 6. a: Axial T1-weighted MR images, and, b: sagittal T1-weighted MR image, with gadolinium enhancement showing a lesion into the right cavernous sinus extending to the pterygomaxillary fossa. Percutaneous biopsy stated for an adenocarcinoma and the patient was treated with chemotherapy.

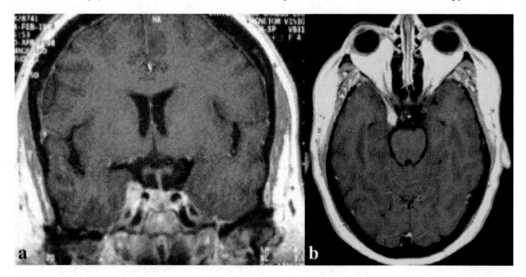

Figure 7. a: Coronal T1-weighted MR image and, b: axial T1-weighted image, with gadolinium enhancement showing a lesion into the right cavernous sinus. This patient was initially referred for diploplia due to a right oculomotor nerve palsy. Percutaneous biopsy stated for an inflammatory pseudotumor. Lesion totally regressed after corticosteroid treatment.

The major inconvenience of the percutaneous biopsy is the difficulty in obtaining a sufficient fragment of pathological processes in order to allow a cytohistopathogical examination. Though, it is necessary to have a close cooperation between the referring neurosurgeon and the cytopathologist. Another problem is the relatively high frequency of presence of meningeal cells in fragments of lesions, which are not meningiomas. This is certainly due to the effraction of the meninges during the biopsy. So it can lead to a wrong diagnosis when no other types of cells are present in the fragment due to a wrong biopsy.

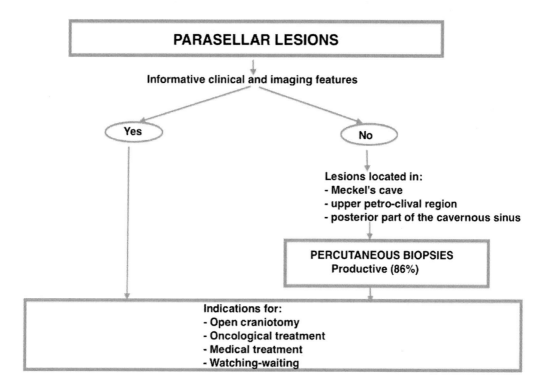

Figure 8. Decision making process for parasellar lesions.

Conclusion

In the argument above, authors recommend performing percutaneous biopsies of the cavernous sinus area lesions through the foramen ovale in all cases of lesions located in the Meckel's cave; posterior part of the cavernous sinus and the upper part of the petroclival region. Especially when neuro-imaging does not provide sufficient information on the histopathological nature of the pathological process. In the function of biopsy results, individual consideration with regards management could be made (Figure 8).

Percutaneous biopsy of lesions through the foramen ovale is an accurate and useful method to orientate the management of parasellar lesions by avoiding unnecessary open craniotomy and its rate of morbi-mortality. The authors recommend performing a percutaneous biopsy in cases of an insufficiently established pathological diagnosis of clinic-radiological findings.

References

[1] Ahn JY, Kwon SO, Shin MS, Joo JY, Kim TS. Chronic granulomatous neuritis in idiopathic trigeminal sensory neuropathy. Report of two cases. *Journal of neurosurgery.* 2002;96(3):585-8. Epub 2002/03/09.

[2] Gottfried ON, Chin S, Davidson HC, Couldwell WT. Trigeminal amyloidoma: case report and review of the literature. *Skull base : official journal of North American Skull Base Society* [et al]. 2007;17(5):317-24. Epub 2008/03/12.

[3] Duma CM, Lunsford LD, Kondziolka D, Harsh GRt, Flickinger JC. Stereotactic radiosurgery of cavernous sinus meningiomas as an addition or alternative to microsurgery. *Neurosurgery.* 1993;32(5):699-704; discussion -5. Epub 1993/05/01.

[4] Al-Mefty O, Smith RR. Surgery of tumors invading the cavernous sinus. *Surg Neurol.* 1988;30(5):370-81. Epub 1988/11/01.

[5] Sindou M, Alaywan M. [Orbital and/or zygomatic removal in an approach to lesions near the cranial base. Surgical technic, anatomic study and analysis of a series of 24 cases]. *Neurochirurgie.* 1990;36(4):225-33. Epub 1990/01/01. La depose orbitaire et/ou zygomatique dans l'abord des lesions proches de la base du crane. Technique chirurgicale, etude anatomique et analyse d'une serie de 24 cas.

[6] Sekhar LN, Pomeranz S, Sen CN. Management of tumours involving the cavernous sinus. *Acta Neurochir Suppl* (Wien). 1991;53:101-12. Epub 1991/01/01.

[7] Inoue T, Rhoton AL, Jr., Theele D, Barry ME. Surgical approaches to the cavernous sinus: a microsurgical study. *Neurosurgery.* 1990;26(6):903-32. Epub 1990/06/01.

[8] Kawase T, van Loveren H, Keller JT, Tew JM. Meningeal architecture of the cavernous sinus: clinical and surgical implications. *Neurosurgery.* 1996;39(3):527-34; discussion 34-6. Epub 1996/09/01.

[9] Hakuba A, Tanaka K, Suzuki T, Nishimura S. A combined orbitozygomatic infratemporal epidural and subdural approach for lesions involving the entire cavernous sinus. *Journal of neurosurgery.* 1989;71(5 Pt 1):699-704. Epub 1989/11/01.

[10] Emery E, Alaywan M, Sindou M. [Respective indications of orbital and/or zygomatic arch removal combined with fronto-pteriono-temporal approaches. 58 cases]. *Neurochirurgie.* 1994;40(6):337-47. Epub 1994/01/01. Indications respectives des deposes orbitaires et/ou zygomatiques en association aux abords fronto-pteriono-temporaux. 58 cas.

[11] Sindou M, Emery E, Acevedo G, Ben-David U. Respective indications for orbital rim, zygomatic arch and orbito-zygomatic osteotomies in the surgical approach to central skull base lesions. Critical, retrospective review in 146 cases. *Acta Neurochir* (Wien). 2001;143(10):967-75. Epub 2001/10/31.

[12] Cusimano MD, Sekhar LN, Sen CN, Pomonis S, Wright DC, Biglan AW, et al. The results of surgery for benign tumors of the cavernous sinus. *Neurosurgery.* 1995;37(1):1-9; discussion -10. Epub 1995/07/01.

[13] Anand VK, House JR, 3rd, al-Mefty O. Management of benign neoplasms invading the cavernous sinus. *Laryngoscope.* 1991;101(5):557-64. Epub 1991/05/01.

[14] Eisenberg MB, Al-Mefty O, DeMonte F, Burson GT. Benign nonmeningeal tumors of the cavernous sinus. *Neurosurgery.* 1999;44(5):949-54; discussion 54-5. Epub 1999/05/08.

[15] Stechison MT, Bernstein M. Percutaneous transfacial needle biopsy of a middle cranial fossa mass: case report and technical note. *Neurosurgery.* 1989;25(6):996-9. Epub 1989/12/01.

[16] Härtel F. Die leitungsanästhese und injektionbehandlung des ganglion gasseri und der trigeminusstämme. *Arch Klin Chir.* 1912;100:627-38.

[17] Kirschner M. Blektrocoagulation des ganglion gasseri. *Zentralbl Chir.* 1932;47:2841-3.

[18] Alvernia J, Wydh E, Simon E, Sindou M, Mertens P. [Microsurgical anatomy of the transoval percutaneous route to the trigeminal cave and the trigeminal ganglion]. *Neurochirurgie.* 2009;55(2):87-91. Epub 2009/03/31. Anatomie microchirurgicale de la voie percutanee transovale vers la cavite trigeminale et le ganglion trigeminal (voie de Hartel).

[19] Alvernia JE, Sindou MP, Dang ND, Maley JH, Mertens P. Percutaneous approach to the foramen ovale: an anatomical study of the extracranial trajectory with the incorrect trajectories to be avoided. *Acta Neurochir* (Wien). 2010;152(6):1043-53. Epub 2010/02/09.

[20] Sindou M, Chavez JM, Saint Pierre G, Jouvet A. Percutaneous biopsy of cavernous sinus tumors through the foramen ovale. *Neurosurgery.* 1997;40(1):106-10; discussion 10-1. Epub 1997/01/01.

[21] Sindou M, Fobe JL, Berthier E, Vial C. Facial motor responses evoked by direct electrical stimulation of the trigeminal root. Localizing value for radiofrequency thermorhizotomy. *Acta Neurochir* (Wien). 1994;128(1-4):57-67. Epub 1994/01/01.

[22] Sweet WH, Wepsic JG. Controlled thermocoagulation of trigeminal ganglion and rootlets for differential destruction of pain fibers. 1. Trigeminal neuralgia. *J Neurosurg.* 1974;40(2):143-56. Epub 1974/02/01.

[23] Messerer M, Dubourg J, Saint-Pierre G, Jouanneau E, Sindou M. Percutaneous biopsy of lesions in the cavernous sinus region through the foramen ovale: diagnostic accuracy and limits in 50 patients. *Journal of neurosurgery.* 2012;116(2):390-8. Epub 2011/11/22.

[24] Soni CR, Kumar G, Sahota P, Miller DC, Litofsky NS. Metastases to Meckel's cave: Report of two cases and comparative analysis of malignant tumors with meningioma and schwannoma of Meckel's cave. *Clin Neurol Neurosurg.* Epub 2010/08/24.

[25] Beck DW, Menezes AH. Lesions in Meckel's cave: variable presentation and pathology. *J Neurosurg.* 1987;67(5):684-9. Epub 1987/11/01.

[26] Sawaya R. *Intracranial metastases.* Current management strategies: Blackwell Futura; 2004.

[27] DeMonte F, Smith HK, al-Mefty O. Outcome of aggressive removal of cavernous sinus meningiomas. *J Neurosurg.* 1994;81(2):245-51. Epub 1994/08/01.

[28] Milker-Zabel S, Zabel-du Bois A, Huber P, Schlegel W, Debus J. Fractionated stereotactic radiation therapy in the management of benign cavernous sinus meningiomas : long-term experience and review of the literature. *Strahlenther Onkol.* 2006;182(11):635-40. Epub 2006/10/31.

[29] Long DM. The treatment of meningiomas in the region of the cavernous sinus. *Childs Nerv Syst.* 2001;17(3):168-72. Epub 2001/04/18.

[30] Coppa ND, Raper DM, Zhang Y, Collins BT, Harter KW, Gagnon GJ, et al. Treatment of malignant tumors of the skull base with multi-session radiosurgery. *J Hematol Oncol.* 2009;2:16. Epub 2009/04/04.

[31] Pollock BE, Stafford SL, Utter A, Giannini C, Schreiner SA. Stereotactic radiosurgery provides equivalent tumor control to Simpson Grade 1 resection for patients with small- to medium-size meningiomas. *Int J Radiat Oncol Biol Phys.* 2003;55(4):1000-5. Epub 2003/02/28.

[32] Yi W, Ohman K, Brannstrom T, Bergenheim AT. Percutaneous biopsy of cavernous sinus tumour via the foramen ovale. *Acta Neurochir* (Wien). 2009;151(4):401-7; discussion 7. Epub 2009/03/07.

[33] Dresel SH, Mackey JK, Lufkin RB, Jabour BA, Desalles AA, Layfield LJ, et al. Meckel cave lesions: percutaneous fine-needle-aspiration biopsy cytology. *Radiology.* 1991;179(2):579-82. Epub 1991/05/01.

[34] Berk C, Honey CR. Percutaneous biopsy through the foramen ovale: a case report. *Stereotact Funct Neurosurg.* 2002;78(1):49-52. Epub 2002/10/17.

[35] Kanpolat Y, Savas A, Bekar A, Berk C. Percutaneous controlled radiofrequency trigeminal rhizotomy for the treatment of idiopathic trigeminal neuralgia: 25-year experience with 1,600 patients. *Neurosurgery.* 2001;48(3):524-32; discussion 32-4. Epub 2001/03/29.

[36] Alvernia JE, Sindou MP, Dang ND, Maley JH, Mertens P. Percutaneous approach to the foramen ovale: an anatomical study of the extracranial trajectory with the incorrect trajectories to be avoided. *Acta Neurochir* (Wien).152(6):1043-53. Epub 2010/02/09.

In: Minimally Invasive Skull Base Surgery
Editor: Moncef Berhouma

ISBN: 978-1-62808-567-9
© 2013 Nova Science Publishers, Inc.

Chapter XVII

Purely Endoscopic Keyhole Supraorbital Approaches for Anterior and Middle Skull Base Tumors

E. Jouanneau[*,1,3,4], M. Berhouma[1,4], T. Jacquesson[1], M. Messerer[1], E. Bogdan[1], A. Gleizal[2,4], G. Raverot[3,4,5] and F. Barral-Clavel[1]

[1] Neurosurgical Department A, Neurological Hospital, Hospices Civils de Lyon, Lyon, France
[2] Oral Maxillofacial Surgery, Hospices Civils de Lyon
[3] Inserm 1028, CNRS UMR5292, Neurosciences Research Center, Neuro-oncology and Neuro-inflammation team
[4] Claude Bernard University, Lyon I, Lyon France
[5] Department of Endocrinology, Hospital Est, Hospices Civils de Lyon, Lyon, France

Abstract

Anterior and middle skull base tumors, mainly meningiomas, are usually operated on using a sub-frontal route with a microscope. With modern radiotherapy, the goal of skull base surgery moves from a radical surgery with high rate of side effect to a functional concept that aims to remove as much as possible of the tumor without compromising the neurological status of patients. Minimally skull base surgery benefits from keyhole and endoscopy techniques. For 3 2 decades, the development of endoscopy helps to imagine innovative approaches for skull base tumors such as the endonasal route. Nonetheless, CSF leak issue and the absence of direct control of the tumor margins may limit the interest of such a route. Keyhole craniotomies have been developed with microscope but vision issue limits their use. Combining advantages of both techniques appears therefore natural and gave birth to intracranial assisted and more recently to fully endoscopic

---

[*] Corresponding author: E Jouanneau, MD, PhD, Department of Neurosurgery A, 59 boulevard Pinel, 69677 Bron cedex, Phone: +33472357495, Fax: 33472357365, E-mail: emmanuel.jouanneau@chu-Lyon.fr.

keyhole surgery. For anterior or middle skull base tumors, Keyhole supraorbital approaches can be done either by a trans-eyebrow or trans-eyelid routes. A step-by-step description of these fully endoscopic alternative routes summarizing advantages and drawbacks compared to others (traditional sub-frontal or more recent endonasal approaches) is reported in this chapter by the authors.

Keywords: Trans eyebrow, trans eyelid, supraorbital keyhole, skull base tumors, endoscopy

Introduction

During the last two decades, skull base surgery has been completed with the development of endoscopic endonasal approaches. After a large experience with endonasal endoscopic surgery for anterior and middle cranial base pathologies, the authors have developed a purely endoscopic approach through a supraorbital keyhole route. The advantages are the absence of temporalis muscle dissection, a very small craniotomy and an insignificant brain retraction as well as excellent cosmetic results.

Trans Eyebrow Route: Technical Note

The technique we are currently using has been previously published [1] and derived from the one described with microscope [2].

Patient Positioning and Operative Room Settings (Fig. 1)

For a right-handed surgeon, midline tumors should be operated on using a right side approach. However, tumor anatomy may oblige to choose a left route mainly when extended to the Sylvian fissure. The patient positioning is the one used when doing a microscopic subfrontal approach (supine position, trunk slightly elevated (30°) to favor venous drainage with the head slightly (15°) turned opposite to the approach). Contrarily to microscope, we do not flex the head as the endoscope provides a direct vision of the skull base. Both the endoscopic column and the neuronavigation face the surgeon, the first on the left and the latter on the right. The nurse is placed on the right and the fellow on the left. Two baskets are set in front of the surgeon for the surgical tools.

Skin Incision and Craniotomy (Fig. 2)

When starting this keyhole surgery, the drape towels can be put in a regular fashion for doing a sub frontal craniotomy if surgical conversion is needed (hemorrhage, insufficient approach, brain edema...). Different eyebrow incisions are possible: inside the eyebrow or at the superior edge of the eyebrow to avoid hair loss issue. We do not have any concern using the first option when limiting coagulation.

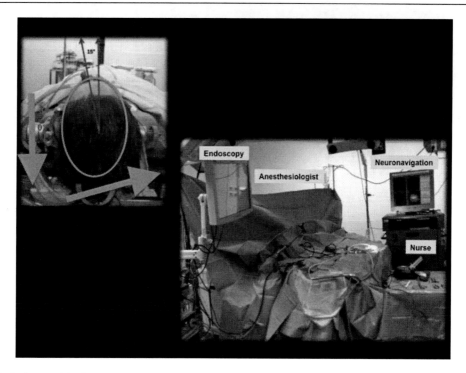

Figure 1. Patient positioning and operative room settings. The patient is supine, the table head 30° raised to favor venous drainage and the head 15° turned left in case of a right approach. The head do not have to be flexed as the endoscope immediately gives a direct view on the skull base. Anesthesiologists are in the back left and the scrub nurse on the right side. Both the endoscopic column and the neuronavigation system face the surgeon, the first on the left and the latter on the right.

The tail of the eyebrow limits the length of incision laterally while medially the supraorbital nerve can be located easily by palpating the orbital notch. An ophthalmologic blade is used to make the incision and the periosteum opened up to the bone while preserving the supraorbital nerve. We use electrical section and/or coagulation as few as possible to protect the eyebrow. The temporal muscle is detached just enough to allow a burr hole below the temporal line and behind the orbital process. With the trans eyebrow approach, the bone hole is usually just up to the Mac Carty point as one cannot pull out enough laterally the skin. Pediatric tools or a drill are used for the hole and the bone cut. The neuronavigation will help to locate the frontal sinus. It is of paramount importance to avoid any aperture of the sinus. Indeed, with such a technique, a frontal cranialization cannot be done properly and any aperture can lead to CSF leak morbidity. In such a situation, a closure with fibrin glue and cement may be advised. A 2 centimeters or 2 cm and half craniotomy is therefore performed with or without an orbital roof removal. Removing the roof is not time consuming and does not have specific morbidity while providing the major advantage of taking away all the bone irregularities that may hide intradural endoscopic exposition. This trick may be replaced by an extradural orbital roof drilling. At this point and before opening the dura mater, we manage medially and above the frontal sinus a small notch with the drill where the endoscope will fit during the surgery.

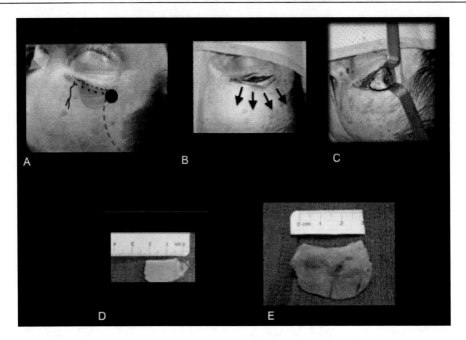

Figure 2. Skin incision and craniotomy. A and B: Two type of incision can be used inside or just few millimeters above the eyebrow. The latter has been reported by some authors in order to avoid hair loss issue. Avoiding as much as possible coagulation, we do not have any trouble and favor the first option to totally hide the incision. C, D and E: the bone hole is made using drill just below the temporalis line and behind the frontal zygomatic processat the Mac carthy point. A frontal or fronto-orbital craniotomy is performed, the latter owing the advantage to suppress all the reliefs that can disturb intradural work without additional morbidity.

Intradural Time (Fig. 3)

The dura mater is opened up tangential to the skull base and temporarily stitched to the draping. In case of orbital roof removing, a special care is given not to apply too much pressure on the ocular globe. A 0°, 4 mm, 18 cm rigid endoscope is therefore placed on the table holder in the previously mentioned notch. A 2.7 mm diameter endoscope can be used as well even if it slightly reduces the vision field. We do not use the four hands working technique, as the craniotomy is really narrow with immediate dangerous relationships (frontal lobe cortex). All the instruments used are those designed for endonasal endoscopic surgery. We do use special flexible coagulation suction (Climdal). Immediately before the dural opening, intravenous osmotic agents are administered in order to help brain relaxation. A lumbar drainage can also be used in selected cases. Opening the ipsilateral Sylvian fissure and basal cisterns allows brain relaxation. A brain retractor is usually placed over the frontal lobe only to avoid brain injury during the instruments introduction. A special care is given to respect the olfactory nerve. Once the endoscope secured in a table self-retaining scope-holder, the subsequent stages are technically identical to microsurgical ones. In contrast with the tridimensional microscopic view, the endoscope provides a panoramic view of the field that can be extended to see around the corners with angled endoscopes. The bidimensional view, sometimes cited as an issue with endoscopic techniques, is corrected with the appreciation of the depth of the two instruments working in the field.

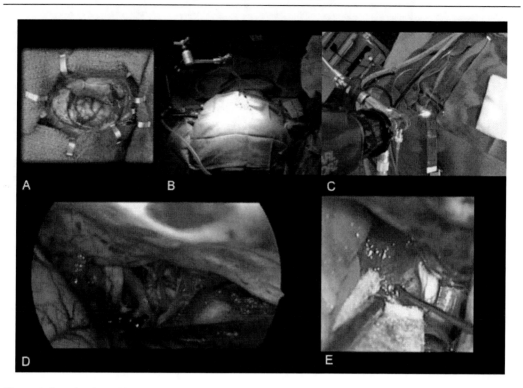

Figure 3. intradural working. A: The dura mater is attached anteriorly. B and C: Once the intradural work begins, the 0° 4 mm short endoscope is secured in its holder and placed in its medial notch specially prepared. A brain retractor is gently placed on the frontal lobe only to avoid brain contusion from in and out instruments' movements. A special care is given to avoid damaging the fragile ispsilateral olfactory nerve. D and E: The intradural work respects the microscopic rules with a large arachnoid aperture to favor CSF aspiration and brain relaxation like in this case.

Closure Step (Fig. 4)

The dura mater is closed using uninterrupted suture reinforced by fibrin glue or equivalent agent. Careful hemostasis has to be made as even with small craniotomies, extradural hematoma may occur. Bone is secured with low profile titanium microplate and screws that may be replaced by non-resorbable wires.

A space for the plate may be prepared by drilling to avoid any cosmetic frontal issue. Two or three plates are used, one lateral placed on the orbital process, one posterior and sometimes one medially close to the sinus. Thereafter, bone defects are filled up with calcium cement for optimal cosmetic results.

Temporal muscle is re-attached to the superior temporal line. The skin is closed in 2 layers including a subcutaneous absorbable 6/0 stitches. A small dressing covers the suture for 2 days and no drainage is required. Ice and external buffering can be used on the scar to limit edema or hematoma during the first two postoperative days.

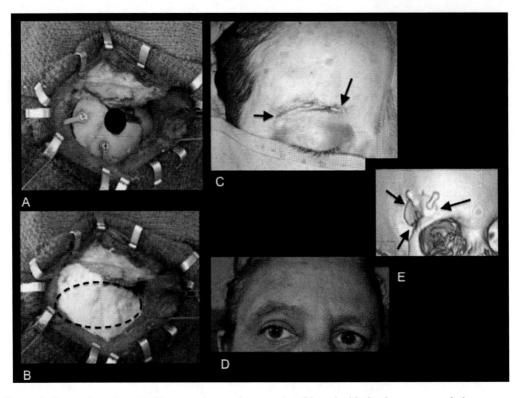

Figure 4. closure time. A and B: Bone craniotomy is secured at this end with titanium screw and plates. Special attention has to be made to hide the material and avoid any uncomfortable reliefs for the patients. Wires may be another option. Cement is used to fill the hole. C: resorbable sutures are used for eyebrow closure with immediate results and D: and the result at 8 days. E: 3D Postoperative CT scan.

Trans-Eyelid Route: Technical Note

Our technique derives from the supraorbital trans-eyelid route as described by Abdelaziz et al. [3] and introduces an intra-dural purely endoscopic work.

Patient Positioning and Operative Room Settings (Fig. 1, Fig. 5)

The setting is similar to the trans-eyebrow approach. This trans-eyelid route will be preferred in case of small eyebrow, huge frontal sinus needing a more lateralized craniotomy and also for anterior tumors like cribriform plate meningiomas. To reach those tumors without being in conflict with the endoscope tube, a tangential route to the bone is needed which is possible only with a more lateral craniotomy. We may also advice to turn a little bit more the head around 20° toward the contralateral side.

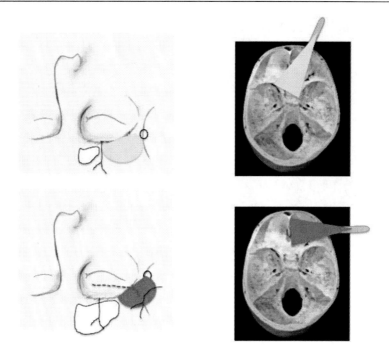

Figure 5. slight differences regarding the craniotomies between trans-eyebrow and trans-eyelid routes. For huge sinus, small eyebrow and anterior tumors, a trans-eyelid route allowing more lateral bone aperture should be preferred.

Skin Incision and Craniotomy (Fig. 6 – Fig. 9)

First of all, two sutures with a Prolen 6/0 are done to close the eye during surgery and avoid cornea ulceration. The upper lid is gently tract and a 2 cm and half incision made in one fold of the eyelid using an ophthalmologic knife. The lateral limit of the incision does not exceed 25 mm from the external canthus to avoid the superior branch of the facial nerve (fig 6). The orbicularis oris muscle and periosteum are properly opened up with a small section tip to reach the orbital roof and zygomatic archfronto-orbital arch (fig 7). The temporal muscle is pushed away laterally to discover the Mac Carty burr hole. Like for the eyebrow approach, the surgeon must absolutely locate the frontal sinus. The bone hole is done at the Mac Carty point with the drill and the craniotomy with the pediatric saw. A protecting retractor is inserted into the orbit without opening the periorbit and the hole is performed until one enters the orbit. Then the drilling exposes the frontal dura mater. The latter is therefore unstuck and the frontal bone cut made until the frontal arch taking great care to leave intact the frontal sinus, with the help of neuronavigation if needed. With small bone scissors or small diamond drill, the orbital roof will be cut protecting the orbit content with a smooth blade. With such an eyelid incision and opening posteriorly the periosteum, we are able to perform a one piece fronto-orbital craniotomy of around 2 cm of height and 2 cm and half large (fig 8). Orbit roof removing is completed as far as possible with the drill. This step is of paramount importance as with such keyholes, the slightest bone crest can disturb vision and tools' progression. Finally, using this eyelid incision, we can perform a wider and more lateral aperture (fig 5).

As for the eyebrow approach, a bone notch will be carried out to manage space for the endoscope.

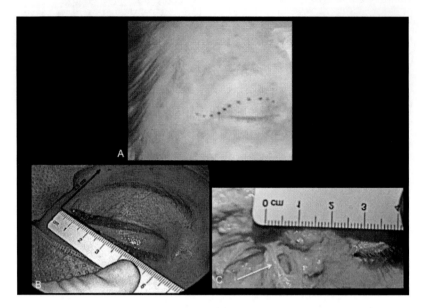

Figure 6. skin incision (A and B operative views, C anatomic data). A: Women with small eyebrow and meningioma of the tuberculum of the sella operated on using a trans-eyelid technique. B and C: eyelid incision that should not exceed 25 mm lateral to the external canthus to avoid any damage to the superior branch of the facial nerve (arrow).

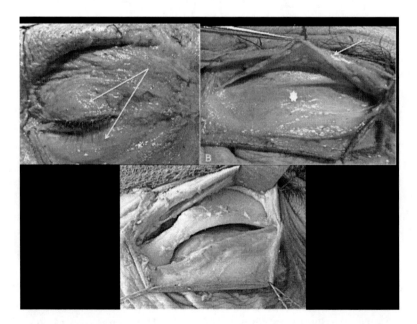

Figure 7. subcutaneous dissection (anatomical data). A: The orbicularis oris muscle is visible on both eyelids (arrow). B: A clear cut is made and the muscle lifted up. The orbital arch is visible with the orbital septum below (*). C: the frontoal-orbital bone has been dissected and prepared for the craniotomy.

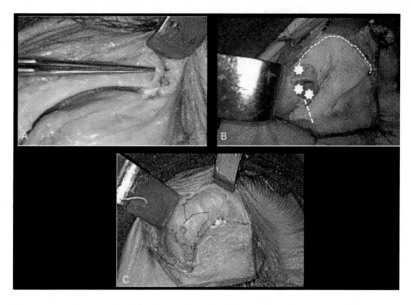

Figure 8: craniotomy steps (anatomical data). A: a bone hole is made on the pterion with the frontal (*) and the temporal (**) dura mater B: the frontal bone cut is drawn C: the orbital roof is removed as much as possible to eliminate disturbing reliefs.

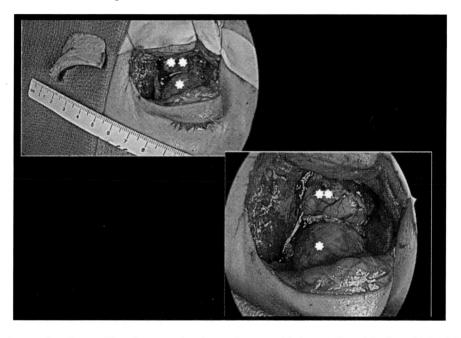

Figure 9. operative pictures. Up, after removing the craniotomy with the eye (*) and the frontal lobe (**). Below, after opening with the dura mater.

Intradural Step

There is no difference between both trans-eyebrow or trans-eyelid techniques regarding the intradural dissection step.

Closure Step (Fig 10)

Once again, no significant difference between the two techniques has to be reported. Dura mater is closed with tight sutures and bone pieces attached with plates or wires taking care to avoid any reliefs. All planes have therefore to be closed with resorbable sutures and we do not need drainage contrary to other authors [3]. A small dressing cover is put in place for 2 days and local application of ice to avoid eyelid edema is advised.

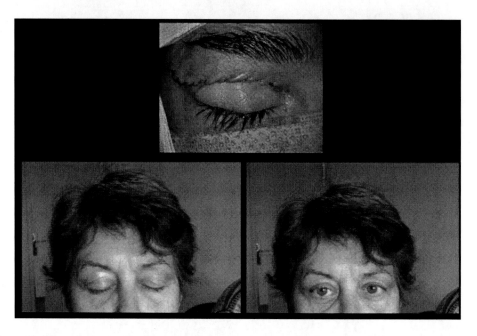

Figure 10. Cosmetic results, Up, immediately after the surgery and below at 3 months.

Respective Indications for Endoscopic Endonasal or Keyhole Supraorbital Surgery for Anterior and Middle Skull Base Tumors

Since the development of modern endoscopy, midline skull base tumors from the cribriform plate to the retrochiasmatic space can be approached using several routes.

The most common route remains the sub fronto-temporal one using microscope. For cribriform lesions, a trans sinusal route can also be used. All these approaches achieve good results but own a non-negligible morbidity with large and poorly esthetic craniotomies, devascularization and atrophy of the temporal muscle and sinuses aperture for the latter. Lastly, by these approaches, the retrochiasmatic space and the infero-medial compartment of the optic nerve cannot be reached without pushing away the optic tract that may be source of damage.

To minimize morbidity as well as for improving the exposition of retrochiasmatic tumors, the endonasal route was initially developed. This route was used first to remove tumors of the tuberculum of the sella using microscope (ectopic pituitary adenomas for instance) [4].

However, the necessary use of a nasal speculum and the axial vision provided by the microscope prevent to use this approach for complex intradural tumors such as large craniopharyngiomas. For 3 2 decades, US and Italian surgical teams (primarily especially Pittsburgh and Naples ones) promote endoscopic endonasal surgery and develop adequate tools first for pituitary surgery and thereafter for intradural skull base tumors [5-12]. Beside tools, the most important contributions concerned the closure techniques with a pedicled flaps for intradural surgery. These techniques dramatically decreased CSF leaks and led these extended endonasal approaches acceptable for patients.

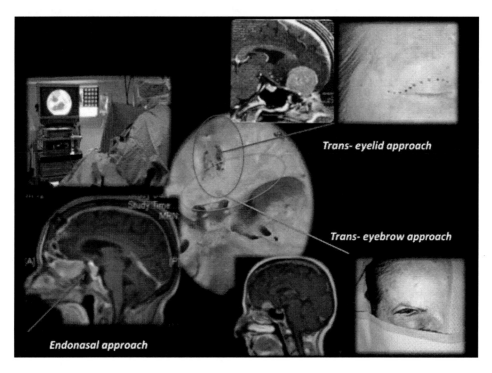

Figure 11. Respective indications for endonasal, trans-eyebrow and trans-eyelid endoscopic approaches for anterior and middle skull base tumors. The endonasal route is the technique of choice for infra and retrochiasmatic tumors. Absence of control of the lateral expansions of tumors and CSF leak issue especially for meningiomas led us to develop cranial keyhole endoscopic approaches. However, a vertical tuberculum sellae and/or short optic nerves with a retrochiasmatic expansion of meningiomas remain good endonasal indications. Trans-eyebrow is a good option in case of regular eyebrow, small frontal sinus, lateral and midline tumors until the cribriform ethmoidal plate. Huge frontal sinus, anterior skull base tumors and small eyebrow indicate a trans-eyelid route.

With an experience of more than 600 cases of pituitary adenomas and 200 extended skull base surgeries, we believe that the best indications for endoscopic endonasal approaches are midline infra or retrochiasmatic tumors as well as clival or craniocervical ones. This route provides also good and safe conditions to approach parasellar tumors cavernous sinus or Meckel's cave tumors growing from medial to lateral pushing away laterally cranial nerves or the internal carotid such as lateral extension of adenomas, chondrosarcomas, chordomas or cavernous sinus or Meckel's cave tumors [11, 13-17]. Approaching such tumors from above may lead to cranial nerves or vascular morbidity and difficulty to completely remove the tumor. The nasal route is therefore an interesting option.

The first issue using the endonasal corridor is for anterior or middle skull tumors that are not purely midline. . Indeed, as soon as the tumor goes over half of the optic nerve or the ICA, it is difficult to control the dissection without pulling out the tumor that may be dangerous for microvessels. This absence of control of the lateral margin of the tumors especially in case of meningiomas with frequent lateral extension going far away along the skull base can increase the rate of subtotal resection and tumor recurrence. This may also be an issue for malignant tumors such as ethmoidal adenocarcinomas.

The second issue is CSF leaks and as a consequence severe infection. In our own experience, this point is particularly relevant for anterior skull base tumors (cribriform and planum or tuberculum of the sella locations).

Therefore, except in case of vertical tuberculum of the sella and/or short optic nerves, we now prefer approaching anterior skull base tumors especially meningiomas from above using supraorbital keyhole endoscopic approaches.

We described herein 2 techniques for supraorbital keyhole approaches, the trans eyebrow and the trans eyelid routes. Originally described with microscope [2,3], we did develop a pure endoscopic technique to improve the vision inside the cranial compartment [1]. Indeed, when the light source is placed outside the keyhole (microscope) the vision is far less good than when the light source is directly brought inside the surgical field (endoscope). One disadvantage of endoscopy is a 2D vision and a necessary training to operate watching a screen TV. For the latter, a surgeon experienced in endonasal endoscopic surgery has no real difficulty to switch from microscope to endoscope. In practice, 3D informations are obtained when working with 2 surgical tools that give you the depth of your surgical field.

These 2 keyhole approaches appear to be very close. However, in our own experience of more than 40 patients with skull base tumors treated either with a trans-eyebrow or with a trans-eyelid incision, slight differences do exist that have clinical implications.

First, when choosing such approaches with facial incisions instead of regular one hidden into the hairline, the cosmetic result must be perfect. Both incisions give similar results in our experience, perhaps with a slight advantage for trans-eyebrow incision really invisible in all our cases. Some authors advice doing the incision few millimeters up to the eyebrow to avoid lesion of the eyebrow but we do not have any problem making the incision inside if you avoid excessive coagulation. Eyelid incision gave us also good cosmetic results but the most lateral part of the incision may be apparent in some patients for months. The skin is also far more fragile in that location and need great care during surgery avoiding excessive traction and coagulation. Temporary eyelid sutures during the surgery are also mandatory to protect the cornea. The choice between the two techniques is finally depending first upon the size of the eyebrow: a short one as in some women will indicate a trans-eyelid approach. Secondly, upon the size of the frontal sinus you need to respect: Indeed, using these small approaches, you cannot perform a good cranialisation. Largely pneumatized sinuses implicate therefore a more lateral approach that cannot be done with a trans-eyebrow incision. Thirdly, of the exact location of the tumor: For cribriform tumors, a trans-eyelid incision with a lateral pteriono-frontal craniotomy is necessary. Head's patients will have also to be rotated a little bit more. Doing so, we are able to work tangentially behind the frontal sinus to reach the most anterior part of the skull base.

LastlyContrarily, if the tumor has a lateral extension in the temporal fossa, a trans-eyebrow with a head minimally rotated or neutral is an optimal positioningwill have to be prefered.

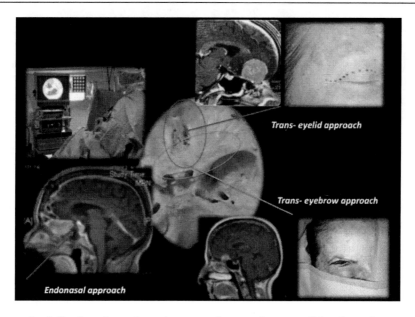

Figure 11. Respective indications for endonasal, trans-eyebrow and trans-eyelid endoscopic approaches for anterior and middle skull base tumors. The endonasal route is the technique of choice for infra and retrochiasmatic tumors. Absence of control of the lateral margins of tumors and CSF leak issue especially for meningiomas led us to develop cranial keyhole endoscopic approaches. However, a vertical tuberculum sellae and/or short optic nerves with a retrochiasmatic extension of meningiomas remain good endonasal indications. Trans-eyebrow is a good option in case of regular eyebrow, small frontal sinus, lateral and midline tumors until the cribriform ethmoidal plate. Huge frontal sinus, anterior skull base tumors and small eyebrow indicate a trans-eyelid route.

Conclusion

In summary, supraorbital keyhole approaches that can be done using either by trans eyebrow or trans eyelid are valid options for small or middle size anterior or middle skull base tumors. Aanteriorly based tumors with a large frontal sinus and small eyebrow must be approached through a trans-eyelid route. More posterior lesions laterally extended with a regular eyebrow are good candidates for trans-eyebrow approaches.

References

[1] Berhouma M, Jacquesson T, Jouanneau E. The fully endoscopic supraorbital trans-eyebrow keyhole approach to the anterior and middle skull base. *Acta Neurochir* (Wien). 2011 Oct;153(10):1949–54.
[2] Reisch R, Perneczky A, Filippi R. Surgical technique of the supraorbital key-hole craniotomy. *Surg Neurol.* 2003 Mar;59(3):223–7.
[3] Abdel Aziz KM, Bhatia S, Tantawy MH, Sekula R, Keller JT, Froelich S, et al. Minimally invasive transpalpebral "eyelid" approach to the anterior cranial base. *Neurosurgery.* 2011 Dec;69(2 Suppl Operative):ons195–206; discussion 206–207.

[4] Oldfield EH, Jane JA Jr. Endoscopic versus microscopic pituitary surgery. *J. Neurol. Neurosurg. Psychiatr.* 2013 Jan 23; Mason RB, Nieman LK, Doppman JL, Oldfield EH. Selective excision of adenomas originating in or extending into the pituitary stalk with preservation of pituitary function. J Neurosurg. 1997 Sep;87(3):343-51.

[5] Castelnuovo P, Dallan I, Pistochini A, Battaglia P, Locatelli D, Bignami M. Endonasal endoscopic repair of Sternberg's canal cerebrospinal fluid leaks. *Laryngoscope.* 2007 Feb;117(2):345–9.

[6] Gardner PA, Prevedello DM, Kassam AB, Snyderman CH, Carrau RL, Mintz AH. The evolution of the endonasal approach for craniopharyngiomas. *J. Neurosurg.* 2008 May;108(5):1043–7.

[7] Jho HD, Carrau RL. Endoscopic endonasal transsphenoidal surgery: experience with 50 patients. *J. Neurosurg.* 1997 Jul;87(1):44–51.

[8] Kassam AB, Gardner P, Snyderman C, Mintz A, Carrau R. Expanded endonasal approach: fully endoscopic, completely transnasal approach to the middle third of the clivus, petrous bone, middle cranial fossa, and infratemporal fossa. *Neurosurg Focus.* 2005 Jul 15;19(1):E6.

[9] Kassam AB, Mintz AH, Gardner PA, Horowitz MB, Carrau RL, Snyderman CH. The expanded endonasal approach for an endoscopic transnasal clipping and aneurysmorrhaphy of a large vertebral artery aneurysm: technical case report. *Neurosurgery.* 2006 Jul;59(1 Suppl 1):ONSE162–165; discussion ONSE162–165.

[10] Liu JK, Das K, Weiss MH, Laws ER, Couldwell WT. The history and evolution of transsphenoidal surgery. *J. Neurosurg.* 2001 Dec;95(6):1083–96.

[11] Schwartz TH, Fraser JF, Brown S, Tabaee A, Kacker A, Anand VK. Endoscopic cranial base surgery: classification of operative approaches. *Neurosurgery.* 2008 May;62(5):991–1002; discussion 1002–1005.

[12] Snyderman CH, Carrau RL, Kassam AB, Zanation A, Prevedello D, Gardner P, et al. Endoscopic skull base surgery: principles of endonasal oncological surgery. *J Surg Oncol.* 2008 Jun 15;97(8):658–64.

[13] Berhouma M, Messerer M, Jouanneau E. [Shifting paradigm in skull base surgery: Roots, current state of the art and future trends of endonasal endoscopic approaches]. *Rev. Neurol.* (Paris). 2012 Feb;168(2):121–34.

[14] Kassam AB, Prevedello DM, Carrau RL, Snyderman CH, Gardner P, Osawa S, et al. The front door to meckel's cave: an anteromedial corridor via expanded endoscopic endonasal approach- technical considerations and clinical series. *Neurosurgery.* 2009 Mar;64(3 Suppl):71–82; discussion 82–83.

[15] Griffith AJ, Terrell JE. Transsphenoid endoscopic management of petrous apex cholesterol granuloma. *Otolaryngol Head Neck Surg.* 1996 Jan;114(1):91–4.

[16] Alfieri A, Jho HD. Endoscopic endonasal approaches to the cavernous sinus: surgical approaches. *Neurosurgery.* 2001 Aug;49(2):354–360; discussion 360–362.

[17] Alfieri A, Jho HD. Endoscopic endonasal cavernous sinus surgery: an anatomic study. *Neurosurgery.* 2001 Apr;48(4):827–836; discussion 836–837.

Chapter XVIII

Mini-Invasive Microvascular Decompression for Posterior Fossa Neurovascular Conflicts

Francesco Acerbi[†*], Morgan Broggi[†], Marco Schiariti, Melina Castiglione, Giovanni Broggi and Paolo Ferroli

Department of Neurosurgery,
Fondazione IRCCS Istituto Neurologico Carlo Besta,
Milano, Italy

Abstract

Introduction: Idiopathic trigeminal neuralgia (TN), glossopharyngeal neuralgia (GN) and hemifacial spasm (HFS) are usually sustained by a neurovascular conflict in the cerebellopontine angle (CPA). Microvascular decompression (MVD) is a non-ablative surgical procedure designed to resolve these neurovascular conflicts.

This chapter will present the surgical technique of minimal invasive MVD, mainly for TN, and it will focus on the value of each visualization technique (pure microscopic, fully endoscopic and microscopic endoscope assisted) for MVD.

Materials and Method: 103 MVD procedures performed between June 2010 and December 2011 were retrospectively reviewed. There were 84 patients affected by TN, 17 by HFS, and 2 patients with TN and GN simultaneously; of these, 7 (7%) were recurrences (6 TN, 1 HFS).

With the patient in supine position, a small mini-invasive elliptical retrosigmoid craniectomy is used to approach the CPA and the interested nerve. After careful exploration of the nerve root entry zone, the offending vessel is identified and moved

---

[*] Corresponding author: Francesco Acerbi, MD PhD. Department of Neurosurgery. Fondazione IRCCS Istituto Neurologico Carlo Besta, Milano, Italy. Email: acerbi.f@istituto-besta.it. Phone: +39 02 23942411. Fax: +39 02 70635017.
[†] These authors equally contributed to the paper.

away. A non compressive technique with either Oxidized regenerated cellulosa or Teflon with or without fibrin glue is always used to keep the vessel in its new position far from the nerve. The visualization techniques used were: pure microscopic in 65 (63%) cases, fully endoscopic in 9 (9%) cases and microscopic endoscopic-assisted in 29 (28%) cases.

Results: A neurovascular conflict was found in 95 (92%) cases, while 8 patients had no intraoperative evidence of neurovascular conflict. The microscope or the endoscope alone was able to show the conflict in all 74 cases. In the 29 microscopic endoscope assisted cases, 11 conflicts previously not clearly visible with the microscope were revealed and a complete conflict resolution was confirmed in 10 cases. Overall, the pain relief rate for TN and GN was excellent in 58 cases (67.5%), good in 15 (17.5%) and poor in 13 (15%), while the degree of spasm resolution in HFS was complete in 13 patients (76%) and decreased in 4 (24%).

Conclusion: Minimal invasive MVD represents the gold standard first line treatment for neurovascular conflict resolutions in the CPA; its aim should be to free the interested nerve from any contact.

MVD under pure microscopic view remains the technique of choice. The endoscope is a useful adjunctive imaging tool in confirming neurovascular conflicts identified by the microscope, revealing conflicts missed by the microscopic survey alone and verifying adequate nerve decompression.

Keywords: Microvascular Decompression; Trigeminal Neuralgia; Retrosigmoid approach; Endoscopy; Minimal invasive neurosurgery

Abbreviations

AICA: Anterior Inferior Cerebellar Artery
BA: Basilar artery
CN: Cranial Nerve
CPA: Cerebellopontine Angle
CSF: Cerebrospinal Fluid
GA: General Anesthesia
GN: Glossopharyngeal Neuralgia
HFS: Hemifacial Spasm
ICG: Indocyanine Green
MEA: Microscopic Endoscopic-Assisted
MEV: Mastoid Emissary Vein
MR: Magnetic Resonance
MS: Multiple Sclerosis
MVD: Microvascular Decompression
OA: Occipital Artery
PICA: Posterior Inferior Cerebellar Artery
REZ: Root Entry Zone
SCA: Superior Cerebellar Artery
SPV: Superior Petrosal Vein
SS: Sigmoid Sinus
TN: typical idiopathic Trigeminal Neuralgia
TREZ: Trigeminal Root Entry Zone

TS: Transverse Sinus
VA: Vertebral Artery

1. Introduction

There is a subgroup of cranial nerves (CN) dysfunctions that are due to a contact or, better a conflict, between a blood vessel and the nerve; these conflicts usually occurs in the posterior fossa cisterns, mainly in the cerebellopontine angle (CPA), between the root entry zone (REZ) of the nerve at the brainstem and its exit out of the posterior fossa [1-4]. Theoretically, apart from the olfactory and the optic nerves, all other CN could be potentially interested by such a phenomena, but the nerves most frequently involved are the Vth, the VIIth and the IXth ones. Thus, the most common clinical disease of this group is by far the so-called "typical-idiopathic Trigeminal Neuralgia" (TN), followed by hemifacial spasm (HFS) and by glossopharyngeal neuralgia (GN).

The incidence of TN is about 2.3 to 4.5 per 100,000 new cases per year. Age at onset is variable, but the incidence increases in the fifth, sixth and seventh decades of life; the female:male ratio is 1.8:1. TN is most frequent in the right side of the face; typically the branches involved are V2 and V3 either alone or together [5-7] Considering HSF, the average annual incidence rate is 0.74 per 100,000 in men and 0.81 per 100,000 in women. The average prevalence rate is 7.4 per 100,000 population in men and 14.5 per 100,000 in women. The left side is more commonly affected. The incidence and prevalence rates were highest in those from 40 to 79 years of age [8, 9]. HFS may be associated with TN, GN and vestibular or cochlear nerve dysfunction [10].

Isolated GN is rare; it occurs at a rate of 1 case in every 70 to 100 cases of TN [1, 11]; concurrence of TN and GN is often possible [12].

1.1. Trigeminal Neuralgia

The clinical features of the so-called "typical-idiopathic" TN have been well documented and are now universally recognized. Eller et al. and Burchiel et al. [13] classified trigeminal neuralgia on the basis of two broad categories: the patient's history and seven specific diagnostic criteria. Eller's and Burchiel's Trigeminal Neuralgia type 1 identifies the typical-idiopathic drug-resistant type, which is by far the most common one.

TN is a chronic pain syndrome whose patients suffer from idiopathic, episodes of spontaneous facial pain. This pain, which is experienced in one or more divisions of the trigeminal nerve, expresses itself as paroxysms of brief and excruciatingly intense bouts of stabbing, electrical shocks that can arise either spontaneously or in response to gentle tactile stimulation of a trigger point on the face or in the oral cavity; they may also be triggered by such natural activities as: chewing, speaking, washing the face, or brushing the teeth [14]. TN is almost always experienced unilaterally although there have been reports describing bilateral signs and symptoms [13]. The pain typically occurs after conspicuously obvious pain-free intervals that can last for weeks, months and years. The neurological examination is almost always normal even though in some cases a slight degree of sensory loss can be observed [15]. The diagnosis of TN is always based on the patient's clinical history [16], supported by *ad hoc* Magnetic Resonance (MR) sequences [17]. In terms of its pathophysiology, the features of classic TN are currently thought to be related to a compression of the trigeminal nerve root, usually by a blood vessel, at or near the REZ [2, 18, 19]. Dandy [20] was the first to propose in 1934 that there might be a causal relationship between pain paroxysms and the compression of the trigeminal nerve. In the late 1950's Gardner sustained that a therapeutic pain relief might be obtainable by decompressing the trigeminal nerve [21, 22]. Later on, the notion that a TN was caused by a microvascular compression gained much support from the work of Jannetta [2], who not only was able to find a compressing vascular contact in a high percentage of TN patients, but was also able to demonstrate that prolonged pain relief could be obtained by a microsurgical decompressive technique, Microvascular Decompression (MVD) [23, 24].

This compression is typically caused by adjacent arterial loops, usually from the superior cerebellar artery (SCA) or the anterior inferior cerebellar artery (AICA) [25]. Sometimes a big and tortuous loop of the basilar artery (BA) might be in conflict with the nerve [26]. Seldom, veins can be in conflict with the nerve too; the complex of the superior petrosal veins (SPV) or Dandy's veins is the most frequently involved. Multiple conflicts (arterial or mixed arterial and venous) may also occur. Secondary TN might be caused by multiple sclerosis (MS) or tumors, arteriovenous malformations, and aneurysms located in this region.

1.2. Hemifacial Spasm

HFS is characterized by an intermittent, involuntary, unilateral and painless spasmodic contraction of muscles of the face innervated by the facial nerve; it may be limited to the upper or lower half of the face only [27]. HFS usually starts with rare contractions of some muscles only (mainly the orbicularis muscle), then it progresses to the entire half of the face and it increases in frequency and intensity of the spasms. Along with palatal myoclonus, HSF continues during sleep [27, 28]. As for TN, neurovascular conflict is the most common cause; for its anatomical location the artery most often responsible for this is the AICA or one of its branches, followed by the posterior inferior crebellar artery (PICA), the SCA, the vertebral artery (VA) or, again, a tortuous dolichoectasic BA; it is rather uncommon to find a venous conflict. Other causes of HFS include tumors, cysts, aneurysms, vascular malformations and MS.

1.3. Glossopharyngeal Neuralgia

GN is a spasmodic, lancinating, and paroxysmal pain that starts in the posterior throat, the tonsil, the base of the tongue or the external ear canal, and can radiates down the throat and to the side of the neck [14, 29]. The type of pain is similar to that experienced with TN. Often pain is triggered by swallowing or yawning. In addition, these painful attacks can be associated with hemodynamic instability resulting from reflexive autonomic outflow that can lead to hypotension, syncope, life-threatening cardiac irregularities and seizures syncopal episodes [30-34]. These additional symptoms might be explained by the simultaneous involvement of the vagus nerve; this fact was first described by Dandy [35] and, not surprisingly, White and Sweet proposed the term of "vago-glossopharyngeal neuralgia [36].

Single arterial conflict, usually from PICA and VA, is the cause in the majority (around 70%) of GN patients; multiple arterial or venous conflicts are possible, but not so frequent [37].

2. Surgical Technique of MVD

Medical and other treatment modalities are available for all these three syndromes.

TN can be managed with oral medications; there are also well effective percutaneous treatments and radiosurgery is an option too. Local injection of botulinum toxin may be effective in treating HFS. GN was historically treated mainly by surgical lesions of (neurotomies, rhizotomies) of the glossopharyngeal and vagal fibers or tractotomy/nucleotomy [38-39]; some of these procedures were used for TN too [40-42]; cocainization of tonsillar pillars and fossa has been used too.

However, as already said, the etiology of primary TN, HFS and GN is usually identified in a neurovascular conflict. Hence, MVD is universally recognised as the first treatment option in all patients affected by one of these syndromes able to undergo general anaesthesia (GA).

In the following paragraphs, the surgical technique of mini-invasive MVD will be presented, focusing on MVD for TN, for obvious epidemiological reasons.

2.1. MVD for TN

2.1.1. Patient Positioning

Step 1: Following the induction of GA, with the patient supine, the head is supported by a Mayfield three points fixation device. The head is then elevated above the level of the heart with the intention of facilitating cranial venous drainage. *Step 2:* The head is gently and carefully rotated towards the contralateral side and a sandbag is then placed under the ipsilateral shoulder in order to support it. *Step 3:* The neck is flexed 10-15° anteriorly with the chin positioned approximately two fingers breadth from the shoulder while keeping the vertex almost parallel to the floor. These steps are aiming to create a working corridor for the surgeon with minimal disturbance from the ipsilateral shoulder. *Step 4:* The vertex is then gently depressed 10° inferiorly to allow for an optimal visualization of the under surface of the tentorium and the upper neurovascular structures of the CPA. Finally, the patient is taped securely onto the table to allow for further rotation during the procedure when necessary (Figure 1). This position provides an optimal trajectory straight to the nerve. Special attention must be paid to the ipsilateral shoulder, which in some cases (especially overweight or obese patients) should be taped and gently pulled caudally, so that it does not decrease vision and working corridor.

2.1.2. Skin Incision and Soft Tissues Dissection

Only a small (5 x 5 cm) area behind the ear is shaved. Identification of anatomical surface landmarks is performed before the skin incision. The course of the Transverse Sinus (TS) is identified and marked along a line connecting the inion with the zygomatic arch at the level of the supramastoid crest passing through the asterion. A second line showing the tip and the body of the mastoid is then drawn. The junction between the TS and the Sigmoid Sinus (SS) is identified and marked at the level of the asterion. After this, the planned retrosigmoid craniectomy is drawn on the skin too (Figure 2).

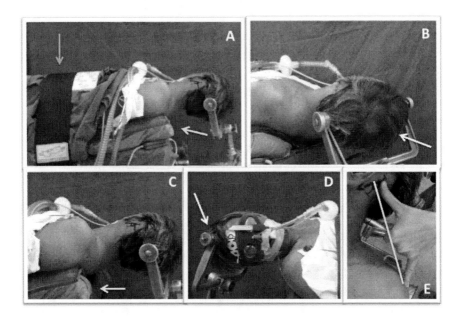

Figure 1. Patient positioning, left side procedure. A General view; note that the patient is taped securely onto the table to allow for further rotation during the procedure when necessary (orange arrow) and the sandbag under the ipsilateral shoulder in order to minimize the neck rotation (yellow arrow). B Postero-lateral view; the vertex is then gently depressed 10° inferiorly to allow for an optimal visualization of the under surface of the tentorium and the upper neurovascular structures of the CPA (see text, white arrow). C Enlargement of A: the neck is not stretched thank to the sandbag placed under the ipsilateral shoulder (yellow arrow). D Anterior view: again, the vertex is then gently depressed 10° inferiorly (white arrow) .E The position should create the necessary working corridor for the surgeon without disturbance from the ipsilateral shoulder (yellow line).

Neuronavigation can be used to confirm correct landmarks identification and proper craniectomy planning, even though it is not mandatory, and we do not use it routinely. Following the usual sterile draping preparation, a vertical skin incision approximately 4 to 6 cm in length is performed about 3 to 5 cm posteriorly from the internal acoustic meatus centred over the pre-planned craniectomy.

The soft tissues are dissected with the aid of a periosteal elevator. The insertion of the splenius capitis and longissimus capitis are dissected for a brief tract with electrocautery and then retracted to expose the retrosigmoid suboccipital bone area. An orthostatic retractor is positioned to maintain adequate exposure of the occipital bone. The occipital artery (OA) is usually encountered between splenius and longissimus capitis. Once the OA is identified, it is then coagulated and cut. This is generally necessary in order to achieve an adequate bone exposure. The mastoid emissary vein (MEV) connects the suboccipital venous plexus to the SS; this landmark is helpful in localizing the SS. Once the MEV is identified, its bone orifice(s) is/are sealed with bone wax. The size and the position of both the OA and the MEV, however, are known to show a considerable variation [43,44] and knowledge of their anatomy is therefore recommended.

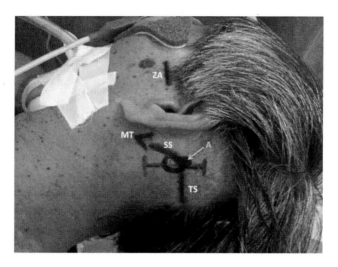

Figure 2. The main anatomical landmarks are drawn on the skin, left side approach; the small retrosigmoid craniectomy and the skin incision are marked. TS: Transverse sinus. SS: Sigmoid sinus. MT: Mastoid tip. ZA: Zygomatic arch. A: Asterion. I: incision. C: craniectomy.

2.1.3. Bone Removal: Craniectomy

The recognition of bone landmarks is reconfirmed prior to start bone drilling. The asterion is defined as the junction of the lambdoid, parietomastoid and occipitomastoid sutures.

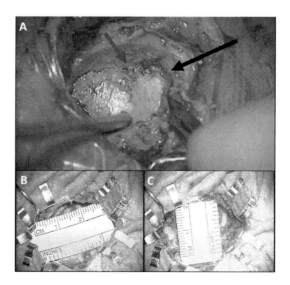

Figure 3. A Left side approach. The craniectomy is completed and the dura mater is exposed. The black arrow points the transverse sinus and the red arrow points the sigmoid sinus. B and C Right side approach, measurements of the craniectomy.

It is most often found in a position over the transverse-sigmoid junction (approximately two thirds of the cases), but in approximately one third of the cases it is found below the transverse-sigmoid junction [45]. Using a high- speed 5-7 mm cutting drill (X-max/E-max2 plus, Anspach, The Anspach Effort, Palm Beach Gardens, FL, USA), a small vertical elliptical craniectomy (3 x 2 cm) is performed just caudally to the asterion (Figure 3).

The opening is elliptically shaped with the anterior and superior borders constituted by the posterior margin of the SS and the inferior margin of the TS respectively and the upper anterior extremity of the ellipse by the sinuses junction (Figure 3). In many cases some posterior mastoid air cells must be opened to obtain an adequate exposure. These must be sealed with muscle and fibrin glue in order to reduce the risk of infection and rhinoliquorrhea.

2.1.4. Exposure of the CPA and Nerve Decompression

It is at this stage of the procedure that the microscope is introduced. The dura mater is opened in a curvilinear fashion following the shape of the craniectomy with its base toward the SS. A short cut towards the TS-SS junction allows for retraction of the free dural edge thus exposing both the tentorium and the posterior surface of petrous ridge and creating a direct supra-latero-cerebellar corridor below the superior petrosal sinus. A cottonoid is introduced and advanced over the cerebellar surface until the arachnoid of the CPA cistern is found and widely opened in the region between the VIIth-VIIIth CN and the SPV complexes. Progressive Cerebrospinal Fluid (CSF) drainage creates the working corridor avoiding any need for cerebellar retraction.

Looking cranially, the trigeminal nerve generally comes into view without any interference from the SPV complex. In some cases the SPV complex can be in the way to the nerve. It is typically formed by three veins converging into a main collector that can be sacrificed just before entering the superior petrosal sinus, leaving a common chamber to allow for anastomotic circulation, that can be studied by indocyanine green (ICG) videoangiography, by the aid of a specifically designed temporary clipping test [46].

At this point, the trigeminal nerve can be easily identified as running just behind and medially deep to the SPV complex. The nerve is explored in all its length from its origin at the brainstem to its exit through the porus trigemini.

The neurovascular conflict is often easy to find although, at times, it can be hidden by the nerve itself. In these particular cases, the use of the endoscope with a Microscopic Endoscopic-Assisted (MEA) technique can be useful. The authors performed also fully endoscopic MVD, but they believe the microscopic technique is easier, faster and safer (see below).

The Trigeminal REZ (TREZ) is the most common site of vascular compression. It is the locus of transition from central to peripheral myelin and the most vulnerable part of the nerve. After the neurovascular conflict is identified (Figure 4), a sharp dissection of the arachnoidal bands that fix the artery into its position allows for moving away the artery. Great attention must be paid to the small perforating arteries that can, on rare occasions, complicate the artery mobilization. At this point, pieces of oxidized regenerated cellulosa (Tabotamp, Codman Ethicon, Johnson and Johnson, New Jersey, USA) or small pieces of polytetrafluoroethylene (Teflon) with or without fibrin glue are strategically placed to keep the artery far from the nerve so that, at the end of the procedure, this is free from any contact. The aim of MVD, in the authors' opinion, is to free the nerve from any contact both from the vessel and from the interposed material.

2.1.5. Surgical Closure

Once the haemostasis is assured, the surgical corridor is then irrigated with warm saline solution. The dura mater is closed by 4-0 vicryl separated stitches. Any residual dural gap is plugged with muscles fragments, or small bovine collagen sheets (Condress, Abiogen Pharma spa, Ospedaletto, PI, Italy). Dural closure can be reinforced with sealants (Duraseal, Covidien, Dublin, Ireland; or Tissue-Patch-Dural, TPD, Tissuemed ltd, Leeds, UK). The closure is completed by positioning over the craniectomy area a further layer of Condress. The operating microscope is moved away and the craniectomy may by covered with an MR compatible titanium plate (Bioplate, Bioplate inc, Los Angeles, CA, USA) to avoid any skin depression or left without any cover. The muscular, subcutaneous and skin layers are closed in the usual manner.

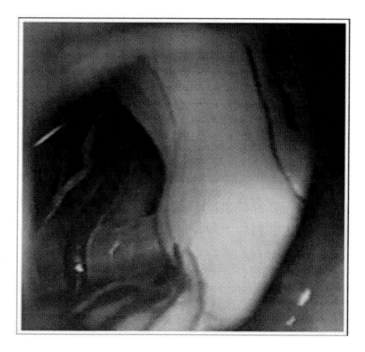

Figure 4. Microscopic view during left side MVD for TN showing the neurovascular conflict between the trigeminal nerve and the SCA. Note how much the nerve is deformed by the artery.

2.2. Specific Consideration for MVD for HFS and GN

Even though the aim of the procedure is the same, i.e. to decompress the nerve from a neurovascular conflict, obviously there are some variations between performing MVD for TN, for HSF or for GN and the reason for this are the different anatomical location and course of the nerves. The major difference certainly consists in the position of the craniectomy.

To better understand these distinctions and for good orientation of the neurovascular structures in the CPA, Rhoton and colleagues [47-50] divided the CPA into three portions (i.e., superior, middle, and inferior) and explained their anatomic relationships with the "rule of 3." This concept includes three foramina (Meckel's cavity, internal acoustic meatus, and jugular foramen), three groups of cranial nerves (trigeminal nerve, facial and vestibulocochlear nerves, and glossopharyngeal, vagal, and accessory nerves), three cerebellar arteries (SCA, AICA and PICA), and three cerebellar surfaces (tentorial surface, petrosal surface, and suboccipital surface). These groups of neurovasculo-osseous structures correspond well with each other, and the rule of 3 allows us to choose an anatomically appropriate route to the destination in the CPA [51-53].

For MVD to treat HFS and GN, the approach is rather similar. The patient's position is the same as for the infratentorial lateral supracerebellar approach described above for TN, except that the head position must be more vertex-down, for better exposure of the inferolateral part of the suboccipital region. The incision is also the same except caudally, where it extends for 3-4 cm longer. Surgical access from a strictly caudal direction, which is one of the keys of this infrafloccular approach, requires the sectioning of more muscle and the skin incision mentioned above to yield an accurate surgical direction and space. The bony opening is made more inferior then for TN, without exposing the TS. It is important to achieve this inferolateral exposure slightly over the medial border of the inferior half of the sigmoid sinus. Arachnoid membrane dissection around the jugular foramen to open the lateral aspect of the cerebellomedullary fissure allows observation of the IXth, X^{th}, and XIth CN and the flocculus without any cerebellar retraction. This is a sufficient exposure for GN. For HFS, there is the need to look upward in a caudorostral direction perpendicular to the VIIIth CN in order to find the REZ of the facial nerve, just medial to the root entry zone of the VIIIth CN in the supraolivary fossette.

3. Visualisation Technique of MVD (Microscopic, Endoscopic and Microscopic Endoscope Assisted)

Since its first descriptions, MVD has been performed with a microsurgical technique. Therefore the largest series available concern the use of the microscope for performing MVD [45, 54-57]. In the last two decades, however, endoscopy has emerged to be a useful imaging tool in many neurosurgical applications either alone or in assistance to the operating microscope [58-61]. Reports documenting the endoscopic anatomy of the CPA have shown the advantage of the endoscope, over the operating microscope, when it came to visualization of the structures in this area [62, 63]. For these reasons, therefore, endoscope-assisted microscopic and full endoscopic approaches have been developed for different pathologies involving this area [64-69]. We actually believe that all three techniques are valuable and that the modern neurosurgeon has to be confident with all of them so to choose the more appropriate one in each situation.

3.1. Methods

In a year and a half period (June 2010-December 2011), 103 MVD were performed at our Institution. There were 58 women and 45 men, the mean age was 62.3 years (range 24-79). TN was the most frequent pathology with 84 out of 103 patients (81.5%) treated, followed by HFS (17 cases, 16.5%); not surprisingly, 2 patients (2%) underwent MVD to treat simultaneously TN and GN. 96 (93%) MVD were first time procedures, while 7 (7%) cases were recurrences (6 patients affected by TN and 1 by HFS).

The visualization techniques used were pure microscopic in 65 (63%) cases and fully endoscopic in 9 (9%) cases; in the remaining 29 (28%) cases a microscopic endoscopic-assisted (MEA) technique was used. The indications for using this technique were: a) To seek for the neurovascular conflict, which could not be found under microscopic view. b) To check for adequate and complete conflict resolution (Figure 5). Intraoperatively, during microscopic cases, the endoscope is always available in the operating room and the surgeon can ask to set it up whenever he/she needs it. This takes on average no more than 5 minutes.

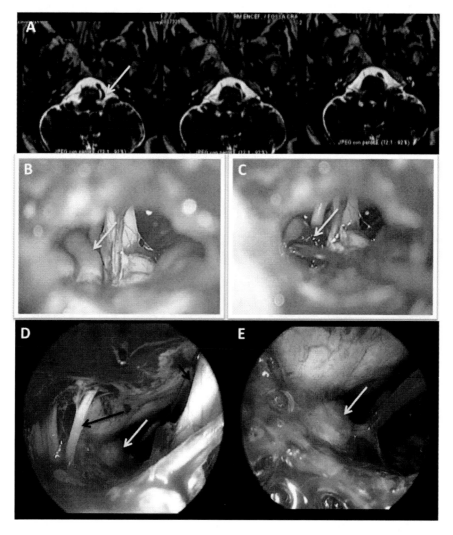

Figure 5. Microscopic endoscopic-assisted technique for left side MVD for recurrent HFS due to dolichoectasic vertebral artery. A Axial MR images showing a big dolichoectasic VA compressing the REZ of the facial nerve at the brainstem (yellow arrow) B Microscopic view, conflict identification (yellow arrow). C Microscopic view, conflict resolution by means of a piece of Tabotamp and fibrin glue (yellow arrow). D Endoscopic view confirming complete conflict resolution: the VA is stuck away (yellow arrow), the VIIth-VIIIth CN complex is free from any contact (short black arrow). On the bottom is visible the IXth-X[th]-XIth CN complex (long black arrow). E Enlargement of D, showing Tabotamp and fibrin glue that keep the artery in its new position (yellow arrow).

The endoscope is usually introduced in the operating field by the surgeon under microscopic view with the assistant looking at the endoscopic screen; when the endoscope reaches the right position in the surgical field, then the assistant holds the endoscope so that the surgeon can use a two hands surgical technique. This is done in order to avoid inadvertent damages to neurovascular structures of the CPA during the introduction of the scope. Conversely, during fully endoscopic cases, the scope is inserted in the CPA directly by the surgeon (with strong endoscopic background and experience), but only after adequate CFS drainage and cerebellar relaxation. In the same way as for the MEA technique, the microscope is always ready in the room should the surgeon needs it to further open the arachnoid of the CPA cistern to gain the necessary corridor for the endoscope to be inserted.

3.2. Results

Overall, a neurovascular conflict was found in 95 (92%) cases, while 8 (8%) patients had no intraoperative evidence of neurovascular conflict. These patients underwent a splitting of the trigeminal fibers, the so-called intradural rhyzotomy. The microscope or the endoscope were able to identify a conflict in all 74 cases when they were used alone (65 fully miscroscopic cases and 9 fully endoscopic cases, Figures 4 and 6)[*].

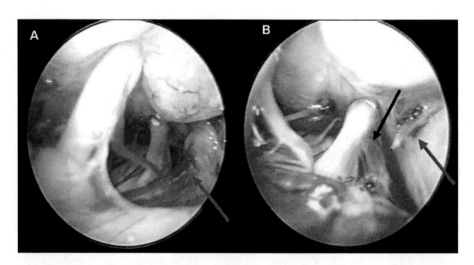

Figure 6. Pure endoscopic approach for a left side MVD for TN. **A** The SPV complex (red arrow) covers the view through the trigeminal nerve. **B** After coagulation and cutting of the vein the conflict (black arrow) becomes more evident.

In the 29 MEA cases, 11 (10.5% overall) conflicts previously not clearly visible with the microscope were revealed and a complete conflict resolution was confirmed in 10 cases (Figure 5). The mean follow-up was 14.2 months (range 6-24 months, between June 2010 and June 2012). There was no mortality in this series. No permanent or transient deficits of trigeminal nerve or of lower cranial nerves functions occurred. Two patients (2%), one operated for TN and one for HFS, experienced permanent partial hearing deficit; they were

---

[*] All patients shown in figures 1 to 6 gave their permission for use of photographs at the time of surgical consent.

both first time procedures. One patient (1%) operated for recurrent TN, presented postoperatively a partial facial weakness, which was mild-moderate immediately after surgery (House & Brackman scale 3) and then almost fully recovered at 6 months after surgery (House & Brackman scale 2). We had 1 CSF leak causing pseudomeningocele, that resolved with 4 days of external spinal drainage. No wound infection nor meningitis occurred. In terms of pain relief for TN and GN, this was classified following the BNI score [70] as: excellent in 58 cases (67.5%), good in 15 (17.5%) and poor 13 (15%), while the degree of spasm resolution in HFS was complete in 13 patients (76%) and decreased in 4 (24%). In the follow-up period there were no recurrences or re-explorations; patients affected by TN who were still complaining of pain were treated either with oral medications or with percutaneous treatments.

4. Discussion and Final Remarks

The earliest enthusiastic report of achieving any beneficial effects by decompressing the trigeminal nerve in patients with TN were made by Gardner and Miklos [21, 22]. But it was not until the mid 1960's that Jannetta and Rand [71] were able to further refine the procedure and popularized the MVD as the first choice in the treatment of TN. Jannetta, who also described the MVD operation in great detail [2, 3, 24, 72] is considered by many, including us, to be the father of this procedure.

4.1. Patient Positioning

Many different positions have been advocated as being ideal for performing MVD. The two most commonly used positions are the sitting position and the lateral position, with variations. Sindou, on the basis of considerable surgical experience with these two positions, has made an exhaustive comparison between the two [73]. His conclusions were that the lateral position allows for a much safer, faster and effective MVD. Concerning the lateral position, Jannetta and Mc Laughlin have indicated a preference for the lateral decubitus, also known as three-quarter prone position [24, 45], while Hitotsumatsu and the Fukushima group have described their preference for a purely lateral park bench position [37, 53]. Our preference, along with other [74], is the supine decubitus (see above). This allows a simple positioning of the patient, even of old age, and makes it possible in the same way to operate with an optimal trajectory straight to the nerves. However, it should be noted that during this step a special attention should be paid to the ipsilateral shoulder, which in some cases (especially overweight or obese patients) should be taped and gently pulled caudally, so that it does not decrease the surgical visualisation and the surgeon's maneuverability.

4.2. Skin Incision and Bone Removal

No matter which type of incision is used (linear or curvilinear), It is critical to avoid injuries to the greater and especially to the lesser occipital nerve, that may cause severe postoperative pain and headache [74]. Most authors recommend a maximum length of a 5 cm skin incision and 3 x 3 cm craniectomy [24, 25, 45]. This is not only because the need for good cosmetic results, but because minimally invasive techniques reduce the chance of adhesion of cervical musculature to exposed dura at the craniectomy site which has been suggested as one of the mechanisms for postoperative headache after MVD [74]. We proposed our minimal invasive approach with an even smaller bone opening, especially because we greatly prefer craniectomy over craniotomy.

Jannetta wrote "The junction between the transverse and the sigmoid sinuses must be visualized before the dural opening" [45]. This concept establishes the ideal place for making the burr hole in MVD for TN. Although it is true that the asterion can be found, in most cases, to lie just above the junction, many authors believe that it's position is extremely variable and instead prefer to follow the MEV as a means of identifying the junction [45, 75]. However, it is known that the site where the MEV flows into the sigmoid sinus can also vary considerably [44]. Over the course of many years, we have found the asterion to be a reliable landmark that results in no major problems, even because with our technique of craniectomy, it is possible to enlarge the bone opening until the TS and SS are enough exposed. The Jannetta group [24, 45] and others [37, 53, 76], have popularized the idea of the "isosceles triangle" where the apex is found to be at the junction. We use a circular craniectomy that is almost elliptical which has just the base of the triangle assuming a curvilinear shape.

The risk of causing a paradoxical rhinoliquorrea at this stage is well known. If it is necessary to open some mastoid air cells to gain the optimal dural exposure it is very important that this opening should immediately be covered with bone wax and eventually, after the dural closure, be sealed with muscle and fibrin glue or sealants.

4.3. Exposure of the Cerebellopontine Angle and Nerve Decompression

Depending on the surgeon's preference, the dura can be opened with or without the use of the microscope.

Sindou opens the dura by turning two small flaps backwards, one along the transverse sinus and the other along the sigmoid one [25]. McLaughlin has suggested the effectiveness of either a curvilinear or a T-shaped dural incision [45]. Our preference is for a curvilinear incision because we believe that a sufficient bone removal allows the same kind of intradural access but, at the same time, with an easier chance of watertight closure. Sampson begins by opening the dura inferomedially so as to initially decompress the cerebellomedullary cystern [37]. We, however, did not notice any differences in CSF drainage between whether we opened the dura cranially or caudally.

Once the dura is opened, however, no cerebellar retraction should be used and we strongly recommend against this, even though it seems to be regularly used by many authors [24, 45, 37, 53, 76]. It is crucially important that the surgeon be patient while gently coming along the lateral cerebellar surface and begins to drain some cc of CSF; once this is accomplished, he or she will then be facing the arachnoid membranes of the CPA cistern. At this point, the simple sharp opening of this arachnoid by means of a micro-scissors allows for an adequate CSF drainage along with a wide CPA exposure.

While performing MVD for TN, as the surgeon goes further inside the CPA, and after having cut the arachnoidal membranes of the VIIth-VIIIth CN complex, the SPV complex should then appear just between the surgeon and the trigeminal nerve. Although Jannetta and his group [24, 45] typically coagulate and then cut the superior petrosal veins when it is necessary, Sindou [25] and Hitotsumatsu [53] have suggested that it might be better to, at least, preserve the main trunk of the complex. Our tendency is to coagulate one or more veins of the SPV complex when it is required and sparing if it is possible the main collector to avoid any complications; more recently, we have used ICG videoangiography in order to evaluate venous flow and possibly predict the presence of anastomotic circulation by the aid of a specifically designed ICG clipping test [46, 77]. Once the affected nerve and its vascular conflict come into view the vessel should be then be moved away and safely retained in this position to avoid any recurrence. Many methods have been advocated for keeping the vessel separated for the nerve. These include: using pieces of muscle, fascia, periosteum, or subcutaneous fat. Our experience is that these strategies usually have little effect. This kind of tissue is usually reabsorbed and/or favors adhesions, fibrosis, and arachnoiditis, as has been observed during reoperations in patients in whom the neurovascular conflict has recurred [78-82]. Most authors dealing with MVD now use synthetic implants, particularly Ivalon sponge [78, 80, 83, 84], Dacron knitted material [73, 82] or Teflon felt [24, 45, 73, 76, 85, 86, 87, 88]. In most of these reports, a small piece of synthetic material has been interposed between the nerve and the vessel. Though they are considered as being biocompatible, Ivalon, Dacron, and Teflon may, in fact, generate granulomas and the consequent risk of distorting or irritating the trigeminal root and, in so doing, produce a recurrence of pain [89-92]. The so-called sling retraction technique [25, 53, 73, 88, 93] has been proposed as an alternative by introducing the concept of a non-compressive technique for MVD. Our group fully trusts this approach and our non-compressive method along with Teflon or a small piece of Tabotamp, with or without fibrin glue positioned in such a way as to keep the vessel away from the decompressed nerve, has proven so far to be quite successful.

4.4. Surgical Closure

Once hemostasis is assured a wide consensus agrees that it is then time to close the dura mater in a watertight manner by using muscle or fascia grafts [24, 37, 45]. In order to reduce the incidence of CSF leakage a small pad of cellulose and Gelfoam can be placed over the durotomy [45] and the dural closure should be reinforced with fibrin glue and cover it with dural substitutes [37]. Towards this goal, we have had success using fibrin glue, Duraseal and, more recently, Tdural [94]. The craniectomy should then be covered so as to prevent the possibility of any unwanted negative cosmetic results. Some authors have performed a cranioplasty with methylmethacrylate [45], while others [24, 37] including sometimes ourselves, cover the craniectomy with a titanium wire mesh star shaped. However, with our small mini-invasive craniectomy, we believe that this is not always necessary.

4.5. Visualization Techniques

There are many reports on MVD using the microscopic visualisation [25, 45, 56, 86] and it is generally felt that this technique is safe and effective for this procedure [95]. We developed a mini-invasive microscopic approach by limiting the length of incision and reducing the diameter of the retrosigmoid craniectomy. In this way, we still had the possibility to obtain a high percentage of pain resolution, we a very low rate of complication, even in older patients [96]. However, sometimes the neurovascular conflict may be hidden by the nerve itself or may be not completely clear under the microscopic view. In this perspective, endoscope may be useful to enlarge the angle of view and better visualize the structures in cerebello-pontine angle, thus possibly identifying an unrecognized conflict or confirm its resolution. Our data demonstrated that 10.5% of new conflicts could be found by the aid of an endoscope-assisted technique. In addition, this visualisation helped in confirming a complete conflict resolution in other 9.7% of cases. However, attention must be paid during endoscope insertion, that we suggest to be performed under microscopic view in order to avoid inadvertent damage to anatomical structures in the CPA (see above).

The possibility of a pure endoscopic technique has been explored in previous papers and it has been suggested that good clinical results in term of post-operative pain control could be reached, with low rate of complications [66, 67]. Based on our experience, we could confirm that the technique is feasible. In the 9 patients submitted between 2010 and 2011 to a pure endoscopic vascular decompression in posterior fossa in our Institution, a conflict could be found and eliminated in all the cases. However, we found the surgery technically demanding, requiring high surgical skills and experience, and certainly a long duration than a pure microscopic or MEA approach. Therefore, we think that at this moment a pure endoscopic technique cannot be demonstrated to be superior to a mini-invasive microscopic approach with endoscopic-assisted visualisation in selected cases.

Conclusion

A mini-invasive microvascular decompression for neurovascular conflicts in posterior fossa is feasible through a limited incision and small craniectomy, with very good post-operative results and low rate of complications. Better anatomical view can be obtained through the insertion of the endoscope that is capable to identify conflicts that were missed with a pure microscopic technique, or confirm conflict resolution. We believe that this tool should be available for every case in order to increase the chance of good post-operative results.

References

[1] P. J. Jannetta, "Cranial rhizopathies", *J.R. Youmans Neurological surgery* 3rd (ed). W.B. Saunders, Philadelphia, pp4169–4182 (1990).

[2] P. J. Jannetta, "Arterial compression of the trigeminal nerve at the pons in patients with trigeminal neuralgia", *J Neurosurg,* Jan;26(1):Suppl: pp159-62 (1967).

[3] P. J. Jannetta, M. Abbasy, J.C. Maroon, F.M. Ramos, M.S. Albin, "Etiology and definitive microsurgical treatment of hemifacial spasm. Operative techniques and results in 47 patients", *J Neurosurg,* Sep;47(3): pp321-8 (1977).

[4] R. K. Laha, P.J. Jannetta, "Glossopharyngeal neuralgia", *J Neurosurg,* Sep;47(3): pp316-20 (1977).

[5] J. M. Zakrzewska, *Trigeminal Neuralgia*. London: WB Saunders, 19 (1995).
[6] H. van Loveren, J.M. Tew Jr, J.T. Keller, M.A. Nurre, "a 10-year experience in the treatment of trigeminal neuralgia. Comparison of percutaneous stereotaxic rhizotomy and posterior fossa exploration", *J Neurosurg,* Dec;57(6): pp757-64 (1982).
[7] J. M. Taha, J.M. Jr. Tew, "Comparison of surgical treatments for trigeminal neuralgia: reevaluation of radiofrequency rhizotomy", *Neurosurgery,* May;38(5): pp865-71 (1996).
[8] R. G. Auger, J.P. Whisnant, "Hemifacial spasm in Rochester and Olmsted County, Minnesota, 1960 to 1984", *Arch Neurol.* Nov;47(11): pp1233-4 (1990).
[9] B. Nilsen, K.D. Le, E. Dietrichs, " Prevalence of hemifacial spasm in Oslo, Norway", *Neurology,* Oct 26;63(8):1532-3 (2004).
[10] M. B. Møller, A.R. Møller, "Loss of auditory function in microvascular decompression for hemifacial spasm. Results in 143 consecutive cases", *J Neurosurg.,* Jul;63(1): pp17-20 (1985).
[11] J. R. Youmans, *Neurological Surgery* 2nd ed. Philadelphia:WB Saunders; (1982).
[12] G. W. Bruyn "Glossopharyngeal neuralgia", *Handbook Clin Neurol.*;4: pp459–473 (1986).
[13] J. L. Eller, A.M. Raslan, K.J. Burchiel, "Trigeminal neuralgia: definition and classification", *Neurosurg Focus,* May 15;18(5):E3. (2005).
[14] G. Broggi, F. Acerbi, M. Broggi, G. Messina, "Surgical therapy for pain", R.G. Ellenborgen, S.A. Abdulrauf, L.N. Sekhar, *Principles of Neurological Surgery* 3rd ed. Elsevier Saunders, pp 737-755 (2012).
[15] W. J. Elias, K.J. Burchiel, "Trigeminal neuralgia and other craniofacial pain syndromes: an overview", *Semin Neurosurg,* 15: pp59–69 (2004).
[16] W. J. Elias, K.J. Burchiel, "Trigeminal neuralgia and other neuropathic pain syndromes of the head and face", *Curr Pain Headache Rep.* Apr;6(2):pp115-24 (2002).
[17] Q. Chun-Cheng, Z. Qing-Shi, Z. Ji-Qing, W. Zhi-Gang, "A single-blinded pilot study assessing neurovascular contact by using high-resolution MR imaging in patients with trigeminal neuralgia". *Eur J Radiol.* Mar;69(3): pp459-63 (2009).
[18] S. J. Haines, P.J. Jannetta, D.S. Zorub, "Microvascular relations of the trigeminal nerve. An anatomical study with clinical correlation", *J Neurosurg.* Mar;52(3): pp381-6 (1980).
[19] P. J. Hamlyn, T.T. King, " Neurovascular compression in trigeminal neuralgia: a clinical and anatomical study", *J Neurosurg.,* Jun;76(6): pp948-54 (1992).
[20] W. E. Dandy, "Concerning the cause of trigeminal neuralgia". *Am J Surg,* 24: pp447–455 (1934).
[21] W. J. Gardner, M.V. Miklos, " Response of trigeminal neuralgia to decompression of sensory root; discussion of cause of trigeminal neuralgia" *J Am Med Assoc.* Aug 8;170(15): pp1773-6 (1959).
[22] W. J. Gardner, "Concerning the mechanism of trigeminal neuralgia and hemifacial spasm", *J Neurosurg.,* Nov;19: pp947-58 (1962).
[23] P. J. Jannetta, "Microsurgical approach to the trigeminal nerve for tic douloureux", *Prog Neurol Surg,* 7: pp180-200 (1976).
[24] P. J. Jannetta, M.R. McLaughlin, K.F. Casey, "Technique of microvascular decompression. Technical note", *Neurosurg Focus,* 18(5):E5 (2005).

[25] M. Sindou, J.M. Leston, E. Decullier, F. Chapuis, "Microvascular decompression for trigeminal neuralgia: the importance of a noncompressive technique--Kaplan-Meier analysis in a consecutive series of 330 patients", *Neurosurgery,* 63(4 Suppl 2): pp341-51 (2008).

[26] X. S. Yang, S.T. Li, J. Zhong, J. Zhu, Q. Du, Q.M. Zhou, W. Jiao, H.X. Guan, "Microvascular decompression on patients with trigeminal neuralgia caused by ectatic vertebrobasilar artery complex: technique notes", *Acta Neurochir* (Wien), 154(5): pp793-7 (2012).

[27] M. S. Greenberg, *Handbook of Neurosurgery*, Greenberg MS 7th ed, Thieme medical publisher, New York, pp542-544 (2010).

[28] J. M. Tew, H.S. Yeh, "Hemifacial spasm" *Neurosurgery* (japan) 2:267-278 (1983).

[29] P. Ferroli, A. Fioravanti, M. Schiariti, G. Tringali, A. Franzini, F. Calbucci, G. Broggi, "Microvascular decompression for glossopharyngeal neuralgia: a long-term retrospectic review of the Milan-Bologna experience in 31 consecutive cases", *Acta Neurochir* (Wien), 151(10): pp1245-50 (2009).

[30] R. E. Weinstein, D. Herec, J.H. Friedman,"Hypotension due to glossopharyngeal neuralgia", *Arch Neuro*, 43(1): pp90-2 (1986).

[31] L. Ferrante, M. Artico, B. Nardacci, B. Fraioli, F. Cosentino, A. Fortuna,"Glossopharyngeal neuralgia with cardiac syncope", *Neurosurgery,* 36(1): pp58-63 (1995).

[32] G. I. Barbash, G. Keren, A.D. Korczyn, N.S. Sharpless, M. Chayen, Y. Copperman, S. Laniado, "Mechanisms of syncope in glossopharyngeal neuralgia", *Electroencephalogr Clin Neurophysio,* 63(3): pp231-5 (1986).

[33] Y. Kong, A. Heyman, M.L. Entman, H.D. McIntosh, "Glossopharyngeal neuralgia associated with bradycardia, syncope, and seizures", *Circulation*, 30: pp109-13 (1964).

[34] J. N. St John, "Glossopharyngeal neuralgia associated with syncope and seizures", *Neurosurgery,* 10(3): pp380-3 (1982).

[35] W. E. Dandy, "Glossopharyngeal neuralgia (tic douloureux): Its diagnosis and treatment", *Arch Surg,* 15: pp198–214 (1927).

[36] J. C. White, W.H. Sweet, "Pain and the neurosurgeon: a 40-year Experience", Charles C Thomas, Springfield, III, pp 265–302 (1969).

[37] J. H. Sampson, P.M. Grossi, K. Asaoka, T. Fukushima, "Microvascular decompression for glossopharyngeal neuralgia: long-term effectiveness and complication avoidance", *Neurosurgery,* 54(4): pp884-90 (2004).

[38] C. Giorgi, G. Broggi, "Surgical treatment of glossopharyngeal neuralgia and pain from cancer of the nasopharynx", *J Neurosurg*, 61: pp952–955 (1984).

[39] F. Isamat, E. Ferran, J.J. Acebes, "Selective percutaneous thermocoagulation rhizotomy in essential glossopharyngeal nevralgia", *J Neurosurg*, 55: pp575–580 (1981).

[40] E. J. Bernard, B.S. Nashold, F. Caputi, J.J. "Moossy Nucleus caudalis DREZ lesions for facial pain", *Br J Neurosurg*, 1: pp81–92 (1987).

[41] B. L. Crue, J.A. Carregal, A. Felsoory, "Percutaneous stereotactic radiofrequency: Trigeminal tractotomy with neurophysiological recordings", *Confin Neurol.,* 34: pp389–397 (1972).

[42] E. Hitchcock," Stereotactic trigeminal tractotomy". *Ann Clin Res,* 2: pp131–135 (1970).

[43] J. E. Alvernia, K. Fraser, G. Lanzino, "The occipital artery: a microanatomical study", *Neurosurgery*, Feb;58(1 Suppl):ONS pp 114-22; discussion ONS pp114-22 (2006).

[44] C. V. Reis, V, Deshmukh, J.M. Zabramski, M. Crusius, P. Desmukh, R.F. Spetzler, M.C. Preul, "Anatomy of the mastoid emissary vein and venous system of the posterior neck region: neurosurgical implications", *Neurosurgery*, Nov;61(5 Suppl 2): pp193-200; discussion pp200-1 (2007).

[45] M. R. McLaughlin, P.J. Jannetta, B.L. Clyde, B.R. Subach, C.H. Comey, D.K. Resnick, "Microvascular decompression of cranial nerves: lessons learned after 4400 operations", *J Neurosurg.* Jan;90(1): pp1-8 (1999).

[46] P. Ferroli, F. Acerbi, G. Tringali, E. Albanese, M. Broggi, A. Franzini, G. Broggi, "Venous sacrifice in neurosurgery: new insights from venous indocyanine green videoangiography", *J Neurosurg.* Jul;115(1): pp18-23. Epub 2011 Apr 8 (2011).

[47] D. G. Hardy, A.L. Jr. Rhoton, "Microsurgical relationship of the superior cerebellar artery and the trigeminal nerve", *J Neurosurg* 49: pp669–678 (1978).

[48] J. R. Lister, A.L. Jr Rhoton, T. Matsushima, D.A. Peace, "Microsurgical anatomy of the posterior inferior cerebellar artery", *Neurosurgery,* 10:170–199 (1982).

[49] R. G. Martin, J.L. Grant, D.A. Peace, C. Theiss, A.L. Jr Rhoton, "Microsurgical relationships of the anterior inferior cerebellar artery and the facial vestibulocochlear nerve complex", *Neurosurgery* 6: pp483–507 (1980).

[50] T. Matsushima, A.L. Jr Rhoton, E.P. de Oliveira, D.A. Peace, "Microsurgical anatomy of the veins of the posterior fossa", *J Neurosurg* 59: pp63–105 (1983).

[51] T. Matsushima, T. Hitotsumatsu, M. Miyazono, T. Inamura, Y. Natori, T. Inoue, "Lateral suboccipital approach based on the anatomy: Its variation and pitfalls, in Oohata K (ed): Surgical Anatomy for Microneurosurgery XIII", Tokyo, *SciMed Publications*, pp 105–116 (2001).

[52] T. Matsushima, T. Inoue, S.O. Suzuki, K. Fujii, M. Fukui, A.L. Jr Rhoton, "Microsurgical anatomy of the cranial nerves and vessels in the cerebellopontine angle, in Yamaura A (ed): Surgical Anatomy for Microneurosurgery IV", Tokyo, *SciMed Publications*, pp 45–55 (1992).

[53] T. Hitotsumatsu, T. Matsushima, T. Inoue, "Microvascular decompression for treatment of trigeminal neuralgia, hemifacial spasm, and glossopharyngeal neuralgia: three surgical approach variations: technical note", *Neurosurgery*, Dec;53(6): pp 1436-41; discussion pp1442-3 (2003).

[54] M. Sindou, G. Acevedo, "Microvascular decompression of the trigeminal nerve, in Spetzler R (ed): Operative Techniques in Neurosurgery", Amsterdam, *Elsevier*, vol 4, pp 110–126 (2001).

[55] M. Sindou, T, Howeidy, G, Acevedo, "Anatomical observations during microvascular decompression for idiopathic trigeminal neuralgia (with correlations between topography of pain and site of the neurovascular conflict). Prospective study in a series of 579 patients", *Acta Neurochir* (Wien) 144: pp 1–13 (2002).

[56] M. Sindou, J. Leston, E. Decullier, F. Chapuis, "Microvascular decompression for primary trigeminal neuralgia: Long-term effectiveness and prognostic factors in a series of 362 consecutive patients with clear-cut neurovascular conflicts who underwent pure decompression", *J Neurosurg* 107: pp 1144–1153 (2007).

[57] M. Sindou, J. Leston, T. Howeidy, E. Decullier, F. Chapuis, "Microvascular decompression for primary trigeminal neuralgia (typical or atypical). Longterm effectiveness on pain; prospective study with survival analysis in a consecutive series of 362 patients", *Acta Neurochir* (Wien) 148: pp1235–1245 (2006).

[58] L.M. Cavallo, A. Messina, P. Cappabianca, F. Esposito, E. de Divitiis, P. Gardner, M. Tschabitscher, "Endoscopic endonasal surgery of the midline skull base: anatomical study and clinical considerations". *Neurosurg Focus*. Jul 15;19(1):E2. (2005)

[59] P. Charalampaki, R. Reisch, A. Ayad, J. Conrad, S. Welschehold, A. Perneczky, C. Wüster, "Endoscopic endonasal pituitary surgery: surgical and outcome analysis of 50 cases". *J. Clin Neurosci.* May;14(5): pp410-5 (2007).

[60] E. de Divitiis, P. Cappabianca, L.M. Cavallo, "Endoscopic transsphenoidal approach: adaptability of the procedure to different sellar lesions", *Neurosurgery,* Sep;51(3): pp699-705; discussion pp705-7 (2002).

[61] A. B. Kassam, P. Gardner, C. Snyderman, A. Mintz, R. Carrau, "Expanded endonasal approach: fully endoscopic, completely transnasal approach to the middle third of the clivus, petrous bone, middle cranial fossa, and infratemporal fossa", *Neurosurg Focus,* Jul 15;19(1):E6 (2005).

[62] G. M. O'Donoghue, P. O'Flynn, "Endoscopic anatomy of the cerebellopontine angle". *Am J Otol.* Mar;14(2): pp122-5 (1993).

[63] P. Cappabianca, L.M. Cavallo, F. Esposito, E. de Divitiis, M. Tschabitscher, "Endoscopic examination of the cerebellar pontine angle", *Clin Neurol Neurosurg.*, Sep;104(4): pp387-91 (2002).

[64] N. Goksu, Y. Bayazit, Y. Kemaloglu, "Endoscopy of the posterior fossa and endoscopic dissection of acoustic neuroma", *Neurosurg Focus.*, Apr 15;6(4):e15 (1999).

[65] R. Jarrahy, G. Berci, H.K. Shahinian, "Endoscope-assisted microvascular decompression of the trigeminal nerve", *Otolaryngol Head Neck Surg.* Sep;123(3): pp218-23 (2000).

[66] R. Jarrahy, J.B. Eby, S.T. Cha, H.K. Shahinian, "Fully endoscopic vascular decompression of the trigeminal nerve", *Minim Invasive Neurosurg.* Mar;45(1): pp32-5 (2002).

[67] M. S. Kabil, J.B. Eby, H.K. Shahinian, "Endoscopic vascular decompression versus microvascular decompression of the trigeminal nerve", *Minim Invasive Neurosurg.,* Aug;48(4): pp207-12 (2005).

[68] H. K. Shahinian, J.B. Eby, M. Ocon, "Fully endoscopic excision of vestibular schwannomas", *Minim Invasive Neurosurg.,* Dec;47(6): pp329-32 (2004).

[69] F. Acerbi, M. Broggi, S.M. Gaini, M. Tschabitscher, "Microsurgical endoscopic-assisted retrosigmoid intradural suprameatal approach: anatomical considerations", *J Neurosurg Sci*, 54(2): pp55-63 (2010).

[70] C. L. Rogers, A.G. Shetter, J.A. Fiedler, K.A. Smith, P.P. Han, B.L. Speiser, "Gamma knife radiosurgery for trigeminal neuralgia: the initial experience of The Barrow Neurological Institute", *Int J. Radiat Oncol Biol Phys*, 47(4): pp1013-1019 (2000).

[71] P. J. Jannetta, R.W. Rand, "Transtentorial retrogasserian rhizotomy in trigeminal neuralgia by microneurosurgical technique", *Bull Los Angeles Neurol Soc.* Jul;31(3): pp93-9 (1966).

[72] P. J. Jannetta, " Treatment of trigeminal neuralgia by suboccipital and transtentorial cranial operations", *Clin Neurosurg,* 24: pp538–549 (1977).

[73] M. Sindou, F. Amrani, P. Mertens, "[Microsurgical vascular decompression in trigeminal neuralgia. Comparison of 2 technical modalities and physiopathologic deductions. A study of 120 cases]". *Neurochirurgie,* 36(1): pp16-25; discussion pp25-6 (1990). [Article in French]

[74] D. A. Silverman, G.B. Hughes, S.E. Kinney, J.H. Lee, "Technical modifications of suboccipital craniectomy for prevention of postoperative headache", *Skull Base*, 14(2):77-84 (2004).

[75] J. D. Day, M. Tschabitscher, "Anatomic position of the asterion", *Neurosurgery*, 42: pp198–199 (1998).

[76] Z. Ma, M. Li, Y. Cao, X. Chen, "Keyhole microsurgery for trigeminal neuralgia, hemifacial spasm and glossopharyngeal neuralgia", *Eur Arch Otorhinolaryngol.* 267(3):449-454 (2010)

[77] P. Ferroli, P. Nakaji, F. Acerbi, E. Albanese, G. Broggi, "Indocyanine green (ICG) temporary clipping test to assess collateral circulation before venous sacrifice", *World Neurosurg.* Jan;75(1): pp 122-5 (2011).

[78] P. J. Jannetta, D.J. Bissonnette, "Management of the failed patient with trigeminal neuralgia", *Clin Neurosurg* 32: pp334–347 (1985).

[79] B. Klun, "Microvascular decompression and partial sensory rhizotomy in the treatment of trigeminal neuralgia: Personal experience with 220 patients", *Neurosurgery* 30:pp49–52 (1992).

[80] S. A. Rath, H.J. Klein, H.P. Richter, "Findings and long-term results of subsequent operations after failed microvascular decompression for trigeminal neuralgia", *Neurosurgery* 39: pp933–940 (1996).

[81] J. Jr Szapiro, M. Sindou, J. Szapiro, "Prognostic factors in microsurgical decompression for trigeminal neuralgia", *Neurosurgery* 17: pp920–929 (1985).

[82] T. Yamaki, K.Hashi, J. Niwa, S. Tanabe, T. Nakagawa, T. Nakamura, T. Uede, T. Tsuruno, "Results of reoperation for failed microvascular decompression", *Acta Neurochir* (Wien) 115: pp1–7 (1992).

[83] R. I. Apfelbaum, "Surgery for tic douloureux". *Clin Neurosurg* 31: pp351–368 (1983).

[84] M. E. Linskey, H.D. Jho, P.J. Jannetta, "Microvascular decompression for trigeminal neuralgia caused by vertebrobasilar compression", *J Neurosurg* 81:pp1–9 (1994).

[85] A. Ammar, C. Lagenaur, P.J. Jannetta, "Neural tissue compatibility of Teflon as an implant material for microvascular decompression", *Neurosurg Rev* 13: pp299–303 (1990).

[86] T. Fukushima, "Posterior cranial fossa microvascular decompression (Jannetta method) for trigeminal neuralgia and facial spasm [in Japanese]". No Shinkei Geka 10: pp1257–1261 (1982).

[87] P. J. Jannetta, "Microvascular decompression of the trigeminal nerve root entry zone: Theoretical considerations, operative anatomy, surgical techniques and results, in Rovit RL, Murali R, Jannetta PJ (eds)2", *Trigeminal Neuralgia*. Baltimore, Williams & Wilkins, pp 201–222 (1990).

[88] M. Sindou, P. Mertens, F. Amrani, "Does microsurgical vascular decompression for trigeminal neuralgia work through a neo-compressive mechanism? Anatomical-surgical evidence for a decompressive effect", *Acta Neurochir Suppl* (Wien) 52: pp127–129 (1991).

[89] D. Y. Cho, C.G.S. Chang, Y.C. Wang, F.H. Wang, C.C. Shen, D.Y. Yang, "Repeat operations in failed microvascular decompression for trigeminal neuralgia", *Neurosurgery* 35: pp665–670 (1994).

[90] C. A. Megerian, N.Y. Busaba, M.J. McKenna, R.G. Ojemann, "Teflon granuloma presenting as an enlarging, gadolinium enhancing, posterior fossa mass with

progressive hearing loss following microvascular decompression", *Am J Otol* 16: pp783–786 (1995).

[91] I. C. Premsagar, T. Moss, H.B. Coakham, "Teflon-induced granuloma following treatment of trigeminal neuralgia by microvascular decompression", *J Neurosurg* 87: pp454–457 (1997).

[92] A. M. Vitali, F.T. Sayer, C.R. Honey, "Recurrent trigeminal neuralgia secondary to Teflon felt", *Acta Neurochir* (Wien) 149: pp719–722 (2007).

[93] T. Matsushima, T. Yamaguchi, T.K. Inoue, K. Matsukado, M. Fukui, "Recurrent trigeminal neuralgia after microvascular decompression using an interposing technique. Teflon felt adhesion and the sling retraction technique", *Acta Neurochir* (Wien) 142(5): pp557-61 (2000).

[94] P. Ferroli, F. Acerbi, M. Broggi, M. Schiariti, E. Albanese, G. Tringali, A. Franzini, G. Broggi, "A Novel Impermeable Adhesive Membrane to Reinforce Dural Closure: A Preliminary Retrospective Study on 119 Consecutive High-Risk Patients", *World Neurosurg*. Nov 1. [Epub ahead of print] (2011)

[95] G. Broggi, M. Broggi, P. Ferroli, A. Franzini, "Surgical technique for trigeminal microvascular decompression", *Acta Neurochir (Wien)*, 154(6): pp1089-1095 (2012)

[96] P. Ferroli, F. Acerbi, M. Tomei, G. Tringali, A. Franzini, G. Broggi, "Advanced age as a contraindication to microvascular decompression for drug-resistant trigeminal neuralgia: evidence or prejudice?" *Neurol Sci*, 31(1):23-28 (2010).

In: Minimally Invasive Skull Base Surgery
Editor: Moncef Berhouma
ISBN: 978-1-62808-567-9
© 2013 Nova Science Publishers, Inc.

Chapter XIX

Radiosurgical Management of Trigeminal Neuralgia

Constantin Tuleasca[1,2,4], Marc Levivier[2,4], Romain Carron[1], Anne Donnet[3] and Jean Regis[1]

[1]Aix-Marseille University, INSERM, UMR and Timone University Hospital, Functional and Stereotactic Neurosurgery Service and Gamma Knife Unit,
Marseille, France
[2]Lausanne University Hospital (CHUV),
Neurosurgery Service and Gamma Knife Center
[3]Neurology Department, Aix-Marseille University
and Timone University Hospital, Marseille, France
[4]University of Lausanne, Faculty of Biology and Medicine, Lausanne, Switzerland

Abstract

Introduction: Trigeminal neuralgia (TN), also known as "tic douloureux", is a serious health problem with a prevalence rate of 4 to 5 per 100.000 people. Medically refractory TN may be treated by microvascular decompression (MVD), percutaneous procedures (glycerol injection, balloon microcompression or thermocoagulation), and Gamma Knife radiosurgery (GKR).

Methods: We describe the historical evolution and concept of GKR for TN, and review the clinical results in the framework of the large experience of pioneering groups. We also present and discuss technical nuances that are of importance for analyzing and understanding this approach.

Results: In 1951, Leksell performed the first radiosurgical procedure for TN by targeting the gasserian ganglion, as identified on X-rays. GKR was implemented in 1968, and gained further interest for the treatment of TN in the early 1990s, with the use of MRI and limitations of other more invasive techniques. In 1993, Rand proposed to move the target from the gasserian ganglion to the cisternal segment of the nerve. Lindquist promoted short time afterwards the idea of targeting the nerve at its emergence from the brainstem (DREZ), with a dose of 70 Gy. We then reported the first 5 cases successfully treated in Marseille with a more anterior target (plexus triangularis target) and a higher dose of 90 Gy. In 1996, the multi-centric trial published by Kondziolka et al. demonstrated the safety-efficacy of GKR with a maximum doses between 70 and 90 Gy. In the only prospective trial published in 2006, we demonstrated a high efficacy/safety rate using an average high dose of 85 Gy and the anterior cisternal target, comparing favorably with the series using a DREZ target and a lower radiation doses.

Long-term results have been reported in two series with 22% of patients pain-free at 7 years (Dhople et al., 2009) and 30% of patients pain-free at 10 years (Kondziolka et al., 2010), respectively, both using the DREZ and quite low radiation doses. The large series of Marseille (Tuleasca et al., work in progress) is the only study reporting long-term results for the anterior cisternal target. From a series of 737 consecutive patients, we analyzed 497 cases with more than one year of follow-up: 45.3% remained pain free at 10 years; the hypoesthesia rate was low, with only 21.1%, being somewhat bothersome in only 8 cases (1.6%) and very bothersome in 3 cases (0.6%).

Nuances in the radiosurgical technique make significant differences in clinical results. For example, the Pittsburgh group demonstrated that increasing the volume of the treated nerve leads to dramatic increase in the risk of toxicity with no clear benefit in efficacy. Our work comparing treatment variables between 2 centers (Marseille and Brussels) demonstrated a dramatic increase of toxicity when using channel blocking. More recently the dose received by the brainstem has been demonstrated to be a negative prognostic factor for trigeminal function injury.

Conclusion: Even if MVD remains the reference technique for the treatment of refractory TN, the current evidence for the long-term safety-efficacy of the GKR is sufficient to propose this method of treatment as a first option. Technical nuances should be taken into account, as they have major impact on the results of this approach.

1. Introduction

Trigeminal neuralgia (TN), also known as "tic douloureux", a name given by the French surgeon Nicholas Andre [1], is a serious health problem with a prevalence rate of 4 to 5 per 100.000 people [2]. Patients typically describe a brutal, intense pain in the face, with "electrical shock-like" characteristics [3].

Several etiologies are considered to be possibly involved in TN, and vascular compression at the emergence from the pons is one of the major pathogenic factors [4-6]. Usually, vessels component of the neurovascular conflict are rather small, such as arteries (superior cerebellar or anterior inferior cerebellar) or, less frequently, prominent veins (petrosal vein, draining veins in the brainstem) [7, 8]. Microvascular decompression (MVD) treats this hypothetic cause, by separating the vascular loop from the trigeminal nerve; MVD is nowadays considered as the reference technique in the drug-resistant TN, with long-term cure rates varying between 69% and 96% [9-18].

Trigeminal pain can be treated with percutaneous techniques also, aiming at the Gasserian ganglion via the foramen ovale, producing a partial lesion on the nerve by different mechanisms of action, that are thermic in thermocoagulation, mechanic in microcompression, and chemical in glycerol injection [19-23].

Nevertheless, surgery as a treatment in idiopathic TN is not a new concept, being one of the most ancient indications in neurosurgery and started far before the medical therapies in this disease (see the first gasserectomy done by Wears in 1885).

The concept of stereotactic radiosurgery was first introduced by the Swedish neurosurgeon Lars Leksell in 1951, when treating a patient suffering from essential TN using a prototype guiding-device linked to a dental X-ray machine for performing a "stereotactic gangliotomy" (24). Gamma Knife radiosurgery (GKR) with 201 Cobalt sources was finally adopted in 1968. Nevertheless, the efficacy of the GKR to treat other type of pathologies, as well as the discovery by Leksell of the glycerol injection, the absence of a high-quality neuroradiological examination and also the appearance of the effective drugs against TN, made GKR to temporarily be abandoned. During the nineties, due to the clear limitations of other surgical therapeutic options, as well as the possibility of accurate direct targeting of the cisternal portion of the nerve, regained interest of GKR for TN [25, 26].

None of these treatment strategies is perfect, but neurosurgeons shall adapt each technique to individual context, as no surgery for TN will provide definitive alievement for all patients [26]. Without any doubt, the heterogeneity of the clinical results in different GKR series (table 1) is suggesting an important impact of the preoperative and perioperative parameters, both part of the dose planning strategy [26].

2. Methods

Between July 1992 and November 2010, 737 patients presenting with intractable trigeminal neuralgia were treated with GKR and prospectively followed-up in the Timone University Hospital in Marseille, France. Our main objective was to evaluate if one single treatment with a target on the cisternal portion of the nerve ("retrogasserian target" or "far-anterior target") alleviates or cures the trigeminal pain, as compared to pre-therapeutic clinical assessment. As secondary objectives, we estimated the changes within the intensity of the pain, the number of attacks before and after the treatment, the neurological exam as well as the influence of the preoperative and perioperative parameters on the safety and efficacy. We excluded patients with TN secondary to multiple sclerosis, those with second GKR treatment and also those presenting with megadolichobasilar artery compression (a special anatomical condition), which are reputed to have more variable response to radiosurgery. Thus, safety and efficacy is reported in this present chapter in 497 patients with medically refractory classical TN, never previously treated by GKR and having a follow-up of at least one year.

All the patients were fulfilling the criteria's of the International Headache Society [3]. Evaluation of the type of the trigeminal pain was made according to the classification proposed by Eller et al. [27] into idiopathic TN1 and TN2. TN1 is described as typical sharp, shooting, electrical shock-like, with pain-free intervals between the attacks, that is present for more than 50% of the time; TN2 is described as an aching, throbbing, or burning pain, present for more than 50% of the time and that is constant in nature (constant background pain being the most significant attribute). Were included only patients fulfilling the criteria of the TN1 type. We would also like to underline the fact that when referring to atypical pain in our series, it is to describe patients with slights characteristics of atipicity, and not with those features from the previous classification, as TN2 is described. Our patients presenting "atypical" pain had still a trigeminal neuralgia and not a different facial pain syndrome, as the term may be misleading. We included no patient that had secondary TN due to a lesion or compressive element on the nerve [3].

During the 18 years of the study, various models of the Gamma Knife were successively used (models B, C, 4C and Perfexion; Elekta Instruments, AB, Sweeden). After application of the Leksell Model G stereotactic frame (Elekta Instruments AB, Sweden) under local anesthesia, all patients underwent stereotactic magnetic resonance imaging (MRI) and computer tomography (CT) for target definition. The MRI sequences used to identify the trigeminal nerve are T2-type CISS without contrast, and contrast-enhanced T1-weighted images. Bone CT always supplements the neuroradiological investigation in order to correct any distortion errors on the MRI images.

A single 4-mm isocenter is positioned in the cisternal portion of the trigeminal nerve at a median distance of 7.6 mm (range 4.5-14) anteriorly to the emergence of the nerve (retrogasserian target, figure 1). The theoretical maximal dose (100%) to be administrated with a 4-mm shot was 90 Gy. This had been adapted to the individual anatomy, taking into account the dose to the brainstem (maximum 15 Gy received by the first 10 mm^3) and also the condition of the patient (lower dose had been administrated in multiple sclerosis cases) [26, 28]. The final median value of the maximum dose delivered in this series was 85 Gy (70 to 90). All the procedures have been performed by the last author (J.R.).

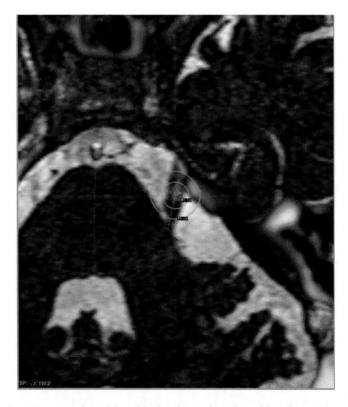

Figure 1. A single 4-mm isocenter is positioned in the cisternal portion of the trigeminal nerve at a median distance of 7.6 mm (range 4.5-14) anteriorly to the emergence of the nerve (retrogasserian target). The theoretical maximal dose (100%) to be administrated with a 4-mm shot is 90 Gy. This has to be adapted to the individual anatomy, taking into account the dose to the brainstem (maximum 15 Gy received by the first 10 mm^3) and also the condition of the patient (e.g. multiple sclerosis).

Patients continued their medication unchanged for one month after GKR and then were able to diminish the drug doses progressively depending on the treatment efficacy. Patients were seen for a neurological examination including facial sensibility, corneal reflex and jaw motility at 3 months, 6 months, and one year after the treatment and then regularly after, once a year.

The study was designed as an open, self-controlled, non-comparative prospective study. Ethics committee (CPPRB1) permission was obtained for this study. Follow-up information was obtained in two ways: direct clinical evaluation and telephone interview by one author (CT) who was not involved in the selection of the cases for treatment.

Outcome measures included initial pain freedom, the onset of the sensory disturbance and the recurrence. The results have been evaluated as follows [29]: Class I: pain free without medication; Class II: pain free with medication; Class III: pain frequency reduction superior to 90%; Class IV: Pain frequency reduction between 50 and 90%; Class V: no significant reduction in pain frequency; Class VI: pain worsening. A recurrence is defined as the change from class I to a lower outcome class. Thus the situation of a patient who had been pain free without medication (Class I) and who then restarted taking specific drugs but who remained pain free on medication (Class II) was considered as a recurrence. A minor recurrence was defined as well tolerated by the patient (lower frequency and intensity of the pain) and not requiring a new surgical therapy. A major recurrence was defined as requiring further surgical procedure [29]. The degree of hypoesthesia is reported using the BNI facial hypoesthesia scale, which uses the following grades: I, no facial numbness; II, mild facial numbness but not bothersome; III, facial numbness that is somewhat bothersome; and IV, facial numbness that is very bothersome. For patients presenting facial sensory dysfunction, we also inquired about their quality of life related to trigeminal neuralgia and whether this sensory problem was bothering them or not. We have asked whether or not they have mastication difficulties.

The latency intervals to become pain free or to develop a recurrence or a sensory disturbance, the date of medication changes, the date of further surgical procedures, were also cautiously monitored.

3. Results

As previously stated, in the present chapter we analyze 497 patients with medically refractory, classical TN, never treated previously by GKR and having a follow-up of at least one year. We present the results by evaluating the widely recognized parameters in the literature: initially pain freedom, sensory disturbance onset and probability of maintaining pain relief.

The median age was 68.3 years (range 28.1-93.2). The median follow-up period was 43.8 months (range 12-174.4). Three hundred and eighty-six, 296, 227, 191, 130, 99, 77, 61, and respectively 24 patients had at least 2, 3, 4, 5, 7, 8, 9, 10 and respectively 12 years of follow-up. Two hundred and twenty-five were men and 272 were women. The median duration of symptoms was 68.3 months (range 1- 531). Twenty- six patients (5.2%) died but were not excluded from the study as they had at least one year of follow- up. Preoperative MRI revealed the presence of a vascular conflict in 278 cases (55.9%).

One hundred and seventy three (34.8%) patients had a prior surgical procedure, of which 102 (20.5%) patients had only one previous intervention, 41 (8.2%) patients had two and 30 (6%) had three or more previous surgeries (as described in table 6). The preoperative surgery technique used was radiofrequency lesion in 99 (19.9%) patients, balloon microcompression in 64 (12.9%), MVD in 45 (9.1%) and glycerol rhizotomy in 6 (1.2%) patients. GKR was the first surgical procedure in 324 patients (65.2%).

The median maximal dose (100%) was 85 Gy (range 70-90). The median distance between the DREZ and the isocenter was of 7.6 mm (range 4.5-14).

Four hundred and fifty-four patients (91.75%) were initially pain-free in a median time of 10 days (range 1-459). The probability of remaining pain-free at 3, 5, 7 and 10 years was 71.8%, 64.9%, 59.7% and 45.3%, respectively.

The hypoesthesia actuarial rate at 5 years was 20.4% and at 7 years reached 21.1% and remained stable till 14 years with a median delay of onset of 12 months (range 1-65). A facial hypoesthesia somewhat (8 cases, 1.6%) or very (3 cases, 0.6%) bothersome was reported in a total of 11 patients (2.2%). Interestingly, the hypoesthesia rate was higher in cases with latter pain free (after 30 days), compared to those alleviated within the first 48 hours, or between 48 hours and 30 days [30].

One hundred and fifty seven (34.4%) patients initially pain free experienced a recurrence with a median delay of 24 months (range 0.62-150.06). The rate of recurrence sufficiently severe to require a new surgery was 67.8% at 10 years. Postoperative hypoesthesia was a positive predictive factor both for maintaining pain relief as well as for recurrence without further surgery.

4. Discussion

4.1. Historical Perspective

Leksell performed the first radiosurgical procedure for TN by targeting the retrogasserian ganglion ("stereotactic radiogangliotomy") in 1951, using X-ray films [24]. The series published by Lindquist in 1991, using this target included 46 patients from whom 18% maintained pain relief at 2 years.

In 1993, Rand proposed to move the target from the ganglion to the cisternal segment of the nerve (the so-called retrogasserian or far-anterior target). We then reported the first 5 cases successfully treated in Marseille with the same target and a high dose of 90 Gy. Lindquist promoted short time afterwards the idea of targeting the so-called DREZ (as used in most of the literature using this target, allowing this terminology might not be appropriate for the trigeminal nerve), at the emergence of the nerve from the brainstem, with a lower dose of 70 Gy [2].

In 1996, the multicentric trial published by Kondziolka et al. demonstrated that the maximal dose plays a major role in the efficacy, without any effect on the safety, in presuming that the target was located in the same place in all the patients [31]. This trial gathered together the first cases from several centers: Los Angeles, Marseille, Pittsburg, Rhode Island and Seattle. It advocated that the maximal radiation dose plays a major role on the efficacy, with no effect on its safety, by identifying short and long-term outcomes, the cut-off of a necessary radiation dose and the risk of complications. More precisely, patients treated with less than 70 Gy were rarely found to become pain-freedom [31]. In the light of this fact, the maximal dose of 70 Gy became a standard for all the centers using the technique used by the Pittsburgh group. The results were encouraging, with 72% of the patients pain-free, 84% with good or excellent results and with only 6% hypoesthesia. Even if this multicentric trial was a pioneering study on the topic, there was clearly heterogeneity of the targets and dose planning strategies between these different groups, which is the main bias of this study, and may have led to some confusion in interpreting the results of further studies [29, 32, 33].

In the only prospective trial published in 2006 [29], we compared favorably our results, using an average high dose of 85 Gy and the retrogasserian target, to that of the series with a lower radiation doses and the so-called DREZ target. As we were preoccupied by the potential risk of irradiating the brainstem with high doses, we decided to use a more anterior cisternal target, at the level of the plexus triangularis (figure 2 and figure 3, A), which is more in agreement with the target classically used for microsurgical rhizotomy and thermocoagulation [34, 35].

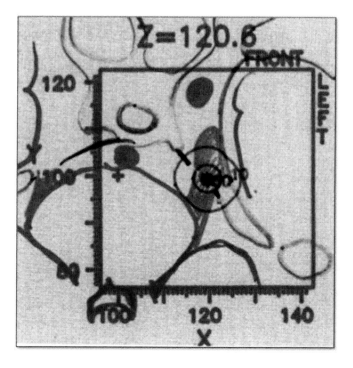

Figure 2. Plexus triangularis target as published in 1995 (100). The 4 mm shot is positioned on the nerve at the level of the trigeminal incisura of the petrous bone apex. The targeting is based both on CISS MR and CT bone window.

4.2. Far Anterior and DREZ Targets

The far anterior target refers to the placement of a unique, 4-mm shot, on the cisternal portion of the trigeminal nerve, at around 7-8 mm or more anteriorly from the brainstem emergence of the nerve (figure 3, B). At the opposite, the so-called DREZ (posterior) target refers at placing the 4-mm shot at the emergence of the trigeminal nerve at the level of the pons, with heterogeneous definitions about the isodose that is supposed to cover the brainstem depending on centers (figure 3, C). No level 1 or 2 evidence exists in favor of anterior versus posterior target. Although this remains a subject of controversy in the literature, our experience oriented us to use an optimal distance between 7.5 and 8 mm, as it showed clear benefit in terms of efficacy, toxicity and long-lasting pain relief [25, 26, 33].

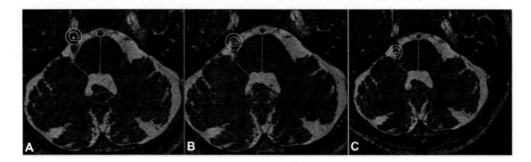

Figure 3. Targeting in radiosurgery for trigeminal neuralgia: the plexus triangularis (figure 3, A); the far anterior target refers to the placement of a unique, 4-mm shot, on the cisternal portion of the trigeminal nerve, at around 7-8 mm or more anteriorly from the brainstem emergence of the nerve (figure 3, B); at the opposite, the DREZ (posterior) target refers at placing the 4-mm shot at the emergence of the trigeminal nerve at the level of the pons, with heterogeneous definitions about the isodose that is supposed to cover the brainstem depending on centers (figure 3, C).

What accounts for the difference between the two anatomical targets? The dose to the DREZ and the dose to the brainstem, more precisely the dose to the trigeminal nerve pathway in the brainstem, are dramatically increased with the so-called DREZ target. The DREZ, also known as the Obersteiner-Redlich zone, is histologically defined as the zone where the peripheral myelination (Schwann cells) leaves places to the central type of myelination (oligodendrocytes). There are two issues that must be kept in mind. One is related to the variability of the limit between the Schwann cells and the glial environment of the nerve as it exits the brainstem, classically located 3 mm from the emergence of the nerve. The second one is related to the fact that this zone can be variable in length and may extend further more in the distal portion of the nerve [13]. As there is no *in vivo* possibility to individually evaluate the extent of the DREZ in the nerve, this makes the term "DREZ target" inappropriate. Beside this anatomical issue, there is also a dose planning-related definition given for the DREZ by Christer Lindquist: "a 4-mm collimator centered on the trigeminal root entry zone treated with a dose of 70 Gy at the center, including the nerve root and the adjacent brainstem within the 50% isodose surface (35 Gy)" (figure 4) (2). This first definition had changed in current clinical practice and the authors are now reporting that no more than the 20% isodose line should irradiate the brainstem (2), or even 30%, if we are considering the Pittsburgh group technique. So, the so-called DREZ target continues to evolve as a concept. A recent paper of Arai et al. placed it "at the midposterior portion of the trigeminal nerve, anterior to the pons" [36], which could be also be considered equivalent to the far anterior target in some cases.

There are many clinical implications related to the anatomical aspect of dose planning. One is that the DREZ target is suggested to yield more toxicity than the far anterior target. The more severe complication of radiosurgery for TN is the "dry-eye", reported by Matsuda et al. [37] and present in 3 of 41 cases of TN treated with 80 Gy at the DREZ target in their series. This complication was significantly related to the irradiated volume of the brainstem [37]. Also, 77 patients out of 104 treated have been found to have hyper intensities on the region of the DREZ on follow-up MRI [37]. We never had this complication in our series using a much higher irradiation dose but the far-anterior target, neither in the prospective trial published in 2006, nor in the results reported in the present chapter [26, 29]. In another paper, including 47 patients in a retrospective study, we recommended a minimal distance of 5 mm and an optimal distance of 8 mm for the placement of a unique shot (38). A recent study made by Park et al. [39] compared both targets (far anterior and the DREZ) and found much more bothersome complication in the DREZ target group (3 cases, 13.1% compared to 0 cases). Other studies also suggested that a target placed at the DREZ or close to it seems to be associated with higher risk of hypoesthesia (including bothersome hypoesthesia) [32, 40, 41].

4.3. Outcome After Radiosurgery: Pain Freedom, Hypoesthesia and Recurrence

There is an important discrepancy between the results that have been published in the literature (table 1). Also, very frequently, different groups report results using heterogeneous methodology. Globally, a good initial outcome on pain relief is present in most of the radiosurgical papers and varies between 35% and 100%; the risk of recurrence ranges between 0 and 46%; the risk of trigeminal nerve dysfunction varies between 0 to 57% [28, 29, 38-40, 42-61].

Pain Freedom

Different concepts are employed to define pain freedom after radiosurgery, as due to its delayed efficacy, these can be interpreted differently according to different study groups: complete or with more than 90% alleviation, with or without medication etc. In our opinion, it is mandatory to clearly and separately report all nuances of this specific outcome. Multiple sclerosis [25, 29, 54, 62-70], previous surgical treatment on the same side [29, 71, 72] or atypical TN (73) are considered to be negative predictors. The presence of a neurovascular compression has been suggested to have no influence by Sheehan [60], a predictor of failure by Shaya [74] and of a success by Brismann [75]. We have recently shown that previous MVD is also a negative predictor for pain freedom [76], but this finding needs further cautious analysis.

Patient related parameters and operative technical nuances give different results. In this sense, the maximal dose is considered related to the initial pain freedom (see table 1). Pollock et al. published a study comparing the 70 Gy maximal dose in 27 patients and the 90 Gy maximal dose in 41 patients [41]. The target was the DREZ. They concluded that a high dose of radiosurgery is associated with higher chances of pain relief [41]. A study published by the Stanford University group advocated that with a maximal dose higher than 75 Gy, patients had a much higher chance of being pain free [77]. Park et al. (39) published recently a retrospective study opposing the DREZ and the retrogasserian target. Patients treated with the retrogasserian target were more likely to become pain free (BNI classes I-IIIb) than those with the DREZ target (93.8% compared to 87%). The time of response was also shorter in the first group than in the second (mean of 4.1 weeks compared to 6.4 weeks). All these findings are in line with our personal experience [78].

Dhople et al. [47], used the DREZ target and a median prescription dose of 75 Gy (range 70-80 Gy), and reported a series of 102 patients with a median follow-up of 5.6 years. They reported an initial pain relief of 81%, new bothersome facial numbness in 6% of the cases and a recurrence rate of 56%. The maintenance of pain relief was of 22% at 7 years. Matsuda et al. [79] reported a series of 104 patients, treated with GKR, a unique 4-mm shot, doses of 80 or 90 Gy and the target at the trigeminal nerve root. They found 98% initial pain relief with 49% new trigeminal nerve dysfunction, appearing between 4 and 68 months after the treatment. Kondziolka et al. [52] published recently a series of 503 patients with 107 cases having more than 5 years of follow-up. A single 4 mm isocenter was used in 99% of the cases and two isocenters in 1%. The target was located at around 3 to 8 mm anterior from the emergence of the trigeminal nerve from the pons. The isocenter was usually situated so that the brainstem surface was irradiated at the 20% isodose line or less. The majority of the patients (92%) received doses of at least 80 Gy with a maximum of 90 Gy. They had 89% initial pain relief with 10.5% hypoesthesia and a recurrence rate of 42.9%. One patient (0.2%) developed deafferentation pain. Only 29% were still controlled with or without medications at 10 years. Verheul et al. [69] reported a series of 450 treatments in 365 patients with a median follow-up of 28 months. They were all treated with a maximal dose of 80 Gy and a unique 4 mm shot at the DREZ; 6% of the patients presented a somewhat bothersome hypoesthesia with only 0.5% very bothersome hypoesthesia. The pain relief at 5 years was of 75%. Loescher et al. [54] reported a series of 72 patients, 58 with essential and 14 with secondary TN (8 with multiple sclerosis). They were all treated with a 4 mm isocenter at the DREZ and a maximal dose of 80 Gy. The initial pain relief in essential trigeminal neuralgia was quite low, 71% at 6 months, with a rather high hypoesthesia rate of 31%.

Hypoesthesia

We advocated earlier in this chapter the role that is played by the anatomical localization of the target in the appearance of hypoesthesia and its degree of severity [37, 39]. The UCLA group reported their results in 126 patients treated between with a 4 mm shot and the target being placed at the emergence of the nerve (DREZ), with 90 Gy at the center (80). The rate of numbness was very high (58.3%), with 19.4% very bothersome, 30.5% subjective dry eyes and 30.5% decreased corneal reflex. In patients with hypoesthesia, Marshall et al. [55] found a significantly higher dose to the DREZ (57.6 compared to 47.3 Gy).

The anatomical location of the target is not the only dose planning strategy factor of major importance. The integrated dose to the nerve (i.e. the volume of the nerve that is irradiated and the total dose received by the nerve) has been reported to correlate to the risk of trigeminal nerve dysfunction.

Flickinger et al. demonstrated that increasing the volume of the treated nerve leads to a dramatic increase of the risk of toxicity (i.e. bothersome hypoesthesia), with no clear benefit on efficacy [49]. The Stanford University team reported the results of 83 patients treated with maximal doses varying between 71.4 and 86.4 Gy, with a special strategy, covering all the nerve (important integrated dose) [77]. Due to the volume of the treated nerve, the hypoesthesia rate was high (74%) with 39% severe numbness. Moreover, the authors demonstrated that a longer length of nerve treatment resulted in higher rates of numbness.

We also focused on this issue and published a paper about using the far-anterior cisternal target and a median dose of 90 Gy, as used in Brussels at that time [38]. We found increased trigeminal nerve injury associated with increased nerve length included in the 50% isodose in patients in which the 90 Gy dose prescription necessitated source plugging. So, the efficacy was similar with the series in Marseille, but the toxicity was much higher (43% of hypoesthesia in Brussels's series instead of 15% in Marseille's series). After a cautious analysis, we showed that in cases of a narrow cistern with too high dosage delivered to the brainstem, the attitude of the 2 teams was not the same. In Marseille, we lowered the maximal dose and then, if still necessary, we would have used shielding (figure 5); in Brussels, the dose of 90 Gy was kept *a priori*, and shielding of the sources was first done in order to reduce the dose to the brainstem. At that time, we maintained that this plugging strategy accounted for the different rates of toxicity (the hypoesthesia rose from 15% with no bothersome hypoesthesia to 50% including 10% bothersome hypoesthesia) [38]. A comparative, retrospective study gathering together the patients treated in Marseille and in Brussels, established that in patients with a large cistern, our methods were similar and results homogenous, with around 20% of trigeminal nerve disturbance. In patients with narrow cistern, the Brussels shielding strategy led to dramatic increase of the mean dose (42.86 Gy compared to 38.01 Gy) and also of the integrated dose (3.28 Gy instead of 2.76 Gy) to the nerve. We concluded that there is no significant benefit with increasing of the volume of the treated nerve [81]. Our attitude was of saying that radiosurgery, unlike percutaneous treatment, should offer very good pain control without any hypoesthesia for the majority of the patients. Maybe also in the case of trigeminal neuralgia radiosurgery could act by a neuromodulator mechanism, which remains a matter of speculation and debate [82].

A correlation between hypoesthesia and pain response does exist and hypoesthesia has been reported in some studies as a positive predictor for pain relief [59, 83]. We have also found similar results in our present series both for the recurrence and the recurrence without further surgery [78]. This postoperative hypoesthesia was not mandatory for long-term pain relief. Park et al. [39] found much more bothersome complication in the DREZ target compared to the far anterior one (3 cases, 13.1% compared to 0 cases). Moreover, there were huge differences regarding the frequency of bothersome facial numbness and dry-eye syndrome (13.1%, 8.7% compared to o%, 0%). Dhople et al. [47] reported new bothersome facial numbness in 6% of the cases. In the series of Matsuda et al. [79], they found 49% of new trigeminal nerve dysfunction, appearing between 4 and 68 months after the treatment. Also, recently, they made a comparison between the posterior and anterior target affirming that the first one is safer. On cautious analysis, the posterior target is closed to the far anterior one (that we also use) and the anterior is actually the historical, plexus triangularis target. Matsuda et al. [37] also reported higher toxicity with higher dose rate. Kondziolka et al. [52] reported 10.5% hypoesthesia. One patient (0.2%) developed deafferentation pain. Only 29% were still controlled with or without medications at 10 years. Verheul et al. [69] found 6% of the patients presenting a somewhat bothersome hypoesthesia, with only 0.5% a very bothersome hypoesthesia. In the series of Loescher et al. [54] the hypoesthesia rate was 31%.

In our series (Tuleasca et al., work in progress), the hypoesthesia actuarial rate at 5 years was 20.4% and at 7 years reached 21.1% and remained stable till 14 years with a median delay of onset of 12 months (range 1-65). A facial hypoesthesia somewhat (8 cases, 1.6%) or very (3 cases, 0.6%) bothersome was reported in a total of only 11 patients (2.2%).

An "evidence-based" review [84] published recently establishes the fact that the hypoesthesia rate currently reported with GKR is not significantly different than after the MVD and is also much less frequent than after percutaneous techniques.

Recurrence

As for pain freedom, recurrence is reported using heterogeneous methodologies, which makes sometimes difficult to appreciate the results on long-term basis. Papers should report pain freedom at precise intervals, such as 1, 3, 5, 7, 10 years instead of the status at the last follow-up. Depending on studies, the recurrence rate varies between 0 to 42%. Also, if two studies reported long-term follow-up concerning the DREZ target [47, 52], there is no such a paper concerning the far anterior one.

Dhople et al. [47] and Kondziolka et al. [52] used the DREZ target and quite low doses and reported a quite steady rate of failures: 22% at 7 years for the first one and 30% for the second one.

We analyzed recently our series of 737 patients operated with GKR, at a far anterior target and using high-doses of irradiation (the median dose of 85 Gy) [30, 76, 78, 85]. The probability of remaining pain relief was of 45.3% at 10 years. Furthermore, the rate of recurrence that was sufficiently severe to require a new surgery was of 67.8% at 10 years. In our series (Tuleasca et al., work in progress), patients with multiple sclerosis related trigeminal neuralgia were found to have much more recurrence than the idiopathic cases.

4.4. Radiosurgery in the Context of Other Surgical Methods

The surgical treatment of medically refractory TN consists of percutaneous ablative techniques, MVD and GKR.

Thermocoagulation, balloon micro-compression and glycerol injection have in common the fact that they act through an ablative mechanism of action and are usually performed under a brief general anesthesia. Very high rates of trigeminal nerve dysfunction, which is classically necessary for complete and prolonged efficacy, are found. They are simple techniques and easy to repeat and readily suitable for the elderly. The efficacy is immediate and they very useful in patients with resistant and devastating pain.

Microvascular decompression is performed under general anesthesia, with craniotomy, and it is established as a technique of choice. It has been first performed on the basis of the observations made by Dandy [86], with a technique developed initially by Gardner and Miklos and perfected by Janetta. Even if accepted frequently as a first line treatment, its rate of failure vary between 15 to 35% [10, 11, 13, 87-89] and its long-term cure rate in 69% to 96% [12, 13, 15, 16, 87, 90]. It has the major advantage of treating the probable cause of disease as well as offering a very low risk of subsequent trigeminal nerve dysfunction [10, 13, 16, 64, 87-89, 91-93]. Major complications include dead, brainstem infarction, intracerebral hematoma, cerebellar edema, hydrocephalus, facial palsy, ipsilateral hearing loss, severe facial numbness, cerebrospinal fluid leak, meningitis, and others [10, 13, 16, 64, 87-89, 91-95], never encountered with GKR. However, if both MVD and GKR have very similar rate of long-term recurrences, the probability of being pain free without medication looks better with MVD [15, 96]. The majority of the authors (among those who have at their disposal the technical and human resources allowing them to perform both MVD and GKR, as well as percutaneous techniques) are agreeing on the fact that the evidence for long-term safety and efficacy of GKR is nowadays sufficient for proposing it as first intention [29, 47, 97, 98]. Advantages and disadvantages of each technique must be loyally exposed to the patient according to the data provided by the reliable peer-reviewed series published during these last 20 years.

No definitive answer can be given to the question of the superiority of one technique on the other. Radiosurgery has the advantage of being the least invasive technique available and is performed under local anesthesia. Additionally, the rate of trigeminal dysfunction is remarkably low and comparable to that occurring with MVD. Postoperative GKR hypoesthesia increases the probability of pain cessation but the majority of the patients experience pain freedom without sustaining any trigeminal nerve dysfunction. It appears as late as 5 years after the treatment, as reported in recent series [52]. Our results confirm an actuarial rate at 5 years of 20.4% and at 7 years of 21.1%, which remained stable until 14 years and had a median delay of onset of 12 months (range 1-65). A facial hypoesthesia somewhat (8 cases, 1.6%) or very (3 cases, 0.6%) bothersome was reported in a total of 11 patients (2.2%). In the meta-analysis of Gronseth et al. [99], the rate of hypoesthesia reported after radiosurgery is similar to the rate of hypoesthesia after MVD, and much lower than the rate of hypoesthesia reported after percutaneous procedures. This observation is suggestive for the fact that radiosurgery may involve neuromodulator mechanisms and not only a pure destructive effect [82]. Technical nuances play a major role and their impact on current clinical practice may explain the large variability of safety and efficacy reported in the literature.

4.5. Curent Indications of Radiosurgery in Trigeminal Neuralgia

According to the current literature, GKR is frequently proposed in everyday practice, including in candidates for MVD [72]. Some authors use the very good safety-efficacy ratio of radiosurgery as an argument for promoting this treatment as a first-line alternative to conventional methods. We still recommend MVD in young patients with clear evidence of neurovascular compression on preoperative MRI as it remains, in our opinion, the gold-standard treatment for this particular group. However, if young patients decline MVD, radiosurgery shall be offered as the alternative choice, on the grounds that it results in a very low rate of numbness, rarely bothersome, for a similar rate of efficacy as compared to percutaneous methods. As in our centers all the main surgical techniques are available and currently practiced, this diminishes the bias in terms of optimal treatment choice for an individual patient at a particular moment at her/his medical history.

In our experience, a high radiation dose (median maximal dose 85 Gy), on a retrogasserian target (at a median 7.6 mm distance from the emergence form the brainstem) and with a unique 4 mm shot offers a high probability of pain-freedom of 91.75, a hypoesthesia rate of 21.1% with only 0.6% being very bothersome, a rate of recurrence of 34.4% and a probability of maintaining pain relief of 45.3% at 10 years. Even if MVD remains the reference technique for the treatment of refractory TN, the current evidence for the long-term safety-efficacy of the GKR is sufficient to propose this method of treatment as a first intervention in medically refractory TN. Technical nuances should be taken into account, as they have major impact on the results of this approach.

Acknowledgments

Supported by Timone University Hospital (Assistance Publique des Hopitaux de Marseille) and Aix-Marseille University

References

[1] Andre N. *Observations pratiques sur les maladies de l'urethre et sur plusieurs faits convulsifs*. Paris: Delaguette; 1976.
[2] Alexander E. *Stereotactic Radiosurgery*. Lunsford D, editor. New York: Mcgraw-Hill 1993.
[3] Headache Classification Subcommittee of the International Headache Society: *The International Classification of Headache Disorders*. . Cephalgia 24 (1 Suppl)2004. p. 9-160.
[4] Haines SJ, Jannetta PJ, Zorub DS. Microvascular relations of the trigeminal nerve. An anatomical study with clinical correlation. *J Neurosurg.* 1980 Mar;52(3):381-6.
[5] Hamlyn PJ, King TT. Neurovascular compression in trigeminal neuralgia: a clinical and anatomical study. *J. Neurosurg.* 1992 Jun;76(6):948-54.
[6] Jannetta PJ. Arterial compression of the trigeminal nerve at the pons in patients with trigeminal neuralgia. *J. Neurosurg.* 1967 Jan;26(1):Suppl:159-62.
[7] Lorenzoni JG, Massager N, David P, Devriendt D, Desmedt F, Brotchi J, et al. Neurovascular compression anatomy and pain outcome in patients with classic trigeminal neuralgia treated by radiosurgery. *Neurosurgery.* 2008 Feb;62(2):368-75; discussion 75-6.
[8] Love S, Coakham HB. Trigeminal neuralgia: pathology and pathogenesis. *Brain.* 2001 Dec;124(Pt 12):2347-60.
[9] Bederson JB, Wilson CB. Evaluation of microvascular decompression and partial sensory rhizotomy in 252 cases of trigeminal neuralgia. *J. Neurosurg.* 1989 Sep;71(3):359-67.
[10] Burchiel KJ. Microvascular decompression for trigeminal neuralgia. *J. Neurosurg.* 2008 Apr;108(4):687-8; discussion 8.
[11] Burchiel KJ, Clarke H, Haglund M, Loeser JD. Long-term efficacy of microvascular decompression in trigeminal neuralgia. *J. Neurosurg.* 1988 Jul;69(1):35-8.
[12] Goya T, Wakisaka S, Kinoshita K. Microvascular decompression for trigeminal neuralgia with special reference to delayed recurrence. *Neurol. Med. Chir* (Tokyo). 1990 Jul;30(7):462-7.
[13] Jannetta PJ. Microvascular decompression of the trigeminal root entry zone. *Theoretical considerations, operative anatomy, surgical technique, and results.* Baltimore: Williams& Wilkins; 1990.
[14] Linskey ME, Jho HD, Jannetta PJ. Microvascular decompression for trigeminal neuralgia caused by vertebrobasilar compression. *J. Neurosurg.* 1994 Jul;81(1):1-9.
[15] Linskey ME, Ratanatharathorn V, Penagaricano J. A prospective cohort study of microvascular decompression and Gamma Knife surgery in patients with trigeminal neuralgia. *J. Neurosurg.* 2008 Dec;109 Suppl:160-72.

[16] Sindou M, Leston J, Decullier E, Chapuis F. Microvascular decompression for primary trigeminal neuralgia: long-term effectiveness and prognostic factors in a series of 362 consecutive patients with clear-cut neurovascular conflicts who underwent pure decompression. *J. Neurosurg.* 2007 Dec;107(6):1144-53.

[17] Sindou M, Leston J, Howeidy T, Decullier E, Chapuis F. Micro-vascular decompression for primary Trigeminal Neuralgia (typical or atypical). Long-term effectiveness on pain; prospective study with survival analysis in a consecutive series of 362 patients. *Acta Neurochir* (Wien). 2006 Dec;148(12):1235-45; discussion 45.

[18] Sindou M AF, Mertens P. Decompression vasculaire microchirurgicale pour nevralgie du trijmeau:comparaison de deux modalites techniques et deductions physiopathologiques. Etude sur 120 cas1990.

[19] Brown JA, McDaniel MD, Weaver MT. Percutaneous trigeminal nerve compression for treatment of trigeminal neuralgia: results in 50 patients. *Neurosurgery*. 1993 Apr;32(4):570-3.

[20] Kanpolat Y, Savas A, Bekar A, Berk C. Percutaneous controlled radiofrequency trigeminal rhizotomy for the treatment of idiopathic trigeminal neuralgia: 25-year experience with 1,600 patients. *Neurosurgery*. 2001 Mar;48(3):524-32; discussion 32-4.

[21] Latchaw JP, Jr., Hardy RW, Jr., Forsythe SB, Cook AF. Trigeminal neuralgia treated by radiofrequency coagulation. *J. Neurosurg.* 1983 Sep;59(3):479-84.

[22] North RB, Kidd DH, Piantadosi S, Carson BS. Percutaneous retrogasserian glycerol rhizotomy. Predictors of success and failure in treatment of trigeminal neuralgia. *J. Neurosurg.* 1990 Jun;72(6):851-6.

[23] Slettebo H, Hirschberg H, Lindegaard KF. Long-term results after percutaneous retrogasserian glycerol rhizotomy in patients with trigeminal neuralgia. *Acta Neurochir* (Wien). 1993;122(3-4):231-5.

[24] Leksell L. Sterotaxic radiosurgery in trigeminal neuralgia. *Acta Chir. Scand.* 1971;137(4):311-4.

[25] Regis J, Tuleasca C. Fifteen years of Gamma Knife surgery for trigeminal neuralgia in the Journal of Neurosurgery: history of a revolution in functional neurosurgery. *J. Neurosurg.* 2011 Dec;115 Suppl:2-7.

[26] Regis J., Tuleasca C., Roussel P., A. D. Radiocirugía en la neuralgia del nervio trigemino. In: Aran, editor. *Radiocirugía Madrid: ARAN*; 2012. p. 357-73.

[27] Eller JL, Raslan AM, Burchiel KJ. Trigeminal neuralgia: definition and classification. *Neurosurg Focus*. 2005;18(5):E3.

[28] Regis J, Arkha Y, Yomo S, Murata N, Roussel P, Donnet A, et al. [Radiosurgery in trigeminal neuralgia: long-term results and influence of operative nuances]. *Neurochirurgie*. 2009 Apr;55(2):213-22.

[29] Regis J, Metellus P, Hayashi M, Roussel P, Donnet A, Bille-Turc F. Prospective controlled trial of gamma knife surgery for essential trigeminal neuralgia. *J. Neurosurg.* 2006 Jun;104(6):913-24.

[30] Tuleasca C, Carron R, Resseguier N, Donnet A, Roussel P, Gaudart J, et al. Patterns of pain-free response in 497 cases of classic trigeminal neuralgia treated with Gamma Knife surgery and followed up for least 1 year. *Journal of Neurosurgery* (Suppl). 2012;117:180-7.

[31] Kondziolka D, Lunsford LD, Flickinger JC, Young RF, Vermeulen S, Duma CM, et al. Stereotactic radiosurgery for trigeminal neuralgia: a multiinstitutional study using the gamma unit. *J. Neurosurg.* 1996 Jun;84(6):940-5.

[32] Regis J. High-dose trigeminal neuralgia radiosurgery associated with increased risk of trigeminal nerve dysfunction. *Neurosurgery.* 2002 Jun;50(6):1401-2; author reply 2-3.

[33] Regis J, Bartolomei F, Metellus P, Rey M, Genton P, Dravet C, et al. Radiosurgery for trigeminal neuralgia and epilepsy. *Neurosurg Clin. N. Am.* 1999 Apr;10(2):359-77.

[34] Frazier CH. Operation for the Radical Cure of Trigeminal Neuralgia: Analysis of Five Hundred Cases. *Ann. Surg.* 1928 Sep;88(3):534-47.

[35] Sindou M, Keravel Y, Abdennebi B, Szapiro J. Traitement Neurochirurgical de la Nevralgie Trigeminale. *Neurochirurgie*1987. p. 89-111.

[36] Arai Y, Kano H, Lunsford LD, Novotny J, Jr., Niranjan A, Flickinger JC, et al. Does the Gamma Knife dose rate affect outcomes in radiosurgery for trigeminal neuralgia? *J. Neurosurg.* 2010 Dec;113 Suppl:168-71.

[37] Matsuda S, Serizawa T, Sato M, Ono J. Gamma knife radiosurgery for trigeminal neuralgia: the dry-eye complication. *J. Neurosurg.* 2002 Dec;97(5 Suppl):525-8.

[38] Massager N, Lorenzoni J, Devriendt D, Desmedt F, Brotchi J, Levivier M. Gamma knife surgery for idiopathic trigeminal neuralgia performed using a far-anterior cisternal target and a high dose of radiation. *J. Neurosurg.* 2004 Apr;100(4):597-605.

[39] Park SH, Hwang SK, Kang DH, Park J, Hwang JH, Sung JK. The retrogasserian zone versus dorsal root entry zone: comparison of two targeting techniques of gamma knife radiosurgery for trigeminal neuralgia. *Acta Neurochir* (Wien). 2010 Jul;152(7):1165-70.

[40] Gorgulho AA, De Salles AA. Impact of radiosurgery on the surgical treatment of trigeminal neuralgia. *Surg Neurol.* 2006 Oct;66(4):350-6.

[41] Pollock BE, Phuong LK, Foote RL, Stafford SL, Gorman DA. High-dose trigeminal neuralgia radiosurgery associated with increased risk of trigeminal nerve dysfunction. *Neurosurgery.* 2001 Jul;49(1):58-62; discussion -4.

[42] Anderson WS, Wang PP, Rigamonti D. Case of microarteriovenous malformation-induced trigeminal neuralgia treated with radiosurgery. *J. Headache Pain.* 2006 Sep;7(4):217-21.

[43] Aubuchon AC, Chan MD, Lovato JF, Balamucki CJ, Ellis TL, Tatter SB, et al. Repeat Gamma Knife Radiosurgery for Trigeminal Neuralgia. *Int. J. Radiat Oncol Biol. Phys.* 2010 Oct 5.

[44] Brisman R. Gamma knife radiosurgery for primary management for trigeminal neuralgia. *J. Neurosurg.* 2000 Dec;93 Suppl 3:159-61.

[45] Brisman R. Repeat gamma knife radiosurgery for trigeminal neuralgia. *Stereotact Funct Neurosurg.* 2003;81(1-4):43-9.

[46] Cheuk AV, Chin LS, Petit JH, Herman JM, Fang HB, Regine WF. Gamma knife surgery for trigeminal neuralgia: outcome, imaging, and brainstem correlates. *Int. J. Radiat. Oncol. Biol. Phys.* 2004 Oct 1;60(2):537-41.

[47] Dhople AA, Adams JR, Maggio WW, Naqvi SA, Regine WF, Kwok Y. Long-term outcomes of Gamma Knife radiosurgery for classic trigeminal neuralgia: implications of treatment and critical review of the literature. Clinical article. *J. Neurosurg.* 2009 Aug;111(2):351-8.

[48] Dos Santos MA, Perez de Salcedo JB, Gutierrez Diaz JA, Nagore G, Calvo FA, Samblas J, et al. Outcome for Patients with Essential Trigeminal Neuralgia Treated

with Linear Accelerator Stereotactic Radiosurgery. *Stereotact Funct Neurosurg.* 2011 May 25;89(4):220-5.

[49] Flickinger JC, Pollock BE, Kondziolka D, Phuong LK, Foote RL, Stafford SL, et al. Does increased nerve length within the treatment volume improve trigeminal neuralgia radiosurgery? A prospective double-blind, randomized study. *Int. J. Radiat. Oncol. Biol. Phys.* 2001 Oct 1;51(2):449-54.

[50] Kano H, Kondziolka D, Yang HC, Zorro O, Lobato-Polo J, Flannery TJ, et al. Outcome predictors after gamma knife radiosurgery for recurrent trigeminal neuralgia. *Neurosurgery.* 2010 Dec;67(6):1637-44; discussion 44-5.

[51] Kondziolka D, Lunsford LD, Flickinger JC. Stereotactic radiosurgery for the treatment of trigeminal neuralgia. *Clin. J. Pain.* 2002 Jan-Feb;18(1):42-7.

[52] Kondziolka D, Zorro O, Lobato-Polo J, Kano H, Flannery TJ, Flickinger JC, et al. Gamma Knife stereotactic radiosurgery for idiopathic trigeminal neuralgia. *J. Neurosurg.* 2010 Apr;112(4):758-65.

[53] Little AS, Shetter AG, Shetter ME, Kakarla UK, Rogers CL. Salvage gamma knife stereotactic radiosurgery for surgically refractory trigeminal neuralgia. *Int. J. Radiat Oncol Biol Phys.* 2009 Jun 1;74(2):522-7.

[54] Loescher AR, Radatz M, Kemeny A, Rowe J. Stereotactic radiosurgery for trigeminal neuralgia: outcomes and complications. *Br. J. Neurosurg.* 2011 Aug 4.

[55] Marshall K, Chan MD, McCoy TP, Aubuchon AC, Bourland JD, McMullen KP, et al. Predictive variables for the successful treatment of trigeminal neuralgia with gamma knife radiosurgery. *Neurosurgery.* 2012 Mar;70(3):566-72; discussion 72-3.

[56] Park KJ, Kondziolka D, Berkowitz O, Kano H, Novotny J, Jr., Niranjan A, et al. Repeat Gamma Knife Radiosurgery for Trigeminal Neuralgia. *Neurosurgery.* 2011 Aug 1.

[57] Pollock BE. Radiosurgery for trigeminal neuralgia: is sensory disturbance required for pain relief? *J. Neurosurg.* 2006 Dec;105 Suppl:103-6.

[58] Pollock BE, Foote RL, Stafford SL, Link MJ, Gorman DA, Schomberg PJ. Results of repeated gamma knife radiosurgery for medically unresponsive trigeminal neuralgia. *J. Neurosurg.* 2000 Dec;93 Suppl 3:162-4.

[59] Rogers CL, Shetter AG, Fiedler JA, Smith KA, Han PP, Speiser BL. Gamma knife radiosurgery for trigeminal neuralgia: the initial experience of The Barrow Neurological Institute. *Int. J. Radiat Oncol Biol Phys.* 2000 Jul 1;47(4):1013-9.

[60] Sheehan JP, Ray DK, Monteith S, Yen CP, Lesnick J, Kersh R, et al. Gamma Knife radiosurgery for trigeminal neuralgia: the impact of magnetic resonance imaging-detected vascular impingement of the affected nerve. *J. Neurosurg.* 2010 Jul;113(1):53-8.

[61] Zheng LG, Xu DS, Kang CS, Zhang ZY, Li YH, Zhang YP, et al. Stereotactic radiosurgery for primary trigeminal neuralgia using the Leksell Gamma unit. *Stereotact Funct Neurosurg.* 2001;76(1):29-35.

[62] Brisman R. Trigeminal neuralgia and multiple sclerosis. Arch Neurol. 1987 Apr;44(4):379-81.

[63] Broggi G, Ferroli P, Franzini A, Pluderi M, La Mantia L, Milanese C. Role of microvascular decompression in trigeminal neuralgia and multiple sclerosis. *Lancet.* 1999 Nov 27;354(9193):1878-9.

[64] Broggi G, Ferroli P, Franzini A, Servello D, Dones I. Microvascular decompression for trigeminal neuralgia: comments on a series of 250 cases, including 10 patients with multiple sclerosis. *J. Neurol Neurosurg Psychiatry*. 2000 Jan;68(1):59-64.

[65] Cruccu G, Biasiotta A, Di Rezze S, Fiorelli M, Galeotti F, Innocenti P, et al. Trigeminal neuralgia and pain related to multiple sclerosis. *Pain*. 2009 Jun;143(3):186-91.

[66] Huang E, Teh BS, Zeck O, Woo SY, Lu HH, Chiu JK, et al. Gamma knife radiosurgery for treatment of trigeminal neuralgia in multiple sclerosis patients. *Stereotact Funct Neurosurg*. 2002;79(1):44-50.

[67] Meaney JF, Watt JW, Eldridge PR, Whitehouse GH, Wells JC, Miles JB. Association between trigeminal neuralgia and multiple sclerosis: role of magnetic resonance imaging. *J. Neurol Neurosurg Psychiatry*. 1995 Sep;59(3):253-9.

[68] Rogers CL, Shetter AG, Ponce FA, Fiedler JA, Smith KA, Speiser BL. Gamma knife radiosurgery for trigeminal neuralgia associated with multiple sclerosis. *J. Neurosurg*. 2002 Dec;97(5 Suppl):529-32.

[69] Verheul JB, Hanssens PE, Lie ST, Leenstra S, Piersma H, Beute GN. Gamma Knife surgery for trigeminal neuralgia: a review of 450 consecutive cases. *J. Neurosurg*. 2010 Dec;113 Suppl:160-7.

[70] Zorro O, Lobato-Polo J, Kano H, Flickinger JC, Lunsford LD, Kondziolka D. Gamma knife radiosurgery for multiple sclerosis-related trigeminal neuralgia. *Neurology*. 2009 Oct 6;73(14):1149-54.

[71] Maesawa S, Salame C, Flickinger JC, Pirris S, Kondziolka D, Lunsford LD. Clinical outcomes after stereotactic radiosurgery for idiopathic trigeminal neuralgia. *J. Neurosurg*. 2001 Jan;94(1):14-20.

[72] Pollock BE, Phuong LK, Gorman DA, Foote RL, Stafford SL. Stereotactic radiosurgery for idiopathic trigeminal neuralgia. *J. Neurosurg*. 2002 Aug;97(2):347-53.

[73] Shetter AG, Rogers CL, Ponce F, Fiedler JA, Smith K, Speiser BL. Gamma knife radiosurgery for recurrent trigeminal neuralgia. *J. Neurosurg*. 2002 Dec;97(5 Suppl):536-8.

[74] Shaya M, Jawahar A, Caldito G, Sin A, Willis BK, Nanda A. Gamma knife radiosurgery for trigeminal neuralgia: a study of predictors of success, efficacy, safety, and outcome at LSUHSC. *Surg Neurol*. 2004 Jun;61(6):529-34; discussion 34-5.

[75] Brisman R, Khandji AG, Mooij RB. Trigeminal Nerve-Blood Vessel Relationship as Revealed by High-resolution Magnetic Resonance Imaging and Its Effect on Pain Relief after Gamma Knife Radiosurgery for Trigeminal Neuralgia. *Neurosurgery*. 2002 Jun;50(6):1261-6, discussion 6-7.

[76] Tuleasca C, Carron R, Resseguier N, Donnet A, Roussel P, Gaudart J, et al. Previous microvascular decompression decrese the chances of pain free in patients treated with Gamma Knife radiosurgery for TIC In: Karger, editor. XXth Congress of the European Society for Stereotactic and Functional Neurosurgery; 26-29 September 2012; Cascai. *Stereotactic and Functional Neurosurgery*, 90 (suppl 1), page 79: Karger; 2012. p. 79.

[77] Lim M, Villavicencio AT, Burneikiene S, Chang SD, Romanelli P, McNeely L, et al. Cyber Knife radiosurgery for idiopathic trigeminal neuralgia. *Neurosurg Focus*. 2005;18(5):E9.

[78] Tuleasca C, Murata N, Donnet A, Regis J, editors. Retrogasserian Gamma Knife radiosurgery for idiopathic trigeminal neuralgia. Congress of the International

Stereotactic Radiosurgical Society (ISRS); 2010 Paris. *Journal of Radiosurgery and SBRT* (Stereotactic Body Radiation Therapy)2011.

[79] Matsuda S, Nagano O, Serizawa T, Higuchi Y, Ono J. Trigeminal nerve dysfunction after Gamma Knife surgery for trigeminal neuralgia: a detailed analysis. *J. Neurosurg.* 2010 Dec;113 Suppl:184-90.

[80] Gorgulho A, Mitchell E, De Salles A. Stereotactic radiosurgery (SRS) for trigeminal neuralgia: clinical, radiological and dosimetric characteristics of patients presenting immediate pain relief. Eight International Stereotatic Radiosurgery Society Congress 2007.

[81] Massager N, Murata N, Tamura M, Devriendt D, Levivier M, Regis J. Influence of nerve radiation dose in the incidence of trigeminal dysfunction after trigeminal neuralgia radiosurgery. *Neurosurgery.* 2007 Apr;60(4):681-7; discussion 7-8.

[82] Regis J, Carron R, Park M. Is radiosurgery a neuromodulation therapy?: A 2009 Fabrikant award lecture. *J. Neurooncol.* 2010 Jun;98(2):155-62.

[83] Chen JC, Greathouse HE, Girvigian MR, Miller MJ, Liu A, Rahimian J. Prognostic factors for radiosurgery treatment of trigeminal neuralgia. *Neurosurgery.* 2008 May;62(5 Suppl):A53-60; discussion A-1.

[84] Cruccu G, Gronseth G, Alksne J, Argoff C, Brainin M, Burchiel K, et al. AAN-EFNS guidelines on trigeminal neuralgia management. *Eur. J. Neurol.* 2008 Oct;15(10):1013-28.

[85] Regis J, Tuleasca C, Resseguier N, Carron R, Donnet A, Roussel P. Long-term outcome of radiosurgery for essential trigeminal neuralgia: a prospective series of 130 consecutive patients with more than 7 years of follow-up. In: Karger, editor. XXth Congress of the European Society for Stereotactic and Functional Neurosurgery; 26-29 September 2012; Cascai. *Stereotactic and Functional Neurosurgery*: Karger; 2012. p. 82.

[86] Dandy WE. The Treatment of Trigeminal Neuralgia by the Cerebellar Route. *Ann. Surg.* 1932 Oct;96(4):787-95.

[87] Kolluri S, Heros RC. Microvascular decompression for trigeminal neuralgia. A five-year follow-up study. *Surg. Neurol.* 1984 Sep;22(3):235-40.

[88] Meaney JM, Miles JB. Microvascular decompression for trigeminal neuralgia. *J. Neurosurg.* 1995 Jul;83(1):183-4.

[89] Sekula RF, Marchan EM, Fletcher LH, Casey KF, Jannetta PJ. Microvascular decompression for trigeminal neuralgia in elderly patients. *J. Neurosurg.* 2008 Apr;108(4):689-91.

[90] Burchiel KJ, Steege TD, Howe JF, Loeser JD. Comparison of percutaneous radiofrequency gangliolysis and microvascular decompression for the surgical management of tic douloureux. *Neurosurgery.* 1981 Aug;9(2):111-9.

[91] Heros RC. Results of microvascular decompression for trigeminal neuralgia. *J. Neurosurg.* 2009 Apr;110(4):617-8; author reply 8-9.

[92] Klun B. Microvascular decompression and partial sensory rhizotomy in the treatment of trigeminal neuralgia: personal experience with 220 patients. *Neurosurgery.* 1992 Jan;30(1):49-52.

[93] Miller JP, Magill ST, Acar F, Burchiel KJ. Predictors of long-term success after microvascular decompression for trigeminal neuralgia. *J. Neurosurg.* 2009 Apr;110(4):620-6.

[94] Barker FG, 2nd, Jannetta PJ, Bissonette DJ, Jho HD. Trigeminal numbness and tic relief after microvascular decompression for typical trigeminal neuralgia. *Neurosurgery*. 1997 Jan;40(1):39-45.

[95] Barker FG, 2nd, Jannetta PJ, Bissonette DJ, Larkins MV, Jho HD. The long-term outcome of microvascular decompression for trigeminal neuralgia. *N. Engl. J. Med.* 1996 Apr 25;334(17):1077-83.

[96] Pollock BE. Comparison of posterior fossa exploration and stereotactic radiosurgery in patients with previously nonsurgically treated idiopathic trigeminal neuralgia. *Neurosurg Focus*. 2005;18(5):E6.

[97] Kondziolka D, Lunsford L, Flickinger J. Gamma knife radiosurgery as the first surgery for trigeminal neuralgia. *Stereotact Funct Neurosurg*. 1998;70 Suppl 1:187-91.

[98] Kondziolka D, Zorro O, Lobato-Polo J, Kano H, Flannery TJ, Flickinger JC, et al. Gamma Knife stereotactic radiosurgery for idiopathic trigeminal neuralgia. *J. Neurosurg.* 2009 Sep 11.

[99] Gronseth G, Cruccu G, Alksne J, Argoff C, Brainin M, Burchiel K, et al. Practice parameter: the diagnostic evaluation and treatment of trigeminal neuralgia (an evidence-based review): report of the Quality Standards Subcommittee of the American Academy of Neurology and the European Federation of Neurological Societies. *Neurology*. 2008 Oct 7;71(15):1183-90.

[100] Regis J, Manera L, Dufour H, Porcheron D, Sedan R, Peragut JC. Effect of the Gamma Knife on trigeminal neuralgia. *Stereotact Funct Neurosurg*. 1995;64 Suppl 1:182-92.

[101] Alexander E, Lindquist C. Special indications: Radiosurgery for functional neurosurgery and epilepsy. In: Alexander E, III, Loeffler J, Lunsford L, editors. *Stereotactic Radiosurgery*. New York: Mc Graw-Hill; 1993. p. 221-5.

Chapter XX

Minimally Invasive Approaches to Cranial Nerves for Microvascular Decompression: Hearing Preservation

Emile Simon* and Marc Sindou
Department of Neurosurgery, University of Lyon,
Hôpital Neurologique P. Wertheimer, Groupement Hospital Est, Lyon, France
Department of Anatomy, University of Lyon, Lyon, France

Abstract

Microvascular decompression (MVD) is used widely in our Department, for trigeminal neuralgia (more than 1850 patients operated since 1972), hemifacial spasm (230 patients) and vago-glossopharyngeal neuralgia (29 patients).

Based on this experience, and with the help of electrophysiological monitoring, we do think that the best way to avoid stretching of the cochlear and vestibular nerves is to pass through a real infratentorial–supracerebellar route for access to the trigeminal nerve, and through an infracerebellar-infraflocular route for access to the facial, vagus and glossopharyngeal nerves.

In this chapter, we focus on technical and practical key-points on «how to avoid complications», especially hearing impairments.

Introduction

Hyperactive dysfunctional syndromes affecting cranial nerves are well known entities, described centuries ago, and their pathophysiology has been subject to much debate ever since. However, the most popular theory is that these diseases are likely to be caused — at least as the main factor — by vascular compression at the root [1-4]. Indeed, over the past

---
* Corresponding author: emile.simon@chu-lyon.fr.

decades a number of publications on imaging data, as well as on intra-operative anatomical observations, have brought evidence that a majority of so-called "primary" trigeminal neuralgias (TN), hemi-facial spasms (HFS) and vago-glossopharyngeal neuralgias are related to a vascular compression of the corresponding root. An even larger number of reports on the valuable long-term results of the microvascular decompression (MVD) procedure on the trigeminal nerves, facial nerves and glossopharyngeal nerves and vagus nerves confirmed the validity of this hypothesis for the corresponding diseases.

As minimally invasive microsurgical technics are nowadays "popular", changes have been brought to minimize risks and side-effects. Main changes consisted of the adoption of the lateral position, the reduction in size (2 cm) of the craniotomy in order to reduce the leakage of CSF and retraction of the cerebellum, intracranial routes designed to avoid damage to the nerves specially the fragile cochleo-vestibular nerve, and decompression without the prothesis touching the root to avoid neocompression [5].

Like the many authors who have worked with the intraoperative monitoring of Brainstem Auditory Evoked Potentials (BAEP) to decrease the risk of hearing loss during posterior fossa surgery [6-17], we have used this monitoring for designing the less aggressive trajectories to approach the various cranial nerves, whilst avoiding stretching the cochlear nerve, for performing MVD.

Intraoperative Monitoring of BAEP

Intraoperative monitoring of BAEP was useful to understand how to avoid hearing loss. The recorded waves (or peaks) corresponding to the auditory tract from cochlea to brainstem are peak I to V. Although the generators of these specific peaks have no unequivocal correlation to a single anatomical structure, for reasons of clinical simplification it can be estimated that peaks I and II are generated in the distal and proximal parts of the cochlear nerve, respectively, and peaks III to V along the brainstem auditory pathways from the cochlear nucleus up to the inferior colliculus.

In previous works from our team [18, 19], dangerous moments of the surgery were identified during MVD for TN and HFS using B.A.E.P monitoring. Changes in electrophysiological phenomena were interpreted during the surgery itself by the Neuro-physiologist, who followed the progress of the work on the monitor of the surgical microscope. Neurophysiologist advised the surgeon whenever latencies and / or amplitudes varied significantly compared to baseline established at the beginning of the intervention after induction of anesthesia and before skin incision.

Correlations between Electrophysiological Modification and Surgical Steps

- Peak I warnings. Preservation of a normal amplitude of peak I is the guaranty of integrity of the cochlea; ischemia resulting from vasospasm in the labyrinthine artery or the AICA parent artery would result in a progressive or sudden decrease in amplitude of peak I (Figure 1). Conversely, an exaggerated amplitude of peak I would signify impairment in the inhibitory descending pathways to cochlea through

the vestibular nerve (olivocochlear efferent system), that is injury or stretching or neocompression by the prothesis of the vestibular nerve.
- Peak V warnings. Damage to the cochlear nerve results in a delay in latency and reduction in amplitude of Wave V. When this delay becomes significant, the neurophysiologist must inform the surgeon so that the procedure be stopped, cause identified and dangerous maneuvers corrected (Figure 2). Main critical situations are: stretching of the VIIIth nerve when retracting cerebellum, manipulation of the labyrinthine artery or the antero-inferior cerebellar artery, direct trauma by instruments or a nearby coagulation, and at end of surgery neocompression of the cochlear nerve by the prosthesis [19].

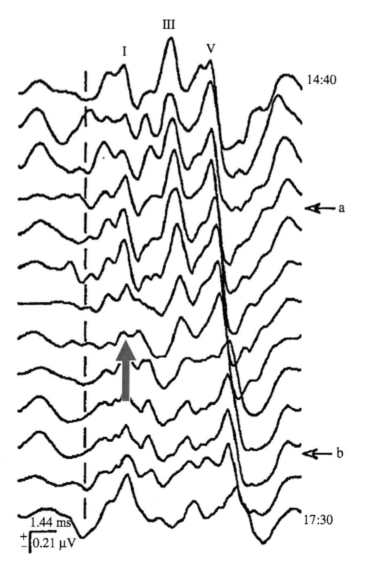

Figure 1. Intraoperative BAEPs monitoring during MVD for HFS. Start (a) and end (b) of surgery. Note that in the mid-course there was a significant but transient decrease in wave I amplitude. This decrease was considered a warning-signal of cochlear dysfunction.

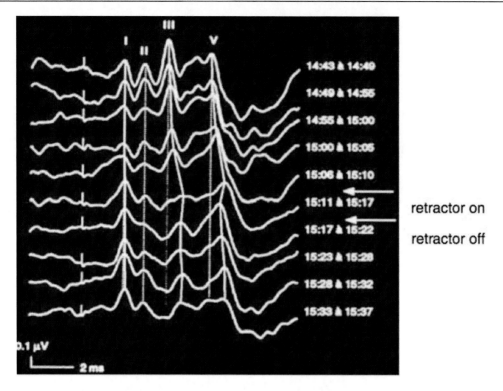

Figure 2. Intraoperative Brainstem Auditory Evoked Potentials (BAEPs) monitoring during Micro-Vascular Decompression (MVD). Note decrease in amplitude and increase in latency of peak V (and also peak III) during manipulations (retractor on), and the return to normal signals when retraction stops (retractor off).

Two levels of warning signals of practical importance to provide information on hearing function during surgery were determined [20]:

a) Delay in latency of Peak V of 0.6 ms. In our series, below this value, no patient had post-operative hearing loss. So below 0.6ms, the surgeon can consider that his or her surgery is within safety limits. Above this value, he or she has to be watchful. This signal may be called the "warning of a real risk".

b) Delay in latency of Peak V of 1 ms. Above this value, in our series, patients had at least some transient hearing loss. Therefore as soon as the delay reaches 1ms, the surgeon has to be advised and to stop the procedure, withdraw the retractor, identify the cause(s) of the VIIIth nerve damage, and urgently make appropriate corrections before resuming surgery. We consider a 1 ms delay as the ultimate "critical warning before irreversibility".

Stability of amplitude and latency of peak V is the warranty of an absence of operative injury, stretching and/or neo-compression of the cochlear nerve.

Importantly, the warning signals have to be interpreted by the neurosurgeon and the neurophysiologist working together in the operating theatre during the time when hearing is at risk. Therefore the neurophysiologist needs to follow the microsurgical steps on the microscope video-screen so that electro-physiological events may be discussed extemporaneously between the two colleagues according to the anatomic/surgical events.

Modification of the Surgical Technique after the (Previously Described) Electrophysiological Observations

The technique was progressively modified due to the experience harvested from the monitoring of BAEP. Initially the patients were operated on in the sitting position, through a retro-mastoid craniotomy 4 cm in diameter.

Access to the trigeminal (for TN) and facial (for HFS) was made via a latero-cerebellar route. Once the conflicting artery identified, detached and transposed, a piece of Teflon or Dacron was interposed between the nerve and the artery.

Changes to the approach [5] consisted of the adoption of the lateral position and the reduction in size (2 cm) of the craniotomy, in order to reduce the leakage of CSF and retraction of the cerebellum. The approach lateral to the cerebellum was replaced by a supero-lateral approach (with conservation of afferent veins to the superior petrosal sinus and respect of the arachnoid membrane covering the facial and the cochleo-vestibular nerves) for the Vth, and an infero-lateral approach for the VIIth.

Study of BAEPs showed that the retraction of the cerebellum resulted in harmful stretching of the auditory nerve. A limited approach through a keyhole craniotomy not only prevents from important CSF depletion, but also from avulsion of the bridging petrosal veins. Preservation of the superior petrosal veins offers in addition a serious guarantee of limiting traction on the eighth nerve and the labyrinthine artery.

Surgical Methods

1. Preoperative Imaging

TN can be diagnosed as primary only after all specific causes (exclusive of vascular compression) have been eliminated by appropriate investigations. Knowing individual pathologic anatomy before surgery is of prime importance to adapt and design the most appropriate approach to every patient. This is possible thanks to modern high-resolution magnetic resonance imaging (MRI) sequences [21, 22]. According to our experience, to reach high level of sensibility and specificity, imaging exploration should combine the following three sequences:

1. 3D-T2 high-resolution (i.e., millimetric) sequence allows obtaining fine and well-balanced images, with the cerebrospinal fluid in intense hypersignal and the other structures of the cerebello-pontine angle in hyposignal, leading to a cisternography-like exploration. The limit is the absence of differentiation of signals between vessels and nerves; therefore sequences showing vessels in hypersignal should be added.
2. 3D-ToF (time of flight) magnetic resonance angiography, if performed with a presaturation filter, visualizes (only) the high-flow vessels, namely, the arteries.
3. 3D-T1 weighted sequence with gadolinium is able to depict all vascular structures, both arteries and veins; these are enhanced by a contrast medium.

Combination of the latter two sequences is indispensable to differentiate veins from arteries (Figure 3 & 4). In a recent prospective study comparing the imaging data and the findings at surgery with a blinded protocol, in 100 consecutive cases of trigeminal neuralgia, we found that the sensitivity of MRI was 96.7% for detecting neurovascular conflicts and specificity 100%. Further, the MRI value for predicting the type of the responsible vessel(s) was 88%, the location along and the site around the root 85.7% and 84.6%, respectively, and most important, the degree of compression was 84.6%; the p value for all these features was less than 0.01 [21, 23].

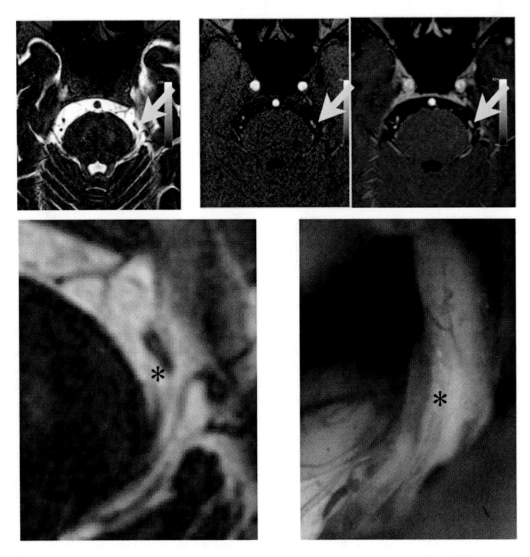

Figure 3. Magnetic Resonance Imaging (3 Tesla, axial, high resolution) of a neurovascular conflict between superior cerebellar artery and the trigeminal nerve on the left. A to C: MRI T2 sequence in high resolution (A), 3D *time-of-flight* (TOF) angiography (B), 3D T1 sequence with gadolinium (C). Conflict (arrow) with a loop of the superior cerebellar artery. D: fusion between T2 and 3D TOF angiography sequence (conflict = asterisk). E: operative findings of the presence of conflict (asterisk). The picture is reversed for didactic reason.

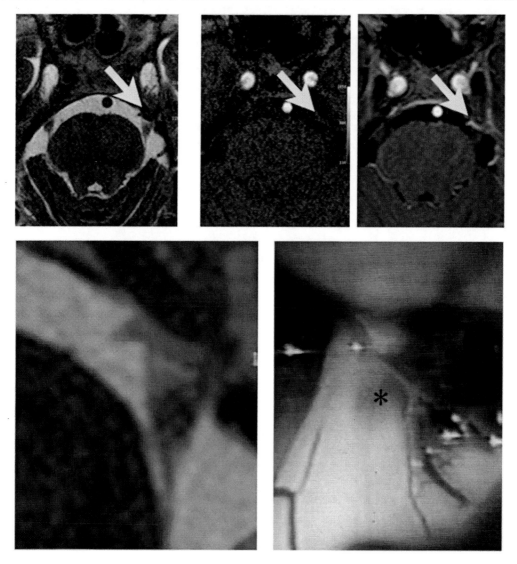

Figure 4. Magnetic Resonance Imaging (3 Tesla, axial, high resolution) of a neurovascular conflict between transverse pontine vein and the trigeminal nerve on the left. A to C: MRI T2 sequence in high resolution (A), 3D time-of-flight (TOF) angiography (B), 3D T1 sequence with gadolinium (C). Note that the conflicting vessel is not visible in 3D TOF angiography (B), but appears on the 3D T1 sequence with gadolinium (C). Conflict (arrow) with lower transverse pontine vein. D: fusion between T2 and T1 sequence with gadolinium 3D. E: operative findings of the presence of venous conflict with a gray area corresponding to an focal demyelination (asterisk).

2. Approach of the Trigeminal Nerve for TN: (Figure 5) [24, 25]

- installation

MVD is performed under general endotracheal anesthesia; the patient is placed in a lateral «"parkbench»" position, with the head - in a three-pin holder - moderately elevated, slightly flexed and rotated 15° toward the contralateral side. Care must be taken to ensure that the shoulder does not make obstacle to the operative microscope handling.

- craniotomy

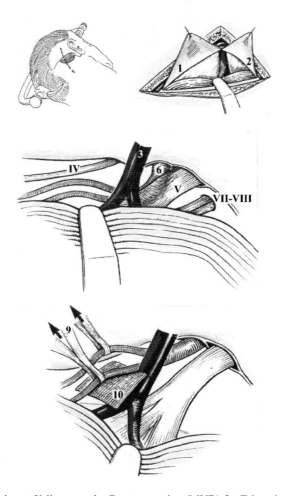

Figure 5. Surgical Technique of Microvascular Decompression (MVD) for Trigeminal Neuralgia with a neurovascular conflict between superior cerebellar artery and the trigeminal nerve on the right. Keyhole retromastoid approach on right side. Upper left view: Position of head, landmarks of skin incision and craniectomy. Upper right view: The dura is opened by turning two small flaps, one being retracted along the transverse sinus [1] and the other along the sigmoid sinus [2]. The self-retaining retractor is placed on the cerebellar hemisphere surface to achieve a supracerebellar-infratentorial approach to the trigeminal nerve. The superior petrosal veins, which drain towards the superior petrosal sinus are respected. Middle view: The arachnoid of the dorso-lateral middle part of the peripeduncular cistern is incised along and below the trochlear nerve [IV]. The SCA [6] and its two division branches are compressing (grade II) the trigeminal root [V] from above. Lower view: The arteries are detached from the nerve and transposed upward just beneath the tentorium. Teflon tapes [9], 3 cm in length and 2 mm in width are passed around the arteries to pull (arrows) and detach then without touching the nerve. Transposition is maintained by wedging the tapes in between the superior aspect of the cerebellum and the inferior surface of the tentorium. Then, to avoid any secondary mal-repositioning of the arteries, a square piece (10 x 10 mm) of semi-rigid prosthesis (made of Dacron or Teflon) [10] is placed on the superior petrosal vein [3], as a console, to hold the arteries.

The "«key-hole"» retromastoid craniotomy is performed just posterior to the base of the mastoid process, so as to expose the dura in the angle formed by the transverse and the sigmoid sinuses. The burr hole must not be made too anteriorly toward the mastoid process as this would endanger the sigmoid sinus, the external wall of which is often reduced to a thin endothelial layer adhesive to the bone. The craniectomy should first expose the transverse sinus and only secondarily the sigmoid sinus, the external wall of the transverse portion of the sinus being stronger and less adhesive to the bone than the sigmoid one. The dura is opened by turning two small flaps, one being retracted along the transverse sinus, and the other along the sigmoid sinus.

- intradural approach

We do think that the best way to avoid stretching of the cochlear and vestibular nerves is to pass through a real infratentorial–supracerebellar route for TN [26]. The arachnoid covering the dorsolateral aspect of the cerebral peduncle is opened from the superior petrosal venous trunk in a medial direction, parallel to the tentorial incisura and 1-2 mm below the trochlear nerve. Care must be taken not to injure the tiny and fragile trochlear nerve, especially with excessive suction. Large opening of the arachnoid around the superior petrosal veins is done. Preservation of the superior petrosal vein and its affluents—the mesencephalic, cerebellar, and pontine afferences—is important, not only to decrease risks of cerebellar swelling and venous infarction but also to protect from excessive retraction and consequently stretching of the VIIth–VIIIth complex. The less the arachnoid in front of those nerves is opened and nerves exposed, the less hearing complications will occur.

- microvascular decompression of the trigeminal nerve

Decompressive surgery should be atraumatic, complete, and non-neocompressive. Because in about one third of the patients there are several compressive vessels and because arachnoid adhesions play a role in abnormal root anatomy [27], the entire trigeminal root has to be dissected free from the trigeminal root entry zone at brainstem to exit from the porus of the Meckel cave, so that it has a straight path with no angulation, kinking, or twisting.

When the superior cerebellar artery is the offending vessel, it is transposed upward and maintained under the tentorium by means of sling(s) 3 to 4 cm in length and 2 to 3 mm in width, made of shredded fibers of Teflon felt; the slings are passed around the artery to exert a pulling effect. To prevent mal-repositioning of the artery, a square piece of semirigid prosthesis made of a piece of Teflon plate is added and placed on the upper petrosal vein to hold the artery.

When the anteroinferior cerebellar artery (AICA) is the offending vessel, a small piece of semirigid Teflon plate and/or a small balled cushion of Teflon fibers interposed between the root entry zone and the loop is used. For the AICA, simple transposition of the loop away from the nerve is often difficult, as AICA frequently has short perforators to the brainstem and gives origin to the labyrinthine artery.

When a vein is found grooving the root, especially the transverse pontine vein cross-compressing the inferior aspect of the root at its exit from the Meckel cave, it is coagulated using fine bipolar forceps and then divided.

In recent works, we studied the practical consequences of having the implants keeping the conflicting vessel(s) apart, in contact or not with the decompressed root. The results strongly support the procedure without the implant touching the root. As a matter of fact, the long-term outcome on pain (Kaplan–Meier curve at 15 years) was significantly better (82% total cure at 15 years, vs. 67%) when the implant was not in contact ($p<0.05$) [28, 29]. These results favor the "pure decompressive effect" of microvascular decompression rather than a "conduction-block mechanism" [30].

- end of procedure

After decompression has been completed, venous hemostasis is checked by asking the anaesthesiologist to perform sustained digital compression at the neck of both jugular veins. Arteriolar hemostasis is checked by irrigating the vessels with droplets of papaverine in saline to suppress all possible spasms due to surgical manipulations.

Then the dura is closed with single stitches. To make closure as watertight as possible, a piece of subcutaneous fascia or of fascia lata is affixed externally to the dura. Additional fatty tissue is packed onto the mastoid cells, if opened.

3. Approach of the Facial Nerve for HFS: (Figure 6) [31]

The method consists of freeing the nerve from the compressive vessel(s), namely the postero-inferior cerebellar artery (PICA), the anterior-inferior cerebellar artery (AICA) and/or the vertebro-basilar artery (VB). In our series of 147 patients PICA, AICA and/or VB were respectively found as the responsible vessel(s) in 61%, 56% and 27% of the cases [31]. Importantly, there can be several compressing vessels in the same patient. Such an eventuality occurred in as many as 40% of the patients. Ignoring this could be the source of surgical failure.

- installation

As for trigeminal neuralgia, MVD is performed under general endotracheal anesthesia, the patient is placed in lateral «parkbench» position, with the head - in a three-pin holder - moderately elevated, flexed laterally toward the contralateral side, and with no head-rotation. Care must be taken to slightly retract the shoulder posteriorly and caudally so that the surgeon can work with one arm on each side of the shoulder of the patient.

- craniotomy

Surgical approach is a "keyhole" retromastoïd craniotomy, in the order of 2 cm in diameter, just posterior to the tip of the mastoid process (Figure 8) so that the facial nerve can be reached from below, passing infero-laterally to the cerebellar flocculus. The burr hole must not be made too anteriorly toward the mastoid process as this would endanger the sigmoid sinus, the external wall of which is often reduced to a thin endothelial layer adhesive to the bone.

- intradural approach

Reaching the facial nerve from below along an infra-floccular route is important for two reasons:

- neurovascular compressions are usually located ventro-caudally at the Root Exit Zone and/or at brainstem.
- a lateral-to-medial retraction of the cerebellar hemisphere would exert stretching of the VIIIth nerve and lead to hearing loss, as demonstrated by intraoperative BEAP recordings [10, 13, 19, 20].

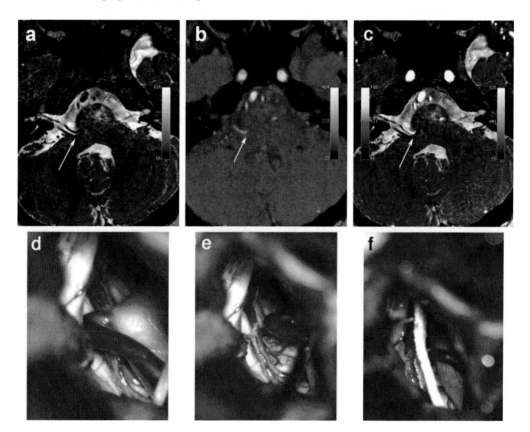

Figure 6. Magnetic Resonance Imaging (3 Tesla, axial, high resolution) of a neurovascular conflict between PICA and the facial nerve on the right. A to C: MRI T2 sequence in high resolution (A), 3D time-of-flight (TOF) angiography (B), fusion between T2 and TOF sequence (C). Identification of neurovascular conflict in a patient with right hemifacial spasm. Responsible artery: loop of PICA at REZ (arrow). D to F: Operative views. An elongated loop of the posterior-inferior cerebellar artery (PICA) is compressing the facial nerve at its exist zone at brainstem (D). After pushing away the offending artery (E), a small plate of Teflon is interposed between the facial nerve and the compressive vessel (F).

After identification of the Xth and IXth nerves before they enter the pars nervosa of the Jugular Foramen, arachnoid is opened, first dorsally to the Xth and IXth nerves, then in front of the choroid plexus emerging from the Foramen of Luschka, and finally at the level of the flocculus where it covers the vestibulo-cochlear nerve.

- microvascular decompression of the facial nerve

Offending loops are usually found just ventral to the choroid plexus, which must be slightly retracted. Mobilization of the compressive artery(ies) must be gentle for vessel(s) and cleavage from the VIIth and VIIIth nerves atraumatic. Care has to be taken to respect the tiny perforating collaterals of VB, PICA or AICA, and manipulation of the labyrinthine artery should be avoided so as not to generate vasospasm and consequently cochlear ischemia.

A good knowledge on the various neurovascular compressive patterns should help to facilitate surgery [32]. The conflicting loop(s) are maintained apart by interposing a small rectangular plate of Teflon (approximately 15 mm × 4 mm in size) between REZ/brainstem on a side and the offending vessel(s) on the other. The piece of Teflon must not exert any "neo-compressive" effect on the facial nerve, and the vestibulo-cochlear nerve as well, and not kink artery(ies). Any excessive vascular manipulation might provoke mechanical vasospasm, and generate ischemia in corresponding territories; regular irrigation of vessels with a few droplets of papaverine in solution is wise precaution (not too much of Papaverine should be used because of its very acid pH). Then dura is closed as tight as possible, completed by affixing a piece of fascia lata on dura and fat tissue on mastoid cells.

4. Approach of the Vagus-Glossopharyngeal Nerves for VGPN: (Figure 7)

According to data from the literature and our own experience, in almost all cases investigated microsurgically, a neurovascular conflict was present, most often with the postero-inferior cerebellar artery (PICA), a megadolicho-vertebral artery sometimes (alone or in combination), a vein (more rarely) at the Roots Entry Zone in the medulla oblongata.

Decompression allowed a long-term good outcome in 80% of cases [33-39]. In our series of MVD for glossopharyngeal neuralgia (23 cases), the hearing has preserved, with the security provided by intraoperative monitoring of BAEP.

- installation

The installation of patient for surgery is the same as for hemifacial spasm.

- craniotomy

The craniotomy is made behind the tip of the mastoid (Figure 8), at the level of the retrocondylar fossa. The dura opening is a small dural flap reflected on the sigmoid sinus.

- intradural approach

The approach to the Roots Entry Zone (REZ) of IXth and X[th] nerves in the lateral sulcus of the medulla oblongata is a path inferolateral to the cerebellar hemisphere and specifically the tonsil. The arachnoid is open from bottom to top, freeing XIth nerve, rootlets of Xth and IXth nerves, and finally the choroid plexus emerging from of the lateral foramen of the fourth ventricle (foramen of Luschka), to the flocculus covering VIIIth nerve. The neurovascular compressions are usually ventral to the REZ of IXth and Xth nerves, contiguous to the lateral side of the medulla oblongata.

- microvascular decompression of the vagus-glossopharyngeal nerve

The conflicting vessel is most often loops of the postero-inferior cerebellar artery (PICA), alone or in combination with vertebrobasilar megadolicho artery. It is extremely important to ensure that all compressive factors have been detected, identified and removed, because the compressive vessels are frequently multiple, often masked by a first conflicting vascular loop. Decompression is performed according to the general principles presented in detail for the trigeminal and hemifacial spasm. Care must be taken to keep the inserted material from compressing the rootlets of IXth and Xth nerves, and to prevent the material to exert traction on VIIIth nerve or even to the floculus that is adherent to it.

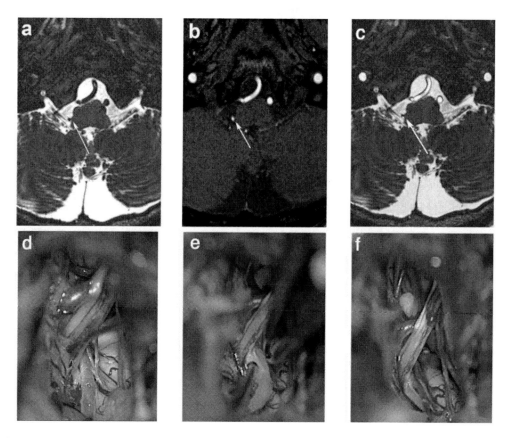

Figure 7. MRI study in a patient with right vago-glossopharyngeal neuralgia. A to C: MRI T2 sequence in high resolution (A), 3D time-of-flight (TOF) angiography (B), fusion between T2 and TOF sequence (C). MRI shows vascular compression of the ninth nerve by the posterior inferior cerebellar artery (PICA) ventrally. D to F: Operative views. Surgical microvascular decompression technique for right vago-glossopharyngeal neuralgia due to compression from the posterior inferior cerebellar artery (PICA). PICA loops compressing ventrally IXth and Xth REZ. After decompression (F), transposition of PICA maintained apart with a prosthesis.

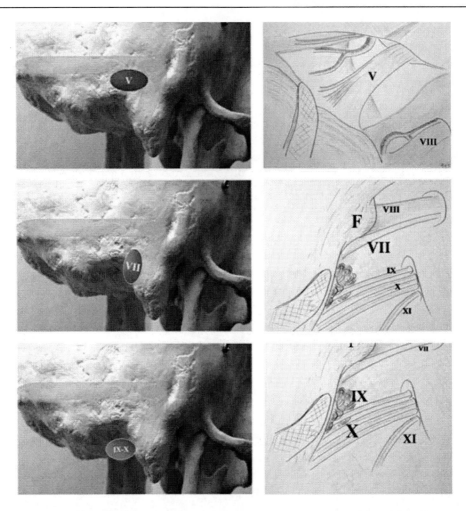

Figure 8. Surgical approach to Trigeminal nerve (A), Facial nerve (B) and vago-glosso-pharyngeal nerve (C). A retromastoid craniectomy of the key-hole type (2cm diameter) is performed: posterior to the base of the mastoid process and below the transverse sinus, to expose the dura at the angle formed by the transverse and the sigmoid sinuses, for TN (upper- left); posterior to the tip of mastoid and adjacent to the sigmoid sinus, to expose the infero-lateral aspect of the cerebellum and corresponding cisterns for VIIth or VIIIth nerve decompression (center left) evenmore inferiorly, at the retrocondylar fossa, to enter the cerebello-pontine – angle infero-laterally to the tonsil for lower cranial nerve decompression (lower-left). To avoid stretching the VIIth–VIIIth nerve complex, the trigeminal nerve is accessed via infratentorial-supracerebellar route along the superior petrosal sinus (upper right), and not by retracting from lateral-to-medial the cerebellum, that would consequently stretch the VIIth–VIIIth nerve complex as well as the labyrinthine artery; the VIIth–VIIIth nerves complexe (center right), is accessed through an infrafloccular trajectory and reached at the caudal aspect of REZ; the lower cranial nerves are approached infero-laterally to the tonsil, to reach their ventral entry/exit zones on lateral aspect of the medulla (lower right).

Discussion

More thanMoreover, the demonstration of direct trauma to the auditory nerve - also infrequent and easily recognizable - the monitoring of BAEP has provedn particularly useful in preventing irreversible deficits secondary to an excessive traction of the VIIth - VIIIth nerves, and/or aggressive manipulation of its vessels. Critical situations are stretching of the

VIIIth nerve when retracting the cerebellum, manipulation of the labyrinthine artery or its parent anteroinferior cerebellar artery, direct trauma by instruments or nearby coagulation, and at the end of the surgery, neocompression of the nerve by the prosthesis interposed between the conflicting vessel(s) and the VIIth–VIIIth complex [19].

However, although useful, intraoperative BAEP monitoring may be withdrawn as soon as the "learning curve" of how to do safe surgery has reached a sufficient level because of the difficulty (and cost) to have BAEP monitoring routinely available.

Unfortunately, most of the Clinical Neurophysiology services may only provide only limited amounts of intraoperative recordings of Evoked Potentials (of all kinds), compared to the more and more candidates (for various diseases) offered by surgeons. This is why it is becoming increasingly necessary to make a selection based on a hierarchy of priorities. In other words, evoked potential monitoring should be reserved in patients who have the greatest risk of neurological impairment, and in the case of MVD, the loss of hearing.

Conclusion

It has been clearly demonstrated with intra-operative BAEP recordings that a lateral-to-medial retraction of the cerebellar hemisphere entails high risk of excessive traction on the VIIth-VIIIth nerve complex, as well as of the labyrinthine artery [40]. The best approach to avoid hearing loss is therefore the infratentorial-supracerebellar route to access the trigeminal nerve, and a route infero-laterally to the cerebellar hemisphere and infra-floccular to reach the facial nerve from below, and also the IXth - X[th] nerves for vago-glossopharyngeal neuralgia.

As soon as the learning curve of the surgical team is "good", the use of intraperative BAEP monitoring for preserving hearing is no longert mandatory anymore, but should be used for specific cases, in particularly in the event of a MVD of the VIIIth nerve for vertigo.

References

[1] Dandy WE. Concerning the cause of trigeminal neuralgia. *Am. J. Surg.* 1934;24:447-55.
[2] Gardner WJ, Miklos MV. Response of trigeminal neuralgia to decompression of sensory root; discussion of cause of trigeminal neuralgia. *Journal of the American Medical Association.* 1959;170(15):1773-6. Epub 1959/08/08.
[3] Gardner WJ. Concerning the mechanism of trigeminal neuralgia and hemifacial spasm. *Journal of neurosurgery.* 1962;19:947-58. Epub 1962/11/01.
[4] Jannetta PJ. Arterial compression of the trigeminal nerve at the pons in patients with trigeminal neuralgia. *Journal of neurosurgery.* 1967;26(1):Suppl:159-62. Epub 1967/01/01.
[5] Sindou M, Amrani F, Mertens P. [Microsurgical vascular decompression in trigeminal neuralgia. Comparison of 2 technical modalities and physiopathologic deductions. A study of 120 cases]. Neuro-Chirurgie. 1990;36(1):16-25; discussion -6. Epub 1990/01/01. Decompression vasculaire microchirurgicale pour nevralgie du trijumeau. Comparaison de deux modalites techniques et deductions physiopathologiques. Etude sur 120 cas.

[6] Chiappa KH. Evoked potentials in clinical medicine. New York: Raven Press; 1983.
[7] Fischer C. Brainstem auditory evoked potential (B.A.E.P.) monitoring in posterior fossa surgery. Neuromonitoring in surgery: Elsevier Science Publishers B.V. ; 1989. p. 191-207.
[8] Fischer C, Ibanez V, Mauguiere F. [Peroperative monitoring of early auditory evoked potentials]. Presse Med. 1985;14(37):1914-8. Epub 1985/11/02. Monitorage per-operatoire des potentiels evoques auditifs precoces.
[9] Friedman WA, Kaplan BJ, Gravenstein D, Rhoton AL, Jr. Intraoperative brain-stem auditory evoked potentials during posterior fossa microvascular decompression. *Journal of neurosurgery*. 1985;62(4):552-7. Epub 1985/04/01.
[10] Grundy BL, Jannetta PJ, Procopio PT, Lina A, Boston JR, Doyle E. Intraoperative monitoring of brain-stem auditory evoked potentials. *Journal of neurosurgery*. 1982;57(5):674-81. Epub 1982/11/01.
[11] Jacobson GP, Tew JM, Jr. Intraoperative evoked potential monitoring. *Journal of clinical neurophysiology : official publication of the American Electroencephalographic Society*. 1987;4(2):145-76. Epub 1987/04/01.
[12] Moller MB, Moller AR. Loss of auditory function in microvascular decompression for hemifacial spasm. Results in 143 consecutive cases. *Journal of neurosurgery*. 1985;63(1):17-20. Epub 1985/07/01.
[13] Moller AR, Moller MB. Does intraoperative monitoring of auditory evoked potentials reduce incidence of hearing loss as a complication of microvascular decompression of cranial nerves? *Neurosurgery*. 1989;24(2):257-63. Epub 1989/02/01.
[14] Radtke RA, Erwin CW, Wilkins RH. Intraoperative brainstem auditory evoked potentials: significant decrease in postoperative morbidity. *Neurology*. 1989;39(2 Pt 1):187-91. Epub 1989/02/01.
[15] Raudzens PA, Shetter AG. Intraoperative monitoring of brain-stem auditory evoked potentials. *Journal of neurosurgery*. 1982;57(3):341-8. Epub 1982/09/01.
[16] Watanabe E, Schramm J, Strauss C, Fahlbusch R. Neurophysiologic monitoring in posterior fossa surgery. II. BAEP-waves I and V and preservation of hearing. *Acta neurochirurgica*. 1989;98(3-4):118-28. Epub 1989/01/01.
[17] Schramm J, Watanabe E, Strauss C, Fahlbusch R. Neurophysiologic monitoring in posterior fossa surgery. I. Technical principles, applicability and limitations. *Acta neurochirurgica*. 1989;98(1-2):9-18. Epub 1989/01/01.
[18] Ciriano D, Sindou M, Fischer C. [Peroperative monitoring of early auditory evoked potentials in microsurgical vascular decompression for trigeminal neuralgia or hemifacial spasm]. Neuro-Chirurgie. 1991;37(5):323-9. Epub 1991/01/01. Apport du monitorage per-operatoire des potentiels evoques auditifs precoces dans la decompression vasculaire microchirurgicale pour nevralgie du trijumeau ou spasme hemifacial.
[19] Sindou M, Fobe JL, Ciriano D, Fischer C. Hearing prognosis and intraoperative guidance of brainstem auditory evoked potential in microvascular decompression. *The Laryngoscope*. 1992;102(6):678-82. Epub 1992/06/01.
[20] Polo G, Fischer C, Sindou MP, Marneffe V. Brainstem auditory evoked potential monitoring during microvascular decompression for hemifacial spasm: intraoperative brainstem auditory evoked potential changes and warning values to prevent hearing

loss--prospective study in a consecutive series of 84 patients. *Neurosurgery.* 2004;54(1):97-104; discussion -6. Epub 2003/12/20.

[21] Leal PR, Froment JC, Sindou M. [Predictive value of MRI for detecting and characterizing vascular compression in cranial nerve hyperactivity syndromes (trigeminal and facial nerves)]. Neuro-Chirurgie. 2009;55(2):174-80. Epub 2009/03/21. Valeur predictive de l'IRM pour la detection et la caracterisation de la compression vasculaire dans les syndromes d'hyperactivite des nerfs craniens (trijumeau et facial).

[22] Hermier M, Leal PR, Salaris SF, Froment JC, Sindou M. [Imaging anatomy of cranial nerves]. *Neuro-Chirurgie.* 2009;55(2):162-73. Epub 2009/03/24. Imagerie anatomique des nerfs craniens.

[23] Leal PR, Hermier M, Froment JC, Souza MA, Cristino-Filho G, Sindou M. Preoperative demonstration of the neurovascular compression characteristics with special emphasis on the degree of compression, using high-resolution magnetic resonance imaging: a prospective study, with comparison to surgical findings, in 100 consecutive patients who underwent microvascular decompression for trigeminal neuralgia. *Acta neurochirurgica.* 2010;152(5):817-25. Epub 2010/01/29.

[24] Sindou M. Microvascular decompression for trigeminal neuralgia. In: Kaye AH, Black PM, editors. *Operative neurosurgery.* London: Churchill-Livingstone; 2000. p. 1595-614.

[25] Sindou M. Microvascular decompression for trigeminal neuralgia. In: Sindou M, editor. Practical Handbook of Neurosurgery From Leading Neurosurgeons. Wien: Springer-Verlag; 2009. p. 333-48.

[26] Sindou M. Operative strategies for minimizing hearing loss associated with microvascular decompression for trigeminal neuralgia. *World neurosurgery.* 2010;74(1):111-2. Epub 2011/02/09.

[27] Sindou M, Howeidy T, Acevedo G. Anatomical observations during microvascular decompression for idiopathic trigeminal neuralgia (with correlations between topography of pain and site of the neurovascular conflict). Prospective study in a series of 579 patients. *Acta neurochirurgica.* 2002;144(1):1-12; discussion -3. Epub 2002/01/25.

[28] Sindou M, Leston J, Decullier E, Chapuis F. Microvascular decompression for primary trigeminal neuralgia: long-term effectiveness and prognostic factors in a series of 362 consecutive patients with clear-cut neurovascular conflicts who underwent pure decompression. *Journal of neurosurgery.* 2007;107(6):1144-53. Epub 2007/12/15.

[29] Sindou M, Leston JM, Decullier E, Chapuis F. Microvascular decompression for trigeminal neuralgia: the importance of a noncompressive technique--Kaplan-Meier analysis in a consecutive series of 330 patients. *Neurosurgery.* 2008;63(4 Suppl 2):341-50; discussion 50-1. Epub 2008/11/15.

[30] Sindou M. Trigeminal neuralgia: a plea for microvascular decompression as the first surgical option. Anatomy should prevail. *Acta neurochirurgica.* 2010;152(2):361-4. Epub 2009/09/17.

[31] Sindou M. Microvascular decompression for hemifacial spasm. In: Sindou M, editor. Practical Handbook of Neurosurgery From Leading Neurosurgeons. Wien: Springer-Verlag; 2009. p. 317-32.

[32] Park JS, Kong DS, Lee JA, Park K. Hemifacial spasm: neurovascular compressive patterns and surgical significance. *Acta neurochirurgica*. 2008;150(3):235-41; discussion 41. Epub 2008/02/26.

[33] Sindou M, Henry JF, Blanchard P. [Idiopathic neuralgia of the glossopharyngeal nerve. Study of a series of 14 cases and review of the literature]. Neuro-Chirurgie. 1991;37(1):18-25. Epub 1991/01/01. Nevralgie essentielle du glossopharyngien. Etude d'une serie de 14 cas et revue de la litterature.

[34] Sindou M, Mertens P. Microsurgical vascular decompression (MVD) in trigeminal and glosso-vago-pharyngeal neuralgias. A twenty year experience. *Acta neurochirurgica Supplementum*. 1993;58:168-70. Epub 1993/01/01.

[35] Olds MJ, Woods CI, Winfield JA. Microvascular decompression in glossopharyngeal neuralgia. *The American journal of otology*. 1995;16(3):326-30. Epub 1995/05/01.

[36] Kondo A. Follow-up results of using microvascular decompression for treatment of glossopharyngeal neuralgia. *Journal of neurosurgery*. 1998;88(2):221-5. Epub 1998/02/06.

[37] Patel A, Kassam A, Horowitz M, Chang YF. Microvascular decompression in the management of glossopharyngeal neuralgia: analysis of 217 cases. *Neurosurgery*. 2002;50(4):705-10; discussion 10-1. Epub 2002/03/21.

[38] Sampson JH, Grossi PM, Asaoka K, Fukushima T. Microvascular decompression for glossopharyngeal neuralgia: long-term effectiveness and complication avoidance. *Neurosurgery*. 2004;54(4):884-9; discussion 9-90. Epub 2004/03/30.

[39] Wakiya K, Fukushima T, Miyazaki S. [Results of microvascular decompression in 16 cases of glossopharyngeal neuralgia]. *Neurologia medico-chirurgica*. 1989;29(12):1113-8. Epub 1989/12/01.

[40] Sindou M, Ciriano D, Fischer C. Lessons from brainstem auditory evoked potential monitoring during micro-vascular decompression for trigeminal neuralgia and hemifacial spasm. In: Schramm J, Moller AR, editors. Intra-operative Neurophysiologic Monitoring in Neurosurgery. Berlin: Springer-Verlag; 1991. p. 293-300.

Chapter XXI

Endoscopic Transchoroidal Fissure Approach to the Posterior Part of the Third Ventricle and Posterior Fossa

Kheireddine A. Bouyoucef[1], Mohamed Si Saber[1], A. Youssef Kada[1], Sofiane Imekraz[1], Rebiha Baba-Ahmed[2] and Michael H. Cotton[3,2]

[1] Department of Neurosurgery, University Hospital of Blida, Algiers, Algeria
[2] Department of neuropathology, University hospital of Lamine Debbaghine, Algiers, Algeria
[3,2] University Hospital of Vaud, Lausanne, Switzerland

Abstract

The third ventricle is one of the most difficult areas in the brain to access surgically. It is almost impossible to reach its cavity without incising some neural and vascular structures. Many microsurgical approaches are described and known by neurosurgeons, namely the trans-callosal (anterior, middle & posterior), trans-cortical trans-ventricular, trans-lamina terminalis, inter-forniceal and trans-sphenoidal.

In recent years, neuro-endoscopic procedures are increasingly used and preferred because of their feasibility and efficacy. The endoscopic approach towards tumors located in the pineal region is particularly a case in point, as this technique permits simultaneous treatment of hydrocephalus frequently encountered as well as tumor biopsy under direct vision, and the possibility of establishing an appropriate therapeutic strategy.

We describe a purely endoscopic technique passing through the choroidal fissure located between the fornix and the thalamus in order to approach the pineal region and posterior fossa tumors which are invisible through the foramen of Monro.

The use of this endoscopic approach avoids much neurological morbidity (memory loss and psychiatric disturbances) due, respectively, to the damage to the body of the fornix and the thalamus.

Out of 1800 neuroendoscopic procedures, 390 were performed for intraventricular lesions, being a total of 174 tumors located in the posterior compartment amongst which 124 were at the posterior pole of the third ventricle and 50 were in the posterior fossa. Out of these were 23 purely endoscopic transchoroidal approaches for tumors located at the pineal region and posterior fossa. No mortality and considerably lower morbidity resulted than from other methods.

Using our approach, the massa intermedia is well seen. This marks the boundary between the processes of the anterior part of the third ventricle which must be approached by the foramen of Monro and those involved the pineal region and the aqueduct of Sylvius arising to the fourth ventricle or brainstem by the choroidal fissure approach.

The purely endoscopic Transchoroidal fissure approach is a safe, effective and minimally invasive surgical technique and constitutes a better alternative for reaching the pineal region and posterior fossa tumors. In this way it is possible at one step to treat any associated hydrocephalus under direct vision, study the CSF and obtain a histological diagnosis. Thus, ventriculo-peritoneal shunt procedures with all their complications are avoided, difficult and at times imprecise stereotactic biopsy become unnecessary, and potentially unnecessary large open surgery is rendered irrelevant.

Introduction

Access routes to lesions of the third ventricle (V3), the deeper brain structure, are multiple. They depend on their location and nature as well as the preference and experience of the neurosurgeon. These approaches may be achieved microsurgically and /or assisted by endoscopy or purely endoscopically (Figure 1). Schematically, the main routes are the following:

- High routes: trans-callosal (anterior, middle or posterior), trans-cortical transventricular, inter-thalamotrigonal, supra-cerebellar
- Transbasal routes: bilateral subfrontal, trans-lamina terminalis, fronto-temporal lateral.
- Low routes: trans-sphenoidal.

Lesions situated in the anterior as well as the posterior part of the third ventricle have heretofore been shunned when considering an endoscopic approach. However, our preference is in fact the endoscopic approach for these lesions in addition to others in more accessible sites.

Whilst access to the tumors located at the anterior part of V3 (colloid cyst) is definitely easier by the foramen of Monro, those located at the posterior part, particularly those behind the interthalamic adhesion, are more difficult to access. Consequently we suggest a posterior access, which makes it possible to utilize the whole posterior space of the third ventricle and to remain plumb with the aqueduct of Sylvius. Furthermore this makes it possible to carry out simultaneous Endoscopic Third Ventriculotomy (ETV) in the case of hydrocephalus secondary to the tumors of the fourth ventricle, and to take biopsies of posterior fossa tumors.

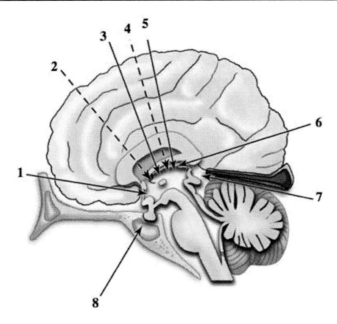

Figure 1. The different approaches to the third ventricle; 1: Subfrontal trans-lamina terminalis; 2: Transcortical transforaminal; 3: Anterior trans-callosal; 4: Transcortical Transchoroidal; 5: Middle trans-callosal; 6: Posterior trans-callosal; 7: Occipital supra-cerebellar; 8: Trans-sphenoidal.

Figure 2. Endoscopic approach to the posterior part of the third ventricle (V3) and fourth ventricle (V4) tumors; a: Endoscopic Transchoroidal or trans-foraminal approach to the brainstem tumors via the ostomy of an endoscopic third ventriculostomy (ETV); b: Endoscopic Transchoroidal approach to intra-axial brainstem and V4 tumors via the aqueduct of Sylvius; c: Endoscopic Transchoroidal approach to tumors of the pineal region.

We describe a purely endoscopic approach for access to the pineal and the posterior fossa tumors (fourth ventricle (V4) & brainstem), which uses a transchoroidal route through the choroidal fissure. (Figures 2 and 3). This route is known to neurosurgeons using microscopic techniques as the inter-thalamo-trigonal approach, and it is this route which we have adapted for pure endoscopic techniques.

Crossing the foramen of Monro in a more traditional approach has the disadvantage of missing a tumor located behind the interthalamic adhesion. Furthermore, to incline the endoscope forward to visualize such a tumor may cause trauma to the choroid plexus, the anterior pillar of the fornix, the interthalamic adhesion, and the peri-aqueductal structures at the periphery of the wound. Injury to the latter may result in bilateral ophthalmoplegia owing to damage to the oculomotor nerve nuclei in the peri-aqueductal gray matter. Injury to the interthalamic adhesion results in collateral effects on memory. These risks are substantially reduced in our approach.

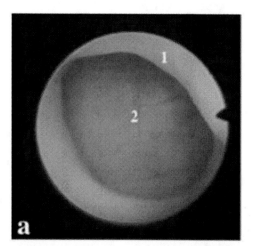 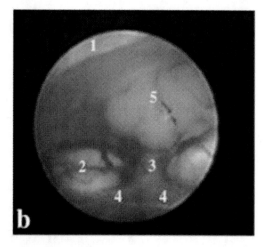

Figure 3. Endoscopic approach to the posterior part of third ventricle (V3) and fourth ventricle (V4) tumors; a: via the cerebral aqueduct often dilated by hydrocephalus, 1: cerebral aqueduct, 2: tumor; b: via the ostomy after treatment of hydrocephalus by ETV, 1: clivus, 2: protuberance, 3: basilar artery, 4: posterior cerebral artery, 5: tumor.

Anatomy

Microsurgical

A perfect knowledge of the richness of neural and vascular structures in this deep region of the brain is mandatory for the neurosurgeon if numerous complications of surgical intervention are to be avoided.

The choroidal fissure is part of the medial wall of the lateral ventricles. It is a C-shaped structure with a ventral concavity (Figures 4 and 5). It begins at the inter-ventricular foramen, continues through the body and atrium, and ends in the temporal horn. Its termination is called the inferior choroidal point [1].

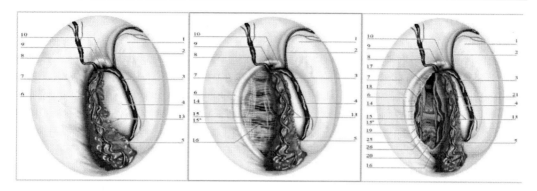

Figure 4. The choroidal fissure; 1: Caudate nucleus; 2: Anterior caudate vein; 3: Thalamo-striate vein; 4: Thalamus; 5: Choroid plexus; 6: Body of the formix; 7: Septum pellucidum; 8: Septal vein; 9: Foramen of Monro; 10: Column of the formix; 11: Tenia thalami; 12: Tenia fornix, 13: Choroidal vessels; 14: Internal cerebral vein; 15: Lateral posterior choroidal artery; 16: Tela choroidea; 17: Optic chiasm; 18: Mamillary bodies; 19: Massa intermedia; 20: Tumor; 21: Infudibular recess; 22: Lateral ventricle; 23: Septum pellucidum; 24: Corpus callosum; 25: Aqueduct; 26: Posterior commissure.

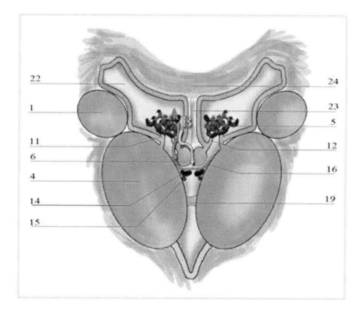

Figure 5. Coronal cut of the choroidal fissure showing the main teniae (fornicis and thalami); 1: Caudate nucleus; 4: Thalamus; 5: Choroid plexus ; 6: Body of the fornix ; 11: Tenia thalami; 12: Tenia fornix; 14: Internal cerebral vein; 15: Medial posterior choroidal artery; 16: Tela choroidea; 19: Massa intermedia; 22: Lateral ventricle; 23: Septum pellucidum; 24: Corpus callosum; A: Transchoroidal approach; B: Interfornical approach.

The choroidal fissure is the site of attachment of the choroid plexus in the lateral ventricles, between the thalamus and the fornix. As the fornix, the choroidal fissure wraps around the thalamus. Therefore, the fornix forms the outer circumference of the choroidal fissure. The choroid plexuses are linked to the thalamus by the taenia choroidea, and to the fornix by the taenia fornicis.

Inside the choroidal fissure run the choroidal arteries. They arise from the posterior cerebral arteries. The medial posterior choroidal arteries vascularize the choroid plexus of the roof of the third ventricle and the body of the lateral ventricle and the lateral posterior choroidal arteries vascularize the choroid plexus of the atrium, body and posterior part of the temporal horn.

The ventricular veins join the deep venous system passing through the choroidal fissure. The lateral group of ventricular veins course through the thalamic (or inner) side of the choroidal fissure, while the medial group of veins course through the forniceal (or outer) side. Veins draining the frontal horn and body of the lateral ventricle empty into the internal cerebral veins (ICV).

The transventricular transchoroidal approaches allow access to the deepest cerebral structures, by minimizing cerebral retraction. Creating an opening through the choroidal fissure along the taenia fornicis is generally preferred because fewer difficulties are encountered than along the taenia choroidea, through which pass many important veins and arteries.

The choroidal fissure (CF) is divided into three portions: rostral (body), dorsal (atrial) and caudal (temporal). The rostral portion of the choroidal fissure is situated in the body of the lateral ventricle, between the body of the fornix and the superior surface of the thalamus. Opening through the CF exposes the roof of the third ventricle and the velum interpositum, a closed space from the foramen of Monro to the area above the pineal body, through which course the internal cerebral veins. The rostral portion of the CF is mainly related to the terminal branches of the medial posterior choroidal arteries and the lateral posterior choroidal artery [2, 3].

Endoscopic

Entry in the right frontal horn of the lateral ventricle reveals first, the choroid plexus which runs from posterior to anterior and lateral to medial, directly reaching the posterior part of the foramen of Monro.

Neuronavigation through a rigid endoscope, with a 5-6mm working channel and 1mm tools, is possible when the ventricles are small. The foramen of Monro is bordered anteriorly by the anterior pillar of the fornix. Seen clockwise from the anterior pillar of the fornix, being the anterior part of the foramen of Monro, at 12 o'clock, are noted in sequence: the bulge of the head of the caudate nucleus, the thalamo-striate vein, the choroid plexus and the superior choroidal vein and finally in the midline, the body of the fornix overlooked at right angle by the septum pellucidum and the anterior septal vein. (this joins the thalamostriate vein at the posterior part of the the foramen of Monro), the superior choroidal vein forming the ICV which will cross the tela choroidea posteriorly to reach the Gallian vein.

The opening of the choroidal fissure between the body of the fornix and the choroid plexus through the taenia fornicis permits direct access to the roof of the third ventricle. The choroid plexus is displaced laterally revealing the tela choroidea (an arachnoid-like membrane) through which are perceived the two internal cerebral veins and the choroidal arteries (latero-posterior and medio-posterior) (Figure 6a), which run through the roof of the third ventricle. Their separation using a balloon (Fogarty No. 4 probe), gives direct access to the posterior chamber revealing anteriorly the inter- thalamic commissure, which marks the boundary between two areas: anteriorly, two mammillary bodies and infundibulum, and posteriorly, the "midbrain", the posterior commissure, the pineal recess, the pineal gland itself and the supra-pineal recess (Figure 6b).

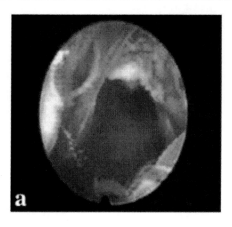 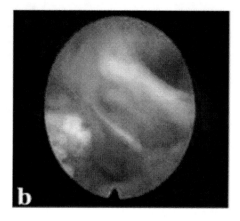

Figure 6. Endoscopic view of the Transchoroidal approach; a: posterior part of the third ventricle (V3) (NB Opening the choroidal fissure allows direct access to the roof of V3 between the internal cerebral veins and choroidal arteries; b: posterior chamber of the third ventricle (V3).

Prolonged irrigation is performed to remove any debris which may obstruct the aqueduct of sylvius.

The ventricles are inflated by Ringer's solution to avoid ingress of air and to prevent ventricular collapse. Furthermore, it is preferable to perform the narrowest possible cortotomy in order to avoid any CSF leak.

Surgical Technique

With the patient in supine position, the head neutral, a 2cm precoronal incision is made 3cm from the midline. A burr hole is performed, followed by cruciform opening of the dura matter.

A rigid endoscope, 18cm long, 6 mm in diameter and provided with a 0° or 30° lens is introduced into the right lateral ventricle, often dilated by obstructive hydrocephalus.

A sample of CSF is taken to look for tumors cells and for the determination of tumor markers.

The first landmark is the choroid plexus which is a guidepost for the foramen of Monro; it is always helpful to traverse the foramen of Monro by performing ETV and to have a look at the posterior part of V3 if the tumor is visible. If it not, a transchoroidal fissure approach is necessary.

Endoscopic Transchoroidal Approach

Under visual control, using the coagulator, the choroid plexus is delicately retracted to the thalamus, by stretching its forniceal attachment (taenia fornicis).

An easy fenestration by a simple introduction of the bipolar coagulator is made, until the tela choroidea, which represents the roof of the V3, is visualized. This is then opened. A Fogarthy catheter n° 4 is used to open the choroidal fissure, avoiding the vascular structures laterally.

Thus, complete access to the posterior chamber of V3, the pineal region and cerebral aqueduct is achieved easily.

Lesions of the pineal region are easily accessible for biopsy or total resection. Whenever possible, those of the posterior fossa will be approached through the cerebral aqueduct (V4 tumors) or through an ostomy (exophytic brainstem tumor).

Endoscopic Interforniceal Approach (Figure 7)

This is a variant of the previous approach. Under endoscopic control surgical dissection veers towards the midline at the junction between the septum and the two bodies of the fornix. Fine dissection permits access to the tela choroidea forming the roof of the V3. Opening it with bipolar coagulation gives wide access to V3 after gradual inflation of a balloon. The danger here is risk of damage to the body of the fornix, ICV and choroidal arteries.

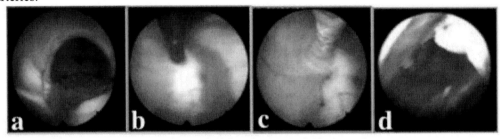

Figure 7. Endoscopic view; a: inside the frontal horn of the right VL showing a dilated foramen of Monro with the main landmarks (septal vein medially, thalamo-striate vein laterally and choroid plexus posteriorly; b:opening the choroidal fissure (taenia fornicis) with a gentle introduction of the coagulator; c:enlargement of the fenestration by balloon inflation; d:opening of the roof of V3 showing the internal cerebral veins and choroidal arteries.

Discussion

A variety of pathologies varying from benign to malignant tumors occurs in the region of the posterior third ventricle and posterior fossa. The complexity of access to this anatomical location makes the management of these lesions a considerable challenge even for the skilled neurosurgeon.

Credit must be awarded to Sir Walter Dandy [4], who pioneered the inter-hemispheric trans-callosal routes to the lateral and third ventricles. In 1921, he described the posterior trans-callosal approach with division of the splenium for removal of pineal tumors, and followed this by reporting a series of third ventricle tumors in 1933. Bush [5] performed the first inter-forniceal approach in 1944. In 1978, Delandsheer & Guiot [6] described a microsurgical inter-thalamotrigonal approach to the third ventricle. In 1998, Wen & al. [7] presented an anatomical study of the choroidal fissure and of the supra- and sub-choroidal surgical routes.

In recent years the endoscopic approach through the foramen of Monro has gained increased popularity for the biopsy and, in selected cases, the removal of tumors of the posterior third ventricle. However, a biopsy of V3 & V4 tumors cannot always be safely accomplished. In fact, in two cases the postoperative result was a complete bilateral ophatalmoplegia that was attributed to peri-aqueductal oculomotor nuclei damage. This was probably due to poor manipulation of the endoscope in relation to the axis of the cerebral aqueduct. To avoid this kind of complication, which was fortunately only transient, we have developed a purely endoscopic inter-thalamotrigonal approach.

Since the first report by Fukushima et al. [8, 9] of an intraventricular method of tumor biopsy, neuro-endoscopy has been increasingly used in the diagnosis and treatment of these tumors. It is widely accepted as an ideal method of diagnosis for intra- or peri-ventricular tumors, which may often cause hydrocephalus [10].

Indeed, in comparison with stereotactic biopsy, endoscopic biopsy has many advantages. By the same minimally invasive approach, through direct vision, it is possible effectively to control tumor bleeding immediately after biopsy, and more importantly, to manage associated hydrocephalus as well as perform CSF analysis (alpha fetoprotein: AFP, human choroinic gonadotrophin: HCG, and others tumor markers). We prefer the use of the rigid endoscope rather than the flexible, because the latter is difficult to handle and its picture quality is inferior.

These procedures eliminate the need for ventriculo-peritoneal shunting and thus avoid its complications. The diagnostic yield by stereotactic biopsy is around 94% [6, 11]. The disadvantages include a relatively higher risk of bleeding, an 8% morbidity, in addition to an increased mortality which has been reported for pineal tumors (1,3- 1,9%) compared to other regions (0,81%) [6, 11].

Open microsurgery, especially using the trans-cortical and trans-callosal routes for intra- or peri-ventricular tumors of the posterior fossa carries severe morbidity [12, 13] and cumulative postoperative mortality. These use the same corridors to reach the third ventricle (para-/interforniceal, supra-/subchoroidal) as endoscopy but their significant morbidity is from 5% to 15% [14]. The most important issue in this approach is the management of the veins and the fornices. Excessive retraction, manipulation or injury (mechanical damage) will cause, respectively, thrombosis and permanent postoperative loss memory. Thus the learning curve for this type of surgery is steeper than that of the endoscopic approach.

Furthermore, in cases of malignant tumors, even extensive surgical resection does not eliminate the need for subsequent adjuvant therapy [15]. A substantial number of neoplasms, including germinomas, malignant non-germinomatous germ cell tumors, pineal parenchymal neoplasms, diffuse brainstem gliomas, and metastases can, however, be successfully managed without surgical intervention, using radio- and/or chemo-therapy [16].

By contrast, the surgeon is sometimes very frustrated to come face to face with a tumor, especially if benign, when he cannot completely remove it. The manufacture of an ultrasonic aspirator adapted to neuro-endoscopic surgery is eagerly sought.

In the department of neurosurgery of the University of Blida (Algeria), the strategies used for the treatment of the tumors located in the pineal region is summarized in Figure 8.

From 1994 to September 2012, 1800 neuroendoscopic procedures were performed in our department. Among which 390 were for intraventricular lesions including 174 located in the posterior compartment (pineal region, posterior fossa, V4 and brainstem). The results are summarized in Tables 1 and 2.

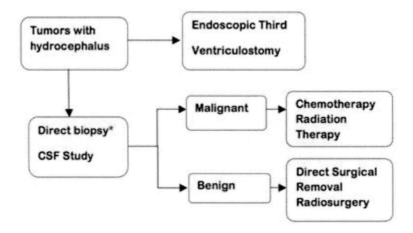

Figure 8. Management decision making for pineal tumors; *: Intra-operaive histological examination is necessary to help the neurosurgeon to take a decision whether to continue surgical intervention or not.

Table 1. Pineal region

Pineal regions tumors	N = 124
Germinomas	26
Pinealoblastomas	13
Pinealocytomas	4
Anaplastic astrocytomas	3
Ependymomas	11
Oligodendrogliomas	1
low grade astrocytomas	24
Not Conclusive	34
Lost	8

Table 2. Posterior fossa region

Posterior fossa tumors	N = 50
Medulloblastomas	16
Astrocytomas III	1
low grade astrocytomas	19
Ependymomas	12
Not conclusive	2

Our results demonstrate the potential value of the endoscopic Transchoroidal and interforniceal approaches for pineal region tumors. With this method, CSF analysis, ETV, biopsy under direct vision and intra-operative histological sampling are all possible at one step.

Furthermore, with the same approach we have performed biopsies for tumors located in V4 (n=36) and for exuberant brainstem tumors through an ostomy (n=6) or aqueduct (n=8).

There was no mortality in our series, whilst morbidity was 4/124 owing to trauma to the interthalamic adhesion but without neurological consequences.

When we tried to reach tumors located in V4 through the foramen of Monro, two patients developed complete ophthalmoplegia, which was fortunately transient. However, the subsequent transchoroidal endoscopic approach (n=23) resulted in no complications.

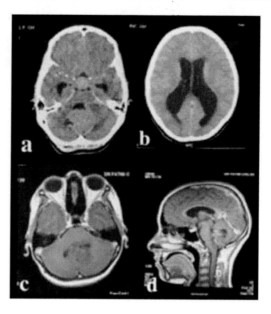

Figure 9. Thirteen year old patient with a medulloblastoma with suprasellar metastasis, having ETV and endoscopie biopsy via the aqueduct; a, b: Axial planes CT-scan imaging showing :a: midline cerebellar mass posterior to V4 ventricle and suprasellar mass associated with, b: hydrocephalus ; c: Axial T1- weighted MR imaging after gadolinium injection showing a midline cerebellar mass posterior to V4 ventricle; d: Sagittal T1-weighted MR imaging after gadolinium injection showing a midline cerebellar mass with posterior compression of the brain stem and the suprasellar metastasis.

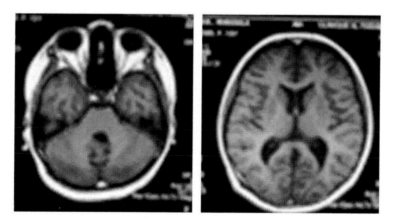

Figure 10. The same patient after 3 years without open surgery, having been treated only by radio- and chemo-therapy, Axial T1 weighted images showing the absence of lesions and complete remission of hydrocephalus.

From the data provided (tables 1 and 2), the number of unfruitful pathological results after biospy might seem high. However, in thse cases, biopsy fragments were deemed inadequate by the pathologist to permit substantiative conclusions. We therefore highlight the paramount importance of maintaining close cooperation between pathologists and neurosurgeons in order to ensure favourable results.

Our algorithm for treatment of tumors of pineal region is shown. We believe that the same algorithm could be used for many high risk tumors located in the posterior fossa such as medulloblastoma, for example.

In fact, since 2005 we have obtained surprisingly good results after only radio- and chemo-theray (without open surgery) with only endoscopic tumor biopsy (Figures 9 and 10).

Conclusion

The endoscopic Transchoroidal approach is an effective, safe and less invasive technique for tumors located in the pineal region and the posterior fossa. Doubtless, such surgery has a steep learning curve, but not necessarily steeper than the traditional open approach. Furthermore, using the same minimally invasive procedure three procedures are possible simultaneously: CSF Study, ETV and tumor biopsy. Thus the risks of stereotaxy and VP shunting are circumvented.

Acknowledgments

For help with scientific editing and, other valuable advice, the authors thank Dr. Mahmoud Messerer of the Departement of neurosurgery of Centre Hospitalier Universitaire Vaudois, Lausanne, Switzerland, and, Dr Julie Dubourg of the Hospices Civils de Lyon, France.

References

[1] Nagata S, Rhoton AL, Jr., Barry M. Microsurgical anatomy of the choroidal fissure. *Surgical neurology*. 1988; 30(1): 3-59. Epub 1988/07/01.

[2] Rhoton AL, Jr., Yamamoto I, Peace DA. Microsurgery of the third ventricle: Part 2. Operative approaches. *Neurosurgery*. 1981; 8(3): 357-73. Epub 1981/03/01.

[3] Hamlat A, Morandi X, Riffaud L, Carsin-Nicol B, Haegelen C, Helal H, et al. Transtemporal-transchoroidal approach and its transamygdala extension to the posterior chiasmatic cistern and diencephalo-mesencephalic lesions. *Acta neurochirurgica*. 2008; 150(4): 317-27; discussion 27-8. Epub 2008/03/04.

[4] Dandy WE. *Benign tumors in the third ventricle of the brain; diagnosis and treatment*: Springfield, Charles C Thomas; 1933.

[5] Busch E. A new approach for the removal of tumors of the third ventricle. *Acta Psychiatr. Scand*. 1944; 19:57-60.

[6] Regis J, Bouillot P, Rouby-Volot F, Figarella-Branger D, Dufour H, Peragut JC. Pineal region tumors and the role of stereotactic biopsy: review of the mortality, morbidity, and diagnostic rates in 370 cases. *Neurosurgery*. 1996; 39(5): 907-12; discussion 12-4. Epub 1996/11/01.

[7] Wen HT, Rhoton AL, Jr., de Oliveira E. Transchoroidal approach to the third ventricle: an anatomic study of the choroidal fissure and its clinical application. *Neurosurgery*. 1998; 42(6): 1205-17; discussion 17-9. Epub 1998/06/19.

[8] Fukushima T. Endoscopic biopsy of intraventricular tumors with the use of a ventriculofiberscope. *Neurosurgery.* 1978; 2(2): 110-3. Epub 1978/03/01.

[9] Fukushima T, Ishijima B, Hirakawa K, Nakamura N, Sano K. Ventriculofiberscope: a new technique for endoscopic diagnosis and operation. Technical note. *Journal of neurosurgery.* 1973; 38(2): 251-6. Epub 1973/02/01.

[10] O'Brien DF, Javadpour M, Collins DR, Spennato P, Mallucci CL. Endoscopic third ventriculostomy: an outcome analysis of primary cases and procedures performed after ventriculoperitoneal shunt malfunction. *Journal of neurosurgery.* 2005; 103(5 Suppl): 393-400. Epub 2005/11/24.

[11] Kreth FW, Schatz CR, Pagenstecher A, Faist M, Volk B, Ostertag CB. Stereotactic management of lesions of the pineal region. *Neurosurgery.* 1996; 39(2): 280-9; discussion 9-91. Epub 1996/08/01.

[12] Ellenbogen RG, Moores LE. Endoscopic management of a pineal and suprasellar germinoma with associated hydrocephalus: technical case report. *Minimally invasive neurosurgery : MIN.* 1997; 40(1): 13-5; discussion 6. Epub 1997/03/01.

[13] Yamini B, Refai D, Rubin CM, Frim DM. Initial endoscopic management of pineal region tumors and associated hydrocephalus: clinical series and literature review. *Journal of neurosurgery.* 2004; 100(5 Suppl Pediatrics): 437-41. Epub 2004/08/04.

[14] Matsutani M, Sano K, Takakura K, Fujimaki T, Nakamura O, Funata N, et al. Primary intracranial germ cell tumors: a clinical analysis of 153 histologically verified cases. *Journal of neurosurgery.* 1997; 86(3): 446-55. Epub 1997/03/01.

[15] Oi S, Kamio M, Joki T, Abe T. Neuroendoscopic anatomy and surgery in pineal region tumors: role of neuroendoscopic procedure in the 'minimally-invasive preferential' management. *Journal of neuro-oncology.* 2001; 54(3): 277-86. Epub 2002/01/05.

[16] Nicholson JC, Punt J, Hale J, Saran F, Calaminus G. Neurosurgical management of paediatric germ cell tumours of the central nervous system--a multi-disciplinary team approach for the new millennium. *British Journal of neurosurgery.* 2002; 16(2): 93-5. Epub 2002/06/06.

Chapter XXII

Optimally Invasive Skull Base Surgery for Large Benign Tumors

Roy Thomas Daniel[1,4,*], *Constantin Tuleasca*[1,4], *Mahmoud Messerer*[1,4], *Laura Negretti*[3], *Mercy George*[2], *Philippe Pasche*[2] *and Marc Levivier*[1,4]

Department of Neurosurgery and Gamma Knife Center[1], ENT and Head and Neck Surgery[2], and Radiation Oncology[3], Lausanne University Hospital
Faculty of Biology and Medicine, University of Lausanne[4], Switzerland

Abstract

Introduction: The management of large benign skull base lesions, such as vestibular schwanommas, meningiomas or pituitary adenomas remains challenging, with microsurgery seen as the main treatment option and complete resection as the primary goal. While previous studies have shown the value of complete resection for good long-term tumor outcome, the postoperative neurological results remain suboptimal. The concept of a "combined approach", as defined below, was developed with a view of preservation or amelioration of neurological outcome as the primary goal while maintaining optimal long-term tumor control. This chapter analyses our series with a review of the treatment outcomes from literature for these tumors.

Material: Between 2010 and 2012, we treated 33 patients, male/female 11/22, with ages between 12.73 and 73.41 years (mean 53.98). There were 11 patients with vestibular schwanommas, 14 with meningiomas (5 petroclival and 9 clinoidal) and 8 with pituitary adenomas. We analyzed the clinical presentation of these patients, as well as audiograms, ophthalmological and endocrinological tests, as indicated. In this context, treatment planning is performed prior to surgery with a combination of techniques namely transcranial microsurgical or endoscopic trans-nasal approach or both, with a view of performing a subtotal tumor resection. The residual part of the tumor, which is frequently

---

[*] Address for correspondence: Prof. Roy Thomas Daniel. Centre Hospitalier Universitaire Vaudois, Rue du Bugnon 46, Lausanne 1011, Switzerland. Tél.: +41 79 556 60 77. Fax: + 41 21 314 26 05. E-mail: roy.daniel@chuv.ch.

adherent to cranial nerves, brainstem or vascular structures, is then treated with Gamma knife radiosurgery. The postoperative outcome with respect to neurological/endocrinological function and tumor control was analyzed.

Results: Patients with vestibular schwanomma underwent subtotal surgery with facial nerve monitoring through a retrosigmoid approach. There were no post-operative facial nerve deficits. Of the 3 patients whom had useful hearing pre-operatively, this improved in 2 and remained stable in 1. After a period of 3 to 6 months, the residual tumor attached to the facial and cochlear nerves was treated with RS. Nine patients with clinoidal meningiomas underwent subtotal resection of the tumor, and one patient also underwent a staged endonasal resection. The cavernous sinus component of the tumor was later treated with RS. The visual status normalized in four and improved in 3. Of the five patients who underwent petroclival meningioma surgery, 3 had House-Brackmann (HB) grade 2 facial function that recovered completely while one continues to have HB grade 4 facial deficit following surgery. Of the 8 patients with pituitary macroadenoma, 2 had GH secreting tumors while the others were non-functional. Seven patients underwent endoscope assisted trans-sphenoidal microsurgery, while one patient had a staged trans-cranial and trans-sphenoidal surgery. While visual status improved/normalized in 4 patients, no patient had worsening of visual or occulomotor function. Two patients had transient diabetes insipidus following surgery. No additional neural dysfunction or tumor progression was noted at latest follow up following RS.

Conclusion: Our data suggest that combined approaches have an excellent clinical outcome with respect to preservation or improvement of neurological function. Microsurgery (transcranial and/or endoscopic transnasal) and radiosurgery are complementary treatment modalities. As long-term results emerge, this multimodal treatment paradigm could become the standard of care for the "optimally invasive" management of difficult skull-base benign tumors.

Introduction

Surgery for large benign skull-base tumors has evolved greatly in the last three decades. Initially deemed inoperable, the description of novel skull-base approaches led many surgeons to attempt a complete resection of these tumors. The use of image guidance, intraoperative electrophysiological monitoring and better techniques of skull-base repair have made these surgeries safer. Despite the large experience accumulated over several decades, the neurological morbidity associated with these aggressive approaches still remains significant and considered inevitable. On the other hand, several large series of radiosurgery for the treatment of small benign skull-base lesions have demonstrated adequate tumor control along with markedly diminished neurological morbidity and improved preservation of functions. In the last decade, endoscopic skull-base surgery has also emerged as an alternate surgical approach for many of these lesions. This chapter deals with our approach towards these lesions, wherein we prefer to combine skull-base microsurgical techniques with endoscopic surgery and radiosurgery in a tailored fashion based on patient and tumor characteristics, with a view to optimize the preservation of neurological functions. This series of patients include only those cases where this multimodal therapy was utilized in a manner where the precise strategy was decided prior to surgery with respect to the combination of techniques used. This chapter deals specifically with the use of this strategy for the treatment of four groups of tumors for which its application has gained interest predominantly, namely vestibular schwannomas, petroclival meningiomas, clinoidal meningiomas, and pituitary adenomas.

The Principle of Combined Approaches

Combined approaches are mostly relevantated to the management of large benign skull-base lesions. In these particularly challenging cases, surgery alone cannot be radical without a high risk for neurological functions, and radiosurgery cannot be used safely as a first line of treatment because the risks of radiation-induced complications are high. Indeed, radiosurgery bears 2 intrinsic risks that are related to the size of the lesion, which in turn and that, limits the treatment of lesions that are "too large". One is related to the fall-off of dose outside of the target area.; W when the volume of the lesion increases, the quality of the dose gradient decreases, and the dose fall-off becomes worse.; T this will increase the so-called area of penumbra around the target, thus increasing the dose outside of the target, and therefore resulting in radiation-induced side-effects. The second is related to the temporary swelling of the tumor that may be generated by radiosurgery in the weeks or months following the treatment. I; in larger lesions, this may have clinical consequences due to an increase in the mass effect. Thus, in large lesions presenting with mass effect, especially when they interface or impinge with critical structures such as the brainstem, or even in small lesions in contact with radiosensitive structures such as the optic apparatus, radiosurgery cannot be used as a primary treatment and combined approaches will be offered, if radical surgery bears too high functional risks.

The surgical techniques used in combined approaches vary with patients and tumor characteristics, and are described below for each of the tumor groups. In our center, radiosurgery is performed with the Gamma Knife. The principle of Gamma Knife surgery (GKS), as it is used as part of combined approaches, is described below. There is some specificity related to the location, anatomy, and characteristics of the skull-base lesions. T; these are emphasized separately for each of the tumor groups addressed in this chapter.

Principle of Gamma Knife Radiosurgery as Part of Combined Approaches

Radiosurgery was invented by the Swedish neurosurgeon Lars Leksell at the beginning of the 1950s and defined as the "delivery of a single, high dose of ionizing radiation to a small and critically located intracranial volume through the intact skull" [1]. Originally, Leksell conceived radiosurgery as a primary tool for functional disorders [1, 2]. In the 1960s, Leksell created the Gamma Knife, a tool for radiosurgery using multiple focusing cobalt-60 sources [2]. Linear accelerators (focused X-rays, as in adapted Linac, dedicated Linac, or robotic devices such as Cyberknife®) and synchrocyclotrons (charged particles) are also used for radiosurgery [3]. Whatever the technique used in radiosurgery, two conditions are mandatory: the first is highest precision and best possible fit between the irradiated volume (called prescription isodose volume, PIV) and the target-volume (TV), in what is called high conformity (% coverage, PIV/TV); the second is minimal irradiation of the normal surrounding tissue, thanks to a steepest dose gradient, in what is called high selectivity [3].

The key limiting factor remains the volume of the target, as a large volume is increasesing the dose delivered outside the target and also the risk of complications, depending on the location and the nature of the lesion.

GKS, as used in our center, is a dedicated neurosurgical stereotactic procedure, combining image guidance with high-precision convergence of multiple gamma rays emitted by 192 sources of Cobalt-60 (Leksell Gamma Knife Perfexion®, Elekta Instruments, AB, Sweden) [3, 4]. In practice, GKS represents a succession of steps closely connected one to each other: stereotactic frame attachment under local anesthesia, acquisition of stereotactic images, target determination and treatment planning, stereotactic irradiation, and clinical and neuroradiological follow-up. The procedure is performed on an ambulatory basis.

Currently, the clinical applications of GKS include benign (pituitary adenomas, meningiomas, vestibular schwannomas, etc …) and malignant (mostly metastases) tumors of the brain and skull-base, vascular malformations (arterio-venous malformations, dural arterio-venous fistulas, cavernomas) and functional disorders (trigeminal neuralgia, epilepsies, movement disorders). The mechanisms of action differ according to the treated condition and the targeting strategy. In the case of tumors, apoptosis may be the major mechanism of cell death [5-7]; in vascular malformations, it induces vessels obliteration by thrombotic endothelial proliferation [8-10]; in functional disorders, GKS is used either to target a specific anatomical point in an anatomical structure (e.g. thalamus [11], anterior limb of the internal capsule [12], trigeminal nerve [13-15] or to target a larger zone, such as an epileptic focus [16], and the mechanism of action may differ [17].

Leksell Gamma Knife Perfexion®, which appeared in 2006, represents a clear technical advance in radiosurgery for the treatment of brain pathologies, allowing very sophisticated dose planning [18]. The advantages consist especially in the possibilities of shaping of the isodoses in very complex manners. It can also treat multiple targets, located close or far from each other, lesions in eloquent areas and lesions with , of with complicated forms. It also saves time in addition to radiation protection which is much improved compared to the previous models and other radiosurgery devices [18]. The technical possibilities of Leksell Gamma Knife Perfexion® are particularly applicable for the post-operative treatment of residual skull-base lesions, which often present with complex shapes and are located very close to highly functional neural structures.

Vestibular Schwanommas

General Considerations

Vestibular schwannomas (VS) are benign tumors that arise from the vestibular branches of the vestibulo-cochlear nerves. These tumors originate from the distal sheath at/or close to the neuroglial–neurilemmal junction that occurs 1 cm away from the pons, commonly at or close to the internal auditory canal (IAC). They are often lobular, well-encapsulated solid tumors, though cystic variants are also seen. The vestibulo-cochlear and facial nerves are flattened and stretched over the tumor surface by progressive tumor growth. Other nerves of the cerebellopontine angle (CPA) like the trigeminal, glossopharyngeal and vagal nerves, along with blood vessels like the anterior inferior and posterior inferior cerebellar arteries may also be apposed or adherent to the tumor capsule, based on the tumor size. The tumor growth proceeds through several stages and thereby represents the different clinical symptoms associated with this tumor. The intracanalicular stage presents with otological symptoms like high-frequency sensori-neural hearing loss, tinnitus, vertigo, and disequilibrium. The extracanalicular stage starts as tumor grows out of the porus acoustics, which causes worsening of the auditory symptoms and the onset of headache and facial hypoesthesia. The cerebellopontine cistern stage is characterized by filling of the CPA with compression of the brainstem and the lower cranial nerves. The last stage is represented bywith hydrocephalus caused by shift and compression of the fourth ventricle [19-22].

Hearing loss is the most frequent clinical symptom of VS, which affects 91% of patients suffering from these tumors. The other common symptoms are related to the vestibular (61%), trigeminal (9%), facial (6%), lower cranial nerves (2.7%), and abducens nerve (1.8%). Patients are classified regarding tumor extension as follows: T1, purely intrameatal; T2, intra- and extrameatal; T3a, filling the cerebellopontine cistern; T3b, reaching the brainstem; T4a, compressing the brainstem; and T4b, severely dislocating the brainstem and compressing the fourth ventricle [23]. The House-Brackmann scale is used to assess facial function: grade I (normal symmetrical function), grade II (slight weakness noticeable only on close inspection, complete eye closure with minimal effort, slight asymmetry of smile with maximal effort, synkinesis barely noticeable, contracture, or spasm absent), grade III (obvious weakness, but not disfiguring, may not be able to lift eyebrow, complete eye closure and strong but asymmetrical mouth movement, obvious, but not disfiguring synkinesis, mass movement or spasm), grade IV (obvious disfiguring weakness, inability to lift brow, incomplete eye closure and asymmetry of mouth with maximal effort, severe synkinesis, mass movement, spasm), grade V (motion barely perceptible, incomplete eye closure, slight movement corner mouth, synkinesis, contracture, and spasm usually absent), grade VI (no movement, loss of tone, no synkinesis, contracture, or spasm) [24]. The auditory function is analyzed by using the Gardner Robertson (GR) classification, as follows: grade I (good-excellent), with pure tone audiogram, 0-30 dB and speech discrimination 70-100 (%); grade II (serviceable), with pure tone audiogram 31-50 dB and speech discrimination 50-69 %; grade III (non-serviceable) with pure tone audiogram 51-90 dB and speech discrimination 5-49 %; grade IV (poor), with pure tone audiogram 91-max dB and speech discrimination 1-4 %; grade V (none), with pure tone audiogram not testable, and speech discrimination 0% ; if PTA and speech do not correlate, a lower class has to be used [25].

Retrosigmoid Approach for VS

The various surgical approaches for these tumors include the retrosigmoid, translabyrinthine and middle fossa approaches. We prefer to use the retrosigmoid approach with the patient in the lateral decubitus position. Electrophysiological monitoring is routinely performed with somatosensory evoqued potentials (SSEPs) and electrical stimulation of cranial nerves V, VII, IX and X along with continuous VIIth nerve EMG. The head is fixed with a 3-point Mayfield clamp with the head in minimal flexion without obstructing venous return. The hair is shaved to simplify the identification of the surface anatomy. A linear slightly curved skin incision is drawn 2 cm behind the pinna, passing through the asterion and terminating 1 cm medial to the mastoid tip. The skin flap is elevated with the periosteum elevator and the neck muscles are divided in line with the skin incision. The burr hole is placed at the asterion and then a sub-occipital craniotomy or craniectomy is carried out. The borders of the lateral part of the transverse sinus and medial border of the sigmoid sinus are exposed with a high-speed drill. Bone wax is used to pack the mastoid air cells and the emissary vein. The dura is opened with a linear incision 5-8 mm from the sigmoid sinus border and curved to the transverse sigmoid junction. Additional oblique incisions at the angles of the primary incision increase surgical view. Several tack-up sutures are placed on the dural edge to increase surgical view and to reduce cerebellar retraction. Cerebrospinal fluid (CSF) is withdrawn from the lateral cerebello-medullary cistern to relax the cerebellum. The tumor is then identified at the IAM and the arachnoidal planes around the poles of the tumor are opened. Stimulation of the posterior capsule of the tumor is performed to assure that the neural elements on the posterior and superior surfaces do not contain an aberrant facial nerve course. The posterior part of the capsule containing the vestibular nerves is incised. Internal debulking is performed with the cavitatory ultrasonic surgical aspirator (CUSA). The posterior lip of the IAM is drilled after opening the petrous dura. This allows the nerves at the IAM to be decompressed. The facial and cochlear nerves are most vulnerable at the lip of the IAM and therefore a thin layer of tumor is left behind on the nerves. The facial nerve is identified anterior to the tumor by stimulation and its course is mapped out. The tumor is progressively decompressed and the capsule is removed except for the portions covering the entire facial nerve. Stimulation of Direct the facial nerve on the anterior capsule with currents (as low as stimulation of 0.1 mA) during its entire course from the brainstem to the IAM is maintained. Stimulation of the nerve through a thin layer of tissue at a level of 1 mA ensures that the thickness of the residual tumor left behind is approximately 5 mm. If cochlear nerve is also to be preserved, a larger tumor residue is intentionally left in place covering the cochlearis nerve.

Alternation of stimulation of the facial nerve directly and through the tumor ensures preservation of facial nerve function while ensuring that the tumor remnant is not too large. Hemostasis is secured, and the drilled petrous bone is sealed with bone wax and covered with a piece of fascia secured with fibrin glue. The dura is closed in a watertight fashion. The mastoid cells are covered with muscle and fibrin glue. Cranioplasty is performed with methylmethacrylate. The pericranium, muscles, fascia, subcutaneous fat and skin are closed in separate layers. A compressive bandage is used after the skin closure [23, 26-29].

Case Series

Eleven patients (6 males and 5 females) with VS had been treated using this combined approach (Ffigure 1). Mean age in this series was 50.3 years (range 24.1- 73.4). These patients presented with hearing loss (7 patients), trigeminal nerve symptoms (4 patients), gait problems (1 patient) and incidentally in three cases. Two patients (18.2%) had a stereotactic fractionated radiotherapy before the surgical intervention, which had failed to ensure tumor control. The lesions were solid in 9 cases (81.8%), and mixed (solid and cystic) in 2 patients (18.2%). Presurgical tumor volume was of a mean of 18.5 cm^3 (range 9.7- 34.9). All patients were operated though a retrosigmoid approach in the lateral decubitus position.

The timing of the GKS was decided on the basis of the residual tumor shape following surgery. This was evaluated on an MRI performed at 3 months after surgery. If the tumor capsule had not closed on itself, an additional 3 months were given before evaluating suitability for GKS. The mean duration between surgery and GKS in this series was 10.5 months (range 4- 22.8 m). The mean tumor volume at GKS treatment was 4.9 cc (range 0.5-12.8). A mean number of 20.7 isocenters was used (range 8-31). Nine patients were treated with 12 Gy and 2 patients with 11 Gy as the dose prescription at the periphery of the tumor. A lower maximal dose was related to a higher tumor volume in cases where cochlear nerve function was also preserved. Dosimetry was made in the sense of the principles of radiosurgery stated above, with improving both conformity and selectivity. In patients with preserved hearing after surgery, special care was made to improve the gradient dose towards the cochlea, as several studies have shown the importance of the dose received by this structure, with a cut-off dose of approximately 4 Gy, so as to offer statistically significant chances of preserving functional hearing).

We did not have any major complications in our series. Two patients needed a second surgical intervention for postoperative meningocele. Two patients needed re-intervention before GKS, because the remnant tumor volume was too large to ensure optimal GKS treatment. This has been done 6.8 and 15 months after the first surgical intervention, respectively.

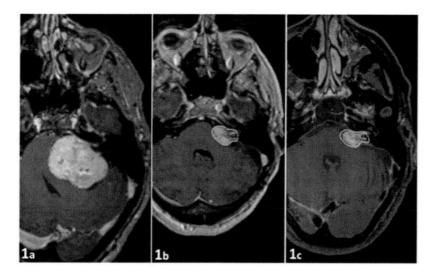

Figure 1. Example of a large vestibular schwannoma treated with combined approach. This 66 year-old female patient presented with hearing loss, tinnitus and gait instability; the MRI showed a large vestibular schwannoma (1a). The tumor was subtotally removed through a retrosigmoid approach. She had normal facial function following surgery. Three months after surgery, she underwent GKS for the residual tumor. The tumor volume before surgery was 24 cc while the target volume at the time of radiosurgery was 2.4 cc. GKS was performed with a dose of 12 Gy at the 50% isodose (1b). The MRI performed 6 months following GKS show slight shrinkage of the tumor, as illustrated by the co-registration and projection of the GKS prescription isodose on the 6-month follow-up MRI (1c).

Postoperative status showed no facial nerve deficits (0%). Four patients with useful preoperative hearing underwent surgery with the aim to preserve cochlear nerve function. Of these patients, the patient who had GR class I before surgery, remained in GR class 1. Two patients improved hearing after surgery, one from GR 5 to GR 3 and the other one with slight improvement, remaining in the same GR 3 class as preoperatively. Mean follow-up after surgery was 15.4 months (range 4- 31.2). One patient, who presented with secondary trigeminal neuralgia before surgery, had transienttory facial hypoesthesia following surgery. No other neurological deficits were encountered. Following GKS, the patients had a mean follow-up of 5.33 months (range 1-13). No new neurological deficits were encountered and hearing remained stable in all patients with preost-surgery preserved hearing.

Discussion

Vestibular schwannomas are known to have a small annual rate of growth [22]. Despite this, recurrences may occur between 7 to 11% when surgical resection is considered to be total [23, 30], and between 7 to 53% is resection is subtotal [31-34]. Following total microsurgical resection of large VSs, facial nerve function preservation (HB I or II) has been achieved in 27 to 58% [23, 30, 35-38] while subtotal resection achieves better preservation rates between 82 to 88% [39, 40]. From recent GKS series data, it is evident that GKS is considered to have an long-term efficacy in controlling tumor growth, of as much as 97-98%, with a very low risk of facial nerve dysfunction, of less than 1% in most of the series [20, 41-44]. Hearing preservation following microsurgical resection in large vestibular schwannomas varies between 0 to 29% (37, 45-50), and in GKS from 38 to 94% [46, 51-53].

Several studies have compared radiosurgery to microsurgery for treating smaller VS. In a retrospective review by Myrseth et al. in 2005 with median follow up of 5.9 years, local control rates between microsurgery and gamma knife were not statistically different (89.2% versus 94.2%) [54]. Facial nerve function and quality of life were both significantly worse in the surgery group versus the GKS group. GKS achieved a facial nerve preservation rate of 94.2% while surgery achieved only 79.8% HB grade I-II [54]. To give further strength to Myrseth's study, Pollock et al. in 2006 performed a prospective study comparing surgery versus radioand knifesurgery in 82 patients with tumors <3cm in size. Median follow up was 42 months and local control rates were not significantly different between the groups (96% versus 100%). Despite these findings, facial nerve function and hearing preservation were both significantly worse in patients treated with surgery versus those treated with Ggamma knife (75% and 5% versus 96% and 3%) [55].

If the "attractive combination" [56] of microsurgery and GKS is used, preservation of facial nerve function occurs between 85.7 to 95% of the cases [40, 57-60]. Regarding combined approaches, Yang et al. [60] reported up to 50% of maintaining serviceable hearing in those patients having preoperative hearing. Our results with 100 % facial nerve preservation (whole series) along with preservation of hearing for all patientsthose with useful pre-operative hearing pre-operatively, confirms the view that combined approaches for large VS achieves much better results than microsurgery used as stand-alone treatment.

Petroclival Meningiomas

General Considerations

Petroclival meningiomas (PCM) account for approximately 3 to 10% of the meningiomas of the posterior fossa [61-63]. Radical removal of these tumors remains challenging, primarily due to the fact that they usually involve multiple regions with significant adherence or invasion of the brainstem, encasement of the basilar artery and its perforators, as well as involvement of cranial nerves V, VII, VIII, IX, X, XI [62, 64]. PCM are typically described as those meningiomas attached to the lateral dura, along the petroclival borderline where the sphenoid, petrous and occipital bones meet [65-67].

Retrosigmoid Approach for PCM

Several skull-base approaches have been described in the literature to resect PCM, namely the trans-labyrinthine, trans-cochlear, total petrosal, retrosigmoid and subtemporal approaches. We prefer to use the standard retrosigmoid approach. The main advantage of this approach is the familiarity with this approach and the fact that when compared to the other approaches it represents a minimally invasive microsurgical strategy that can adequately ensure a planned subtotal resection while preserving neurological function. The retrosigmoid approach has been described in the section on vestibular schwanomma. For the intraoperative electrophysiological monitoring, the occulomotor cranial nerves are also monitored in addition to evoked potentials and monitoring of VII-IX cranial nerves. After dural opening, the tumor is frequently seen anterior to the vestibulo-cochlear and facial nerves. The tumor is accessed through windows between the V,- VII-VIII and the IX-X-XI nerves complex. The tumor is disconnected from its attachment to the petrous bone and tentorium based on its location. Internal decompression of the tumor is performed with CUSA making sure that the nerves are not damaged by manipulation. The continuous EMG and intermittent electrical stimulation of the nerves posterior adjacent to the tumor proves to be invaluable in preserving neural function. The extent of tumor removal depends on the amount of the clival component, invasion of the cavernous sinus and adherence to the brainstem pia and basilar artery and its branches. The parts of the tumor that are deemed risky to remove are left behind for later treatment with GKS. Specific issues related to GKS treatment are definition of the target in "en-plaque" lesions, extension to the tentorial edge and involvement of occulomotor cranial nerves, potential cavernous sinus invasion and persistent brainstem mass effect.

Case Series

We treated four patients who presented with PCM (Ffigure 2). Three were females and one was male. Mean age in this series was 60.27 years (range 42- 73.3). Mean follow-up was 7.6 months (range 5.1- 13.2). All the lesions were located on the left side. They presented with hearing loss (3 patients), hemiparesis (one patient), VIth cranial nerve paresis (one patient); one was an incidental finding. The mean tumor volume before surgery was of 17.1 cc (range 11.2-24.2). The histology was WHO I meningotheliomatous meningioma in all patients. Postoperative clinical examination showed a House-Brackmann grade 2 facial function in 3 cases, which completely recovered. One patient continues to have a grade 4 facial weakness. This patient also had a transient VIth nerve paresis (that completely recovered) and trigeminal hypoesthesia in the V1 distribution. No other complications were encountered. In one case, surgical re-intervention was attempted due to a large residual tumor volume, which was not favorable for GKS; this second surgery did not achieve a significant debulking due to severe adherences to the brainstem and the lower cranial nerves. The patient was later treated with staged GKS. The mean duration time between surgery and GKS was 5.6 months (range 4.3-7). The mean tumor volume at the moment of GKS was 9.9 cc (range 5.1- 14.2). A mean number of 26 isocenters was used (range 21-29). Three patients (75%) were treated with 12 Gy and only one (25%) with 14 Gy at the periphery, at the 50% isodose. The 12 Gy peripheral dose is preferred, especially when there is a contact between the residual tumor and the brainstem.

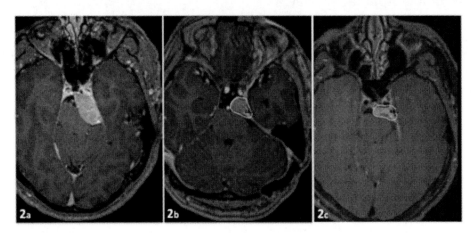

Figure 2. Example of a petro-clival meningioma treated with combined approach. This 42 year-old male patient presented with an incidentally discovered petroclival meningioma. A "wait-and-see" approach was first adopted. Subsequent imaging showed tumor growth and surgery was performed at this stage (2a). He underwent subtotal excision of the tumor. In the immediate postoperative period, the patient had a House-Brackmann grade III facial function along with V1 hypoesthesia. Subsequent clinical follow-up showed an improvement of the facial function to House-Brackmann II. The residual part of the tumor was treated with GKS. The tumor volume before surgery was 11.2 cc; the target volume at the time of radiosurgery was 5.1 cc; the GKS was performed with a dose of 12 Gy at the 50% isodose (2b). Subsequent shrinkage of the residual tumor is illustrated by the co-registration and projection of the GKS prescription isodose on the 6-month follow-up MRI (2c).

Special care was than given to optimize the dose gradient at the interface between the tumor and this neural structures. Two patients had staged GKS. In one case, the remnant tumor volume after surgery was 13.6 cc; an important progression has been confirmed at the moment of GKS into the ethmoidal sinus; the delay between surgery and GKS in this case was of 5.1 months; in the first stage, we treated the evolutive part, along with the invasion into the cavernous sinus and the temporal fossa; a second stage was planned 6 months later, to the posterior part of the tumor, at the interface with the brainstem. No complications have been noted following GKS.

Discussion

Cushing was the first to describe PCM as "the most challenging and formidable meningiomas to treat" [68]. For a long period of time, they were often regarded as "inoperable". A first successful total removal case was reported by Mayberg et al. [69]; before the 1970s, the risk of mortality could be as high as 50% [70-72]. Despite advances in microsurgery, neuroimaging (especially high-quality MRI) and electrophysiological monitoring, they are still associated with high morbidity and mortality [61-65, 67, 73, 74]. Selection of the optimal strategy for these tumors still remain a matter of debate in the current literature [85], and depends on several factors like natural history, patient's age, tumor size, symptoms, quality of life and also the possibility of using radiosurgery as first intention treatment. For large to giant PCMs, microsurgery is generally considered as the primary treatment [75]. When dealing with small to medium lesions, several studies have shown the value of subtotal resection followed by adjuvant therapies, or first intention radiosurgery [63, 75, 76]. GKS has gained increasing acceptance as a therapy in the armamentarium for PCM, in newly diagnosed, subtotally resected or recurrent tumors. Radiosurgery achieves good tumor control in 80% to 100% of the skull-base meningiomas, with low morbidity [75-79]. There have also been few studies reporting results of radiosurgery as a complementary treatment after initial surgical tumor debulking [80, 81]. Surgical mortality in PCM has been reported to range between 0 to 10% in previous series [62, 64-66, 70, 82-85]. Major morbidity varies between 8 to 45%, with cranial nerve deficits between 20 to 70% [62, 64-66, 70, 82-85]. A recent study by Flannery et al. (86) on a large series of 168 patients treated with GKS reported a tumor control of 91% at 5 years and 86% at 10 years (90% after excluding the higher grades). Our series of combined treatment of PCM had no mortality or major morbidity, and the cranial nerve function preservation rates compare favorably with those other series from literature.

Clinoidal Meningiomas

General Considerations

The clinoidal meningiomas (CM) are a subcategory of the sphenoid wing meningiomas that develop at the junction of the anterior and middle cranial fossae. The incidence of CM ranges between 34 to 43.9% of all meningiomas [87-89]. The surgical treatment of the CM needs to deal with the preservation of the carotid artery and its branches, preservation or recovery of visual function, limiting occulomotor cranial nerve palsy and achieving a total resection or adequate tumor control. Visual involvement usually occurs either in relation to direct compression of the optic nerve or in relation to ischemia or demyelination of the optic nerve [90]. These tumors may grow towards the midline and involve the jugum sphenoidale, the diaphragm sellae, the pituitary gland stalk (and infundibulum) and the clivus. They can extend to the cavernous sinus [91] or invade the optic foramen or orbital cavity. They can infiltrate the bone or be associated with cause hyperostosis [89, 91-93].

One of the still confounding issues today is related to their classification, which may differ from one study to another. In 1938, they were classified by Cushing and Eisenhardt [94] into outer (pterional), middle (alar) and inner (deep, clinoidal). Al-Mefty [87, 88, 95] proposed a classification that takes into account not only the point of origin of the meningioma, but also its relationship with the internal carotid artery (ICA), as follows: group I, with implantation on the lower part of the clinoid and that develop in the carotid cistern and incase ICA; group II, originating from the superior or lateral aspect of the anterior clinoid process and as they grow, may enter in contact with ICA; group III, that originating from the optic foramen in which the arachnoid membrane is present between the vessels and the tumor, but may be lacking between tumor and optic nerve. The presence of the arachnoid membrane is of high importance, as it offers a cleavage plane and allows the vessels to be totally freed from the tumor. Finally, Risi et al. [93] proposed a classification regarding the direction of tumor development, into pure clinoid, clinoid with lateral extension and clinoid with cavernous sinus extension.

From a surgical perspective, 2 challenges arise when ensuring optimal strategies of these lesions: the surgical approach and the management of the intracavernous extension of the tumor, which is reported to be as high as 44.1% in several series [87, 93, 96, 97]. The advances in skull-base surgery, like the orbito-zygomatic osteotomy, the epidural approach to the clinoid, and access to the cavernous sinus, have contributed to a large extent to make surgery for these tumors feasible while substantially reducing the risk of vascular injury or neurological deficits.

Pterional Approach for CM

Either a fronto-lateral or a pterional approach is performed, with an orbito-zygomatic osteotomy in addition if necessary, based on tumor location. In the fronto-lateral approach, a burr-hole is placed just posterior to the anterior temporal line. This approach provides a more medial view of the clinoid and suprasellar region, which allows early identification of the optic nerve. For the pterional approach, the single burr-hole is placed at the same location, but the craniotomy is performed more posteriorly, exposing the sphenoid ridge, sylvian fissure and the frontal and temporal lobes. The lesser wing of the sphenoid is drilled-off up to the superior orbital fissure. The superior orbital fissure is unroofed to expose the periorbital fascia, which is opened in cases of tumour infiltration of the optic foramen or intraorbital structures. For cases where there hyperostosis around the optic canal, a partial clinoidectomy is performed prior to extradural unroofing of the optic canal. The dura of the region of the tuberculum sellae and chiasmatic sulcus is inspected for evidence of meningeal tumor infiltration and to identify the optic nerve intradurally. The falciform ligament, around the optic nerve is opened for a short distance towards the orbital apex in cases of tumor in this region. While dealing with clinoidal meningiomas, great care needs to be taken because the carotid and middle cerebral arteries may be embedded in the tumor and in some cases there can be involvement of the vessel wall by the tumor. The tumor is then visualized in the intradural compartment after opening the sylvian fissure and separating the frontal and temporal lobe. The tumor capsule is opened and it is internally decompressed with bipolar coagulation, micro-scissors, and ultrasonic aspiration. After the tumor has been dissected off the vessels, the optic nerve, chiasm or the pituitary stalk, the capsule is removed in pieces. In case of vasospasm following dissection along cerebral arteries, local administration of nimodipine or papaverine sponges can be done to prevent cerebral ischemia. After resection of the tumor, the dural attachment is resected, including the hyperostotic bone of the lesser sphenoid wing or the anterior clinoidal process, if this is possible without endangering nerves or vessels. The portion of the tumor invading the cavernous sinus is not aggressively removed. In these cases a small fat graft is interposed between the optic nerve, chiasm, pituitary stalk and the residual cavernous sinus component of the tumor. This form of "chiasmopexy" or "hypophysopexy" allows for optimal dose treatment with GKS. The dura is closed primarily and for the parts of the skull-base dural defects, these are reconstructed with fascial grafts and fibrin glue. This ensures that there remains no communication with the ethmoidal or sphenoidal sinuses. Bone flaps are repositioned and the muscle and skin flaps are closed in layers over a drain.

From a radiosurgical point of view, a safe optimal distance between the tumor and the optic nerve is mandatory, in order to deliver an optimal radiation dose and to avoid toxicity due to irradiation close to the optic pathways. A good conformity and selectivity of the tumor with respect to the optic pathways is crucial to achieve a marginal maximal dose between 12 and 14 Gy to the tumor, with a an optimal dose gradient in the direction of the optic pathway. This achieves a high-rate of tumor control, with very low risk of optic nerve toxicity. In general, the optic pathway is considered to tolerate up to 8 Gy without risk (maximum of 10 Gy in selected situations for very small volumes) [81, 98-100].

Extended Endoscopic Endonasal Approach to Parasellar Regions

A complete preoperative evaluation is performed with high-resolution brain computed tomography (CT) scans in bone window, CT-angiography and magnetic resonance imaging. Converse to endoscopic endonasal approach to pituitary tumors, this extended approach often uses a bi-nostril approach in order to improve the anatomical exposure, the endoscope being placed in the nostril contralateral to the lesion and surgical tools being placed within both nostrils [101]. The whole endoscopic procedure until the sellar floor opening is performed with a hand-held short 0°, 30° or 45° endoscope (diameter: 4mm; length: 18 cm). The superior and middle turbinates ipsilateral to the lesion are resected. The middle turbinectomy allows both enlarging the nasal corridor to facilitate access to surgical tools and facilitating the identification of major anatomical structures: the vidian nerve and the sphenopalatine artery. Anatomical landmarks are confirmed with the neuronavigation system with CT scan / MRI fusion and Doppler. The contralateral middle turbinate is out fractured. These turbinectomies allow constructing a nasoseptal flap pedicled on the sphenopalatine artery and its nasal branches. The middle turbinate is kept intact until the closure time. Both the flap and this middle turbinate will be used for the cranial base defect reconstruction at the closure time. The nasoseptal flap is elevated with its intact blood supply and placed within the nasopharynx for protection. Opening the maxillary sinus and the posterior ethmoidal cells ipsilateral to the lesion consists in a far lateral approach to the sphenoid recess. The nasal corridor is completed by the realization of posterior nasal septectomy and large bilateral sphenoidotomies. Then, a wide opening of the sphenoid sinus is required to have access to the anterior skull base to the sphenoid floor and beyond to the ispsilateral sphenoid recess consisting in a transpterygoid approach. To be noted, after the opening of the sphenoid sinus, a transcavernous sinus approach, consisting in the exposure of the pituitary dura by removing bone over the sella turcica, may be performed allowing an exposure of the cavernous sinus more directly than the traditional approach to the pituitary. This approach requires the use of Doppler technology to identify the internal carotid artery (ICA). However, we prefer to be not so aggressive and would leave intracavernous tumor in place for later treatment with GKS.

The endonasal route to more lateral lesions is performed through a transmaxillary corridor, which give access to pterygopalatine and infratempral fossa and Meckel's cave. The transpterygoidal approach often combined with transethmoidal and transsphenoidal approaches depending of the targuet to be treated (from optic nerve, orbital apex, cavernous sinus, to infratemporal and pterygopalatien fossa. A large antrostomy allows exposuree and removal of e the entire posterior wall of the maxillary sinus. The first step is to clip the sphenopalatine or maxillary arteries in the pterygomaxillary fossa, to prevent local bleeding. The anatomical landsmarks are the floor of the sphenoid sinus in which run the vidian nerf the maxillary nerve (V2) which is identifiedy aton the floor of the orbit.al floor.. The vidian foramen located in the floor of the sphenoide sinus is a major anatomical landmark allowing identificationying of the petrous internal carotid artery (ICA) and the V2 reaching to the Meckel's cave. Drilling the vidian canal is realized to isolate the ICA and bone removal is realized in order to achieve proximal and distal ICA control. The V2 is followed through the foramen rotundum into the middle cranial fossa. To reach the Meckel's cave lesion, the opening of the dura mater is realized within the space between the distal ICA medially, the V2 laterally, the proximal petrous ICA inferiorly and the abducens nerve superiorly. In case of huge tumors, the anterior temporal fossa is also opened using the anteromedial (between V1 and V2) and anterolateral (between V2 and V3) middle fossa triangles landmarks. To reach lesions located in the petrous apex region, it is necessary to drill medially the bone and along a vertical plane parallel to the ICA. When the bone is thin enough, small pieces are removed and the lesion can then be reached.

The tumor removal step is the same as in microsurgical procedure. The closure time consists ofin plugging with fat the region that communicates with the intradural space. intradurally protecting neurovascular strucures. Extradurally, a resorbable dural substitute is used along with an overlay of a as well as nasoseptal flap and/ theor the middle turbinate, reinforced with fibrin with glue. Plugging of the sphenoid sinus is made with fat.

Case Series

Nine patients (all females) presented with a CM (figures 3 and 4). The mean age in this series was 56 years (range 35- 71 years). Tumors were located on the right side in 2 patients and on the left side in 7 patients. The mean tumor volume before surgery was 20.2 cc (range 4.4- 55.1). Visual loss was the most common symptom (8 cases) of which 2 patients had complete unilateral blindness at the time of surgery. All underwent a transcranial surgery through a frontolateral/ pterional approach, while in one patient an additional endoscopic endonasal tumor resection was also performed. Eight tumors were WHO I meningioma and one tumor was an atypical grade II meningioma. The visual status normalized in four cases (44.4%) and improved in 3 (33.3%). One patient had visual deterioration following surgery, which was clearly related to the chiasmopexy and, following of reduction in size of the fat graft at a second surgery, the visual function completely recovered. One case each of transient IIIrd and VIth nerve palsy were noted. No other worsening of pre operative neural dysfunction was seen. No major complications were thus encountered in this series. There were no wound problems or CSF dural leaks.

Five patients have already been treated with GKS while four await for GKS as part of the combined approach. The mean duration between microsurgery and GKS was 5.2 months (range 3.2- 8.4). The mean tumor volume at the moment of GKS was 4.4 cc (range 0.386- 11.1).

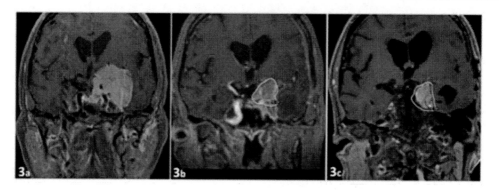

Figure 3. Example of a large clinoidal meningioma treated with combined approach. This 71 year-old female patient presented with markedly reduced visual acuity on the left side and with VIth nerve palsy. The preoperative MRI scan shows a large clinoidal tumor with encasement of the supraclinoid and cavernous sinus carotid artery (3a). She underwent a resection of the tumor though a pterional craniotomy and the part of the tumor encasing the carotid was left behind for GKS. Postoperatively, she completely regained her vision and had no diplopia. The tumor volume before surgery was 39 cc and the target volume for GKS was 7.5 cc. GKS was performed with a dose of 12 Gy at the 50% isodose (3b). Subsequent shrinkage of the residual tumor is illustrated by the co-registration and projection of the GKS prescription isodose on the 6-month follow-up MRI (2c).

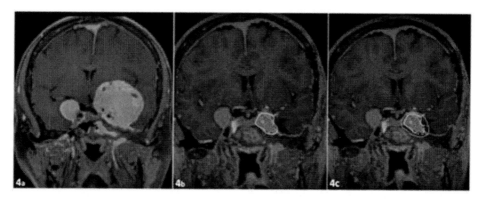

Figure 4. Example of a large clinoidal meningioma treated with combined approach, in a patient with bilateral clinoidal meningiomas. This 47 year-old female patient presented with visual loss on the left side. MR images showed bilateral anterior clinoid meningiomas (4a). On the left side, there was a large clinoidal meningioma markedly displacing the optic nerve and the chiasm and encasing the intradural carotid artery. There was also a component of the tumor in the cavernous sinus and in the sphenoid sinus extending laterally around the orbital apex. The intracranial part of the tumor was approached through a pterional craniotomy. The entire intradural part of the tumor was completely removed including the part encasing the carotid artery and its bifurcation. Following surgery, she completely recovered the vision on the left side. AAt a second stage surgery was performed 3 months aiming at the extracranial part of the tumor, which was removed through an endoscopic endonasal approach. GKS for the cavernous sinus component was performed 6 months later. The tumor volume before surgery was 55.1 cc and the target volume at the time of radiosurgery was 1 cc. GKS was performed with a dose of 14 Gy at the 50% isodose (4b). The maximal dose to the optic chiasm and the left optic nerve was 3.6 Gy and 7.6 Gy respectively. Fig 3c shows the 8 Gy isodose line (as reference for the maximal dose allowed to the optic pathways) in green while the optic chiasm is delineated in pink showing its distance from the treated volume. The smaller tumor on the right side was completely excised through a suprabrow mini-craniotomy 3 months later.

A mean number of 21.8 isocenters was used (range 17-28). Four patients were treated with a marginal dose of 12 Gy and one patient (11.1%) with 14 Gy at the periphery. Four patients await GKS treatment. In our experience, with the current GKS technique and the use of Perfexion®, a minimal marginal dose of 12 Gy is almost always achievable, provided that there is a small distance (even as low as 1-2 mm) between the tumor and the optic pathway. One patient had transitory IVth nerve palsy accompanied by decreased visual acuity and left trigeminal neuralgia (V1 and V2 distribution), 6 weeks after the GKS. This was clearly related to tumor size increase/swelling related to GKS. She was treated with steroids and within 6 weeks all neurological symptoms had completely recovered. In fact, subsequent imaging showed a marked reduction in tumor size with the intracavernous ICA becoming more visible. The mean interval between surgery and GKS was intentionally reduced because of the close proximity of the residual tumor to the optic apparatus. The mean follow-up was 8.1 months (range 1.8-17.5).

Discussion

Complete resection of these types of tumors remains challenging, especially in large or giant cases [102]. In the current literature, the clinical status has been reported to be stable or improved following treatment in 50 to 78% of the cases [88, 92, 93, 103]. In our series, the visual status normalized or improved in 7 of the 9 patients. This has to be seen in the light of the postoperative worsening that has been published in literature, reported to range between 10.7% and 29% [88, 92, 93, 102, 103]. For occulomotor cranial nerve function, we had no long-lasting deficits in any patient, though trans-cavernous surgery is particularly known to produce these deficits in up to 50% based on reports from literature (104, 105). Major morbidity is a well-recognized concern following radical resection of these tumors. Hemiparesis or hemiplegia has been reported to be as high as 13.6% in giant clinoid meningiomas [88, 92, 93, 102, 103]. Mortality has been reported as a complication in 5.9%, but as high as 42.8% in some reports [88, 91-93, 102, 103]. We have had no mortality or major morbidity in our series. Following surgery alone, rates of tumor recurrence occur in up to 17.6% in subtotal resection [88, 89, 92, 102, 103] after macroscopically complete tumor removal to vary between 0 to 26% in several recent series [91, 96, 103, 106, 107]. As the size of the tumor has an important impact on surgical outcome, combined approaches are of huge value in diminishing morbidity and mortality related to surgery, while offering the possibility for GKS to be made in optimal conditions. Gamma Knife surgery has already documented its safety-efficacy in several studies regarding good control of the meningiomas within the cavernous sinus [79, 81, 98, 100, 108-111]. In this context, combined approaches with planned subtotal resection and GKS on the remnant tumor are crucial in obtaining tumor control while achieving a good neurological outcome.

Pituitary Adenomas

General Considerations

Pituitary adenomas (PA) arise from the anterior pituitary gland and are classified based on their size into microadenoma (< 1 cm) or macroadenomas [112]. They represent approximately 10% of intracranial tumors and are most common in the 3rd and 4th decades of life, affecting equally both sexes. Pituitary tumors usually present either due to endocrinological disturbance, or due to mass effect. A small number presents with pituitary apoplexy. Classically, pituitary tumors are divided into two groups: functional (or secreting), and non-functional (non-secretory or else secreting products such as gonadotropin that do not cause endocrinological symptoms). The latter usually do not present until of sufficient size to cause neurologic deficits by mass effect, whereas the former frequently present earlier with symptoms caused by physiologic effects of excess hormones like prolactin, growth hormone (acromegaly) or ACTH (Cushing's disease) [112-116]. Panhypopituitarism may be caused by large tumors of either variety (usually the non-functional type) as a result of compression of the pituitary gland. Except for prolactinomas [117-123], the primary treatment for all other symptomatic PA is surgery [112-116, 124-128]. The endocrinological and oncological outcome following surgery primarily depends on the extensions of the tumor. The majority of the sellar and suprasellar tumors can be treated with surgery only. When they invade the parasellar structures or involve multiple compartments, like subfrontal or temporal fossa extensions, additional surgery, medical therapy or radiosurgery are necessary to achieve tumor control and/or endocrinological remission. This is especially true for secreting PA where even a small residue in the cavernous sinus will preclude remission. The combined approach is therefore specifically indicated in this subgroup of patients.

Endoscopic Endonasal Transsphenoidal Approach for Pituitary Adenomas

Generally the right nostril is chosen for the approach. All necessary instruments are inserted through the chosen nasal cavity parallel to the endoscope. Exceptionally, in case of large tumors, both nasal cavities can be used along with a four-hand technique. The whole endoscopic procedure until the sellar floor opening is performed with a hand-held short 0° (or 30°) angled endoscope (diameter: 4mm; length; 18 cm) allowing panoramic visualization of the anterior wall of the sella turcica as well as the optic and carotid protuberances. When this endoscope is introduced in the nasal cavity, it is necessary to identify the superior and middle turbinates to push them laterally aside. Sometimes, in cases of turbinate hypertrophy such as in acromegalic patients, it will be necessary to remove the middle turbinate in order to allow the endoscope clearing. Then the coagulation and the opening of the mucosa from the sphenoidal ostium to the choana at the base of the vomer is performed. A large sphenoidotomy is performed and the vomer is pushed away until the apparition of the contralateral ostium allowing a better anatomical visualization. The dura mater is opened after the realization of a small bone flap from one cavernous sinus to the other and from the anterior skull base to the clivus. The sellar dura is opened in rectangular shape and the tumor is visualized using a long 0° endoscope (diameter: 4 mm, length: 30 cm). The tumor is resected piecemeal and the pituitary gland is visualized and kept intact along with the arachnoid pouches. At the end of the procedure, the sellar and suprasellar regions are explored using 30° or 45° endoscopes pushed up through the sella turcica to detect potential residue within suprasellar and lateral regions. In order to detect potential cerebrospinal fluid leakage, the anesthesiologist raises the pCo2 to increase the intracranial pressure. The closure is performed by filling the tumor cavity and the durotomy defect with a fat graft. This achieves dural closure as well as serves as a means of distancing the chiasm and pituitary gland from a residual cavernous sinus portion of the tumor [129-133]. This allows the use of high doses of GKS especially for the treatment of secreting PA in the cavernous sinus. Nasal packing is only rarely used.

Case Series

Eight patients (3 males and 5 females) were treated with PA utilizing the combined approach (figure 5). The mean age in this series was 52.3 (range 13-72 years). The presenting symptom was visual deficit in 7 of 8 patients, and one patient presented with symptoms of acromegaly. All patients underwent endoscope-assisted trans-sphenoidal microsurgery, while one patient had a staged trans-cranial resection of tumor in addition. The mean tumor volume before surgery was 16.2 cc (range 5.2-27.1 cc).

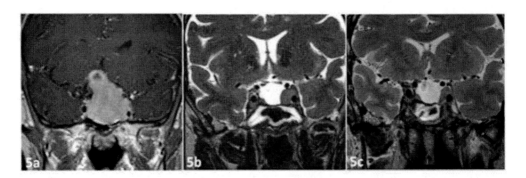

Figure 5. Example of a large pituitary adenoma treated with combined approach. This 13 year-old female patient presented with marked reduction of visual acuity especially on the right side and with a homonymous hemianopia. Endocrine evaluation revealed a non-functional tumor with partial anterior pituitary dysfunction. MRI with gadolinium enhancement showed a large pituitary adenoma, Hardy's grade D, Knosp IV, with bilateral invasion of the cavernous sinus (5a). She underwent a staged trans-nasal trans-sphenoidal endoscopic resection of the tumor. The vision on the left side partly improved while the visual field deficit of the right eye completely recovered. She needed hormonal substitution for the hypopituitarism. A residual tumor located bilaterally in the cavernous sinus was visible on T2-weighted coronal MRI (5b). The tumor volume before surgery was 20.3 cc while the target volume at the time of radiosurgery was 1.1 cc. GKS was performed with a dose of 12 Gy at the 50% isodose for both right and left targets, while the maximal dose to the optic pathway was 8.5 Gy. Six months following GKS, there is a marked reduction of the residual tumor (5c).

Postoperatively, visual status improved/ normalized in all patients while no patient had worsening of visual or occulomotor function. Two patients had transient diabetes insipidus following surgery. Three patients had postoperative anterior pituitary hormone insufficiency, which subsequently improved.

GKS was performed at a mean duration of 7 months (range 2.8-12.5) following surgery. The mean tumor volume at GKS planning was 2.8 cc (range 0.9- 6.4 cc). The residual tumor was located in the right cavernous sinus (4 patients), the left cavernous sinus as well as in the right infrachiasmatic sella (one patient), bilaterally in the right and left cavernous sinus (one patient), in right cavernous sinus with a posterior part in contact with the brainstem and a lateral extension in the temporal fosa (one patient). A mean number of 16.9 isocenters was used (range 4-25). The mean peripheral dose at the 50% isodose, was 24 Gy (range 18-30) in secreting and 13 Gy (range 12-14) in non-secreting PA, respectively. The maximal dose received by the optic pathways was 7.9 Gy (range 5.5-11.7). Current literature admits a maximal dose between 8 and 10 Gy. Only one patient overpassed the maximal limit of 10 Gy, but the dose received by the 1 mm^3 and 10mm^3 of the optic apparatus was 9.4 and 8 Gy, respectively.

Two patients (50%) had staged radiosurgery. One of them had a bilateral residual tumor located in the cavernous sinus. Thus, we decided to perform a staged radiosurgery, so as to avoid the overlap of dose at the level of the chiasm but also pituitary gland and stalk. The other patient had a tumor which was extending in the right cavernous sinus and also at the level of the petro-clival junction; this patient had lost vision on the right side so the main objective was to avoid toxicity for the chiasm and the left optic nerve; thus, we avoided any dose overlap that could endanger the vision of the left eye. No additional deficit after GKS was encountered. The mean follow-up period for this group was 18.7 months (range 5.8-30.7).

Discussion

Surgical approaches to pituitary lesions was initially transcranial until the transsphenoidal approach became the gold standard with the introduction of operative microscopey and fluoroscopy by Guiot et al. (134) and Hardy [135]. Current classification of PA divides it into secreting (or functioning) PA and non-secreting (or non-functioning) PA based on endocrinological presentation, immunohistochemistry and electron microscopy. Surgery is required for the great majority of secreting PA except for prolactinomas, where medical treatment is efficacious in the majority of patients. For non- secreting PA, surgery is required when there are visual and/or endocrinological impairments. This surgery is performed either by a microscopic or endoscopic approach. The main advantage of the endoscopic approach is improved visualization of the sellar and suprasellar anatomy, improved gross total removal [136, 137], reduction of nasal swelling, avoidance of the use of fluoroscopy (138) and shorter hospital stay [139]. Its major disadvantages are the lack of binocular vision and the narrow nasal corridor.

With regards to postoperative results, gross total removal is achieved for non-secreting PA in 35.5 % to 74 % [137, 140, 141] in microscopic series versus in 56% to 93% [133, 137, 142] in endoscopic series. For secreting PA, remission is obtained with surgery alone in 75% to 85% of non-invasive adenomas (143). For ACTH-secreting adenomas and GH-secreting adenomas, remission is respectively obtained in 81 % and 85% with endoscopic approach versus 78% and 67% with microscopic approach [144]. Long term recurrence rates of hormonal hypersecretion vary from 7 to 25% [144]. Visual dysfunction may be present at the time of surgical management. With both microscopic and endoscopic approach, all patients with preoperative visual impairment recover enough vision to resume a normal life, unless a long preoperative evolution has already caused an optic atrophy. The visual outcome reported in literature show more than 92 % of normalization or improvement with no experience of worsening across several large series [137, 144-146].

Nasal morbidities are reported in 0.7% to 7% of cases and are either major bleedings secondary to sphenopalatine artery lesion or minor epistaxis [133, 144, 145, 147]. Sphenoid sinus complications occur in about 2% and 9% in endoscopic and microsurgical series respectively. Other rare complications like fracture of the sphenoid bone, injury of the optic nerves or lesions of the carotid artery have also been reported [148]. The most common complication of transsphenoidal surgery is obviously the occurrence of CSF leaks reported in 1.2% to 6% of endoscopic series and 0.9% to 3 % of microsurgical series [133, 144, 145, 147]. This risk is increased in cases of suprasellar and/or parasellar extensions. Meningitis has a comparable risk to intracranial open skull base surgery with a rate of about 1.8%, increasing with persistence of CSF leaks and postoperative external ventricular or lumbar drains. [149]

Transsphenoidal procedures are at risk of endocrinological impairment such as transient or permanent diabetes insipidus and hypopituitarism. Permanent diabetes insipidus is reported in 1% to 5% and 0.9% to 7.6% in endoscopic and microscopic series respectively [133, 144, 145, 147, 150, 151]. Postoperative hypopituitarism may be avoided by careful tumor resection but occurs in around 14% of cases [133].

Particular attention needs to be given for PA that invades the cavernous sinus, existing in around 10% of cases [152]. Indeed, injury to the internal carotid artery typically occurs during aggressive surgery of macroadenomas extending in the cavernous sinus. These injuries are rare (0% to 0.68%) but may be responsible for mortality [153]. Intracranial hemorrhage is very rare and results from internal carotid lesion, carotid-cavernous fistulas or pseudo-aneurysm, all as a result of aggressive surgery within the cavernous sinus. The overall mortality rate following pituitary adenoma surgery is reported to be between 0% to 0.68% and 0% to 0.9% in endoscopic and microsurgical series respectively [133, 144, 145, 147]. Delayed vascular complications following such surgery can also occur several weeks or even years after such procedure [154]. Trans-cavernous surgery also has a high risk of cranial nerve morbidity [104, 155, 156]. It is mainly these major complications that can be largely avoided following the combined approach where the cavernous sinus portion (in our series) is treated with radiosurgery.

Regarding GKS, tThe two major risks of complications in PA are pituitary insufficiency and visual loss. In non-secreting PA, pituitary insufficiency varies between 0 to 25% and visual complications between 0 to 4.8% [157-162]. In secreting PA, pituitary insufficiency varies between 0 to 19% in acromegaly [163-166], between 0 to 16% in Cushing disease [167-169] and between 0 to 49% in prolactinomas [164, 170, 171]. The most important predictors are the treatment maximal dose and the target volume. Recent papers have shown the importance of the dose received by the pituitary stalk (cut-off at 4.1 Gy) [172] and by the pituitary gland (cut-off at 15 Gy) [173] as predictors of pituitary insufficiency. This is the most frequent complication, usually appearing within 12 months, but it can also appear as a late side-efffeect (up to 100 months) [174]; andit is more frequent after previous radiation therapy. Visual complications can be avoided when using a dose of no more than 8 Gy to the optic apparatus.

As the optic pathway is the most sensitive intracranial structure, the risk increases in cases of previous radiation therapy. A safety distance is necessary to ensure optimal GKS planning. In our unit this is achieved by the use of peroperative spacer (fat or muscle tissue) and is of valuable help to avoid such type of complications. Still used in some centers and also in specific cases, conventional radiation therapy caries a risk of endocrine dysfunction of around 50%, with also a risk of vascular complications (less than 5%), optic neuropathies (1 to 2%), secondary malignancy (1-3%), neuropsychological deficit (less than 1%) and radiation necrosis (less than 1%) [161, 163, 165, 175-177].

Conclusion

Early results from our series of patients who underwent combined treatment for large benign skull base tumors show that there is an excellent clinical outcome with respect to preservation/improvement of neurological functions. Microsurgery (trans-cranial and/or endoscopic trans-nasal) and GKS are to be viewed as complementary treatment modalities. For achieving optimal results, there needs to be a perfect collaboration between the microsurgeon and the GKS surgeon. Preoperative planning of the strategy is crucial to assign parts of the tumor to individual treatment modalities. The inherent philosophy of this treatment along with its safety-efficacy evaluation depends largely on a good understanding of both therapeutic steps and modalities, each of which is of high importance. While the microsurgeon can avoid operating on the part of the tumor that is at highest risk, the GKS surgeon needs to appreciate that surgery will modify local conditions and thereby make GKS treatment planning more difficult. Under these conditions, the contact between the two neurosurgeons is crucial for both the pre-operative strategy and the GKS planning. As long-term results emerge, this multimodal treatment paradigm could become the standard of care for the management of difficult benign skull-base tumors.

References

[1] Leksell L. The stereotaxic method and radiosurgery of the brain. *Acta Chir. Scand.* 1951 Dec 13; 102(4): 316-9.
[2] Leksell L. Cerebral radiosurgery. I. Gammathalanotomy in two cases of intractable pain. *Acta Chir. Scand.* 1968; 134(8): 585-95.
[3] Levivier M, Gevaert T, Negretti L. Gamma Knife, CyberKnife, TomoTherapy: gadgets or useful tools? *Curr. Opin. Neurol.* 2011 Dec; 24(6): 616-25.
[4] Lindquist C, Paddick I. The Leksell Gamma Knife Perfexion and comparisons with its predecessors. *Neurosurgery.* 2008 Feb; 62 Suppl 2: 721-32.
[5] Szeifert GT, Atteberry DS, Kondziolka D, Levivier M, Lunsford LD. Cerebral metastases pathology after radiosurgery: a multicenter study. *Cancer.* 2006 Jun 15; 106(12): 2672-81.
[6] Szeifert GT, Kondziolka D, Atteberry DS, Salmon I, Rorive S, Levivier M, et al. Radiosurgical pathology of brain tumors: metastases, schwannomas, meningiomas, astrocytomas, hemangioblastomas. *Prog. Neurol. Surg.* 2007; 20: 91-105.
[7] Szeifert GT, Kondziolka D, Levivier M, Lunsford LD. Histopathology of brain metastases after radiosurgery. *Prog. Neurol. Surg.* 2012; 25: 30-8.
[8] Szeifert GT, Major O, Kemeny AA. Ultrastructural changes in arteriovenous malformations after gamma knife surgery: an electron microscopic study. *J. Neurosurg.* 2005 Jan; 102 Suppl: 289-92.
[9] Szeifert GT, Salmon I, Baleriaux D, Brotchi J, Levivier M. Immunohistochemical analysis of a cerebral arteriovenous malformation obliterated by radiosurgery and presenting with re-bleeding. Case report. *Neurol. Res.* 2003 Oct; 25(7): 718-21.

[10] Szeifert GT, Timperley WR, Forster DM, Kemeny AA. Histopathological changes in cerebral arteriovenous malformations following Gamma Knife radiosurgery. *Prog. Neurol. Surg.* 2007; 20: 212-9.

[11] Ohye C, Higuchi Y, Shibazaki T, Hashimoto T, Koyama T, Hirai T, et al. Gammaknife thalamotomy for Parkinson's disease and essential tremor: A prospective multicenter study. *Neurosurgery*. 2011 Sep 12.

[12] Kondziolka D, Flickinger JC, Hudak R. Results following gamma knife radiosurgical anterior capsulotomies for obsessive compulsive disorder. *Neurosurgery*. 2011 Jan; 68(1): 28-32; discussion 23-3.

[13] Kondziolka D, Zorro O, Lobato-Polo J, Kano H, Flannery TJ, Flickinger JC, et al. Gamma Knife stereotactic radiosurgery for idiopathic trigeminal neuralgia. *J. Neurosurg.* 2010 Apr; 112(4): 758-65.

[14] Massager N, Lorenzoni J, Devriendt D, Desmedt F, Brotchi J, Levivier M. Gamma knife surgery for idiopathic trigeminal neuralgia performed using a far-anterior cisternal target and a high dose of radiation. *J. Neurosurg.* 2004 Apr; 100(4): 597-605.

[15] Regis J, Metellus P, Hayashi M, Roussel P, Donnet A, Bille-Turc F. Prospective controlled trial of gamma knife surgery for essential trigeminal neuralgia. *J. Neurosurg.* 2006 Jun; 104(6): 913-24.

[16] Bartolomei F, Hayashi M, Tamura M, Rey M, Fischer C, Chauvel P, et al. Long-term efficacy of gamma knife radiosurgery in mesial temporal lobe epilepsy. *Neurology*. 2008 May 6; 70(19): 1658-63.

[17] Regis J, Carron R, Park M. Is radiosurgery a neuromodulation therapy? : A 2009 Fabrikant award lecture. *J. Neurooncol.* 2010 Jun; 98(2): 155-62.

[18] Gevaert T, Levivier M, Lacornerie T, Verellen D, Engels B, Reynaert N, et al. Dosimetric comparison of different treatment modalities for stereotactic radiosurgery of arteriovenous malformations and acoustic neuromas. *Radiother. Oncol.* 2012 Aug 9.

[19] Greenberg MS. Acoustic neuroma. In: Thieme, editor. *Handbook of neurosurgery*. New York: Thieme 2006. p. 429-38.

[20] Régis J, Roche HP, Delsanti C, Thomassin J-M, Ouaknine M, Gabert K, et al. Modern Management of vestibular schwannomas. In: AG K, editor. *Radiosurgery and Pathological Fundamentals*. Basel: Karger; 2007. p. 129-41.

[21] Stangerup SE, Caye-Thomasen P. Epidemiology and natural history of vestibular schwannomas. *Otolaryngol. Clin. North Am.* 2012 Apr; 45(2): 257-68, vii.

[22] Stangerup SE, Caye-Thomasen P, Tos M, Thomsen J. The natural history of vestibular schwannoma. *Otol. Neurotol.* 2006 Jun; 27(4): 547-52.

[23] Samii M, Matthies C. Management of 1000 vestibular schwannomas (acoustic neuromas): surgical management and results with an emphasis on complications and how to avoid them. *Neurosurgery*. 1997 Jan; 40(1): 11-21; discussion -3.

[24] House JW, Brackmann DE. Facial nerve grading system. *Otolaryngol. Head Neck Surg.* 1985 Apr; 93(2): 146-7.

[25] Gardner G, Robertson JH. Hearing preservation in unilateral acoustic neuroma surgery. *Ann. Otol. Rhinol. Laryngol.* 1988 Jan-Feb; 97(1): 55-66.

[26] Camins MB, Oppenheim JS. Acoustic neuromas: surgical anatomy of the suboccipital approach. In: Surgeons TAAoN, editor. *Neurosurgical Operative Atlas* 1991.

[27] Cushing H. *Tumors of the nervus acusticus and the syndrome of the cerebellopontine angle* New York: Hafner; 1963.

[28] Dandy W. *An operation for the total removal of cerebellopontine (acoustic) tumors.* 1925. p. 129-48.

[29] Elhammady MS, Telischi FF, Morcos JJ. Retrosigmoid approach: indications, techniques, and results. *Otolaryngol. Clin. North Am.* 2012 Apr; 45(2): 375-97, ix.

[30] Mamikoglu B, Wiet RJ, Esquivel CR. Translabyrinthine approach for the management of large and giant vestibular schwannomas. *Otol. Neurotol.* 2002 Mar; 23(2): 224-7.

[31] Ramina R, Coelho Neto M, Bordignon KC, Mattei T, Clemente R, Pires Aguiar PH. Treatment of large and giant residual and recurrent vestibular schwannomas. *Skull. Base.* 2007 Mar; 17(2): 109-17.

[32] Kameyama S, Tanaka R, Honda Y, Hasegawa A, Yamazaki H, Kawaguchi T. The long-term growth rate of residual acoustic neurinomas. *Acta Neurochir.* (Wien). 1994; 129(3-4): 127-30.

[33] Kameyama S, Tanaka R, Kawaguchi T, Honda Y, Yamazaki H, Hasegawa A. Long-term follow-up of the residual intracanalicular tumours after subtotal removal of acoustic neurinomas. *Acta Neurochir.* (Wien). 1996; 138(2): 206-9.

[34] Godefroy WP, van der Mey AG, de Bruine FT, Hoekstra ER, Malessy MJ. Surgery for large vestibular schwannoma: residual tumor and outcome. *Otol. Neurotol.* 2009 Aug; 30(5): 629-34.

[35] Briggs RJ, Luxford WM, Atkins JS, Jr., Hitselberger WE. Translabyrinthine removal of large acoustic neuromas. *Neurosurgery.* 1994 May; 34(5): 785-90; discussion 90-1.

[36] Jung S, Kang SS, Kim TS, Kim HJ, Jeong SK, Kim SC, et al. Current surgical results of retrosigmoid approach in extralarge vestibular schwannomas. *Surg. Neurol.* 2000 Apr; 53(4): 370-7; discussion 7-8.

[37] Samii M, Gerganov V, Samii A. Improved preservation of hearing and facial nerve function in vestibular schwannoma surgery via the retrosigmoid approach in a series of 200 patients. *J. Neurosurg.* 2006 Oct; 105(4): 527-35.

[38] Zhang X, Fei Z, Chen YJ, Fu LA, Zhang JN, Liu WP, et al. Facial nerve function after excision of large acoustic neuromas via the suboccipital retrosigmoid approach. *J. Clin. Neurosci.* 2005 May; 12(4): 405-8.

[39] Lownie SP, Drake CG. Radical intracapsular removal of acoustic neurinomas. Long-term follow-up review of 11 patients. *J. Neurosurg.* 1991 Mar; 74(3): 422-5.

[40] Park CK, Jung HW, Kim JE, Son YJ, Paek SH, Kim DG. Therapeutic strategy for large vestibular schwannomas. *J. Neurooncol.* 2006 Apr; 77(2): 167-71.

[41] Gabert K, Regis J, Delsanti C, Roche PH, Facon F, Tamura M, et al. [Preserving hearing function after Gamma Knife radiosurgery for unilateral vestibular schwannoma]. *Neurochirurgie.* 2004 Jun; 50(2-3 Pt 2): 350-7.

[42] Kondziolka D, Lunsford LD, McLaughlin MR, Flickinger JC. Long-term outcomes after radiosurgery for acoustic neuromas. *N. Engl. J. Med.* 1998 Nov 12; 339(20): 1426-33.

[43] Regis J, Tamura M, Delsanti C, Roche PH, Pellet W, Thomassin JM. Hearing preservation in patients with unilateral vestibular schwannoma after gamma knife surgery. *Prog. Neurol. Surg.* 2008; 21:142-51.

[44] Yomo S, Carron R, Thomassin JM, Roche PH, Regis J. Longitudinal analysis of hearing before and after radiosurgery for vestibular schwannoma. *J. Neurosurg.* 2012 Aug 31.

[45] Samii M, Gerganov VM, Samii A. Functional outcome after complete surgical removal of giant vestibular schwannomas. *J. Neurosurg.* 2010 Apr; 112(4): 860-7.
[46] Wiet RJ, Mamikoglu B, Odom L, Hoistad DL. Long-term results of the first 500 cases of acoustic neuroma surgery. *Otolaryngol. Head Neck Surg.* 2001 Jun; 124(6): 645-51.
[47] Fischer G, Fischer C. [Preservation of hearing in acoustic surgery]. *Bull. Mem. Acad. R Med. Belg.* 1995; 150(10-11): 420-7; discussion 7-9.
[48] Fischer G, Fischer C, Remond J. Hearing preservation in acoustic neurinoma surgery. *J. Neurosurg.* 1992 Jun; 76(6): 910-7.
[49] Fischer G, Morgon A, Fischer C, Bret P, Massini B, Kzaiz M, et al. [Complete excision of acoustic neurinoma. Preservation of the facial nerve and hearing]. *Neurochirurgie.* 1987; 33(3): 169-83.
[50] Hecht CS, Honrubia VF, Wiet RJ, Sims HS. Hearing preservation after acoustic neuroma resection with tumor size used as a clinical prognosticator. *Laryngoscope.* 1997 Aug; 107(8): 1122-6.
[51] Wang EM, Pan L, Wang BJ, Zhang N, Dong YF, Dai JZ, et al. [Gamma knife for elderly patients with large vestibular schwannomas: 11-year follow-up]. *Zhonghua Yi Xue Za Zhi.* 2009 May 5; 89(17): 1189-91.
[52] Inoue HK. Low-dose radiosurgery for large vestibular schwannomas: long-term results of functional preservation. *J. Neurosurg.* 2005 Jan; 102 Suppl: 111-3.
[53] Litvack ZN, Noren G, Chougule PB, Zheng Z. Preservation of functional hearing after gamma knife surgery for vestibular schwannoma. *Neurosurg. Focus.* 2003 May 15; 14(5): e3.
[54] Myrseth E, Moller P, Pedersen PH, Vassbotn FS, Wentzel-Larsen T, Lund-Johansen M. Vestibular schwannomas: clinical results and quality of life after microsurgery or gamma knife radiosurgery. *Neurosurgery.* 2005 May; 56(5): 927-35; discussion -35.
[55] Pollock BE, Driscoll CL, Foote RL, Link MJ, Gorman DA, Bauch CD, et al. Patient outcomes after vestibular schwannoma management: a prospective comparison of microsurgical resection and stereotactic radiosurgery. *Neurosurgery.* 2006 Jul; 59(1): 77-85; discussion 77-85.
[56] Tonn JC. Microneurosurgery and radiosurgery--an attractive combination. *Acta Neurochir. Suppl.* 2004; 91:103-8.
[57] Fuentes S, Arkha Y, Pech-Gourg G, Grisoli F, Dufour H, Regis J. Management of large vestibular schwannomas by combined surgical resection and gamma knife radiosurgery. *Prog. Neurol. Surg.* 2008; 21: 79-82.
[58] Iwai Y, Yamanaka K, Ishiguro T. Surgery combined with radiosurgery of large acoustic neuromas. *Surg. Neurol.* 2003 Apr; 59(4): 283-9; discussion 9-91.
[59] van de Langenberg R, Hanssens PE, Verheul JB, van Overbeeke JJ, Nelemans PJ, Dohmen AJ, et al. Management of large vestibular schwannoma. Part II. Primary Gamma Knife surgery: radiological and clinical aspects. *J. Neurosurg.* 2011 Nov; 115(5): 885-93.
[60] Yang SY, Kim DG, Chung HT, Park SH, Paek SH, Jung HW. Evluation of tumor response after gamma knife radiosurgery for residual vestibular schwannomas based on MRI morphology features. *J. Neurol. Neurosurg. Psychiatry.* 2008; 79:431-6.
[61] Claus EB, Bondy ML, Schildkraut JM, Wiemels JL, Wrensch M, Black PM. Epidemiology of intracranial meningioma. *Neurosurgery.* 2005 Dec; 57(6): 1088-95; discussion -95.

[62] Natarajan SK, Sekhar LN, Schessel D, Morita A. Petroclival meningiomas: multimodality treatment and outcomes at long-term follow-up. *Neurosurgery*. 2007 Jun; 60(6): 965-79; discussion 79-81.

[63] Van Havenbergh T, Carvalho G, Tatagiba M, Plets C, Samii M. Natural history of petroclival meningiomas. *Neurosurgery*. 2003 Jan; 52(1): 55-62; discussion -4.

[64] Seifert V. Clinical management of petroclival meningiomas and the eternal quest for preservation of quality of life: personal experiences over a period of 20 years. *Acta Neurochir.* (Wien). 2010 Jul; 152(7): 1099-116.

[65] Al-Mefty O, Fox JL, Smith RR. Petrosal approach for petroclival meningiomas. *Neurosurgery*. 1988 Mar; 22(3): 510-7.

[66] Bricolo AP, Turazzi S, Talacchi A, Cristofori L. Microsurgical removal of petroclival meningiomas: a report of 33 patients. *Neurosurgery*. 1992 Nov; 31(5): 813-28; discussion 28.

[67] Kawase T, Shiobara R, Toya S. Middle fossa transpetrosal-transtentorial approaches for petroclival meningiomas. Selective pyramid resection and radicality. *Acta Neurochir.* (Wien). 1994; 129(3-4): 113-20.

[68] Cushing H. The meningioma (dural endothelioma): their source and favoured seats of origin-cavendish lecture. *Brain*. 1922;45:282-316.

[69] Mayberg MR, Symon L. Meningiomas of the clivus and apical petrous bone. Report of 35 cases. *J. Neurosurg*. 1986 Aug; 65(2): 160-7.

[70] Samii M, Ammirati M, Mahran A, Bini W, Sepehrnia A. Surgery of petroclival meningiomas: report of 24 cases. *Neurosurgery*. 1989 Jan; 24(1): 12-7.

[71] Campbell E, Whitfield RD. Posterior fossa meningiomas. *J. Neurosurg*. 1948 Mar; 5(2): 131-53.

[72] Cherington M, Schneck SA. Clivus meningiomas. Neurology. 1966 Jan;16(1):86-92.

[73] Castellano F, Ruggiero G. Meningiomas of the posterior fossa. *Acta Radiol. Suppl.* 1953; 104: 1-177.

[74] Cho CW, Al-Mefty O. Combined petrosal approach to petroclival meningiomas. *Neurosurgery*. 2002 Sep; 51(3): 708-16; discussion 16-8.

[75] Roche PH, Pellet W, Fuentes S, Thomassin JM, Regis J. Gamma knife radiosurgical management of petroclival meningiomas results and indications. *Acta Neurochir.* (Wien). 2003 Oct; 145(10): 883-8; discussion 8.

[76] Subach BR, Lunsford LD, Kondziolka D, Maitz AH, Flickinger JC. Management of petroclival meningiomas by stereotactic radiosurgery. *Neurosurgery*. 1998 Mar; 42(3): 437-43; discussion 43-5.

[77] Duma CM, Lunsford LD, Kondziolka D, Harsh GRt, Flickinger JC. Stereotactic radiosurgery of cavernous sinus meningiomas as an addition or alternative to microsurgery. *Neurosurgery*. 1993 May; 32(5): 699-704; discussion -5.

[78] Ganz JC, Backlund EO, Thorsen FA. The results of Gamma Knife surgery of meningiomas, related to size of tumor and dose. *Stereotact. Funct. Neurosurg*. 1993; 61 Suppl 1:23-9.

[79] Iwai Y, Yamanaka K, Nakajima H. Two-staged gamma knife radiosurgery for the treatment of large petroclival and cavernous sinus meningiomas. *Surg. Neurol.* 2001 Nov; 56(5): 308-14.

[80] Kayama T. [Contemporary treatment against petroclival meningioma]. *No Shinkei Geka*. 1998 Jan; 26(1): 8-17.

[81] Pollock BE, Stafford SL, Link MJ. Gamma knife radiosurgery for skull base meningiomas. *Neurosurg. Clin. N. Am.* 2000 Oct; 11(4): 659-66.

[82] Bambakidis NC, Kakarla UK, Kim LJ, Nakaji P, Porter RW, Daspit CP, et al. Evolution of surgical approaches in the treatment of petroclival meningiomas: a retrospective review. *Neurosurgery.* 2008 Jun; 62(6 Suppl 3): 1182-91.

[83] Couldwell WT, Fukushima T, Giannotta SL, Weiss MH. Petroclival meningiomas: surgical experience in 109 cases. *J. Neurosurg.* 1996 Jan; 84(1):20-8.

[84] Park CK, Jung HW, Kim JE, Paek SH, Kim DG. The selection of the optimal therapeutic strategy for petroclival meningiomas. *Surg. Neurol.* 2006 Aug; 66(2): 160-5; discussion 5-6.

[85] Yasargil MG. Meningiomas of the basal posterior cranial fossa. *Adv. Tech. Stand. Neurosurg.* 1980. p. 1-115.

[86] Flannery TJ, Kano H, Lunsford LD, Sirin S, Tormenti M, Niranjan A, et al. Long-term control of petroclival meningiomas through radiosurgery. *J. Neurosurg.* 2010 May; 112(5): 957-64.

[87] Al-Mefty O. Clinoidal meningiomas. *J. Neurosurg.* 1990 Dec; 73(6): 840-9.

[88] al-Mefty O, Ayoubi S. Clinoidal meningiomas. *Acta Neurochir.* Suppl. (Wien). 1991; 53:92-7.

[89] Cophignon J, Lucena J, Clay C, Marchac D. Limits to radical treatment of sphenoorbital meningiomas. *Acta Neurochir.* Suppl (Wien). 1979; 28(2): 375-80.

[90] Pamir MN, Belirgen M, Ozduman K, Kilic T, Ozek M. Anterior clinoidal meningiomas: analysis of 43 consecutive surgically treated cases. *Acta Neurochir.* (Wien). 2008 Jul; 150(7): 625-35; discussion 35-6.

[91] Bonnal J, Thibaut A, Brotchi J, Born J. Invading meningiomas of the sphenoid ridge. *J. Neurosurg.* 1980 Nov;53(5):587-99.

[92] Fohanno D, Bitar A. Sphenoidal ridge meningioma. *Adv. Tech. Stand. Neurosurg.* 1986; 14: 137-74.

[93] Risi P, Uske A, de Tribolet N. Meningiomas involving the anterior clinoid process. *Br. J. Neurosurg.* 1994; 8(3): 295-305.

[94] Cushing H, Eisenhardt L. Meningiomas of the sphenoidal ridge. In: Springfield, editor. *Meningiomas: Their Classification, Regional Behavior, Life History and Surgical End Results* 1938. p. 298-319.

[95] Al-Mefty O, Smith RR. Surgery of tumors invading the cavernous sinus. *Surg. Neurol.* 1988 Nov; 30(5): 370-81.

[96] Goel A, Gupta S, Desai K. New grading system to predict resectability of anterior clinoid meningiomas. *Neurol. Med. Chir.* (Tokyo). 2000 Dec; 40(12): 610-6; discussion 6-7.

[97] Lee JH, Jeun SS, Evans J, Kosmorsky G. Surgical management of clinoidal meningiomas. *Neurosurgery.* 2001 May; 48(5): 1012-9; discussion 9-21.

[98] Kondziolka D, Lunsford LD, Coffey RJ, Flickinger JC. Gamma knife radiosurgery of meningiomas. *Stereotact. Funct. Neurosurg.* 1991; 57(1-2): 11-21.

[99] Metellus P, Regis J, Muracciole X, Fuentes S, Dufour H, Nanni I, et al. Evaluation of fractionated radiotherapy and gamma knife radiosurgery in cavernous sinus meningiomas: treatment strategy. *Neurosurgery.* 2005 Nov; 57(5): 873-86; discussion -86.

[100] Roche PH, Regis J, Dufour H, Fournier HD, Delsanti C, Pellet W, et al. Gamma knife radiosurgery in the management of cavernous sinus meningiomas. *J. Neurosurg.* 2000 Dec; 93 Suppl 3: 68-73.

[101] Berhouma M, Messerer M, Jouanneau E. Occam's razor in minimally invasive pituitary surgery: tailoring the endoscopic endonasal uninostril trans-sphenoidal approach to sella turcica. *Acta Neurochir.* (Wien). 2012 Oct 10.

[102] Attia M, Umansky F, Paldor I, Dotan S, Shoshan Y, Spektor S. Giant anterior clinoidal meningiomas: surgical technique and outcomes. *J. Neurosurg.* 2012 Aug 17.

[103] Puzzilli F, Ruggeri A, Mastronardi L, Agrillo A, Ferrante L. Anterior clinoidal meningiomas: report of a series of 33 patients operated on through the pterional approach. *Neuro Oncol.* 1999 Jul; 1(3): 188-95.

[104] Dolenc V. Microsurgical removal of large sphenoidal bone meningiomas. *Acta Neurochir.* Suppl (Wien). 1979; 28(2): 391-6.

[105] Dolenc VV, Pregelj R, Slokan S, Skrbec M. Anterior communicating artery aneurysm associated with tuberculum sellae meningioma--case report. *Neurol. Med. Chir.* (Tokyo). 1998 Aug; 38(8): 485-8.

[106] Abdel-Aziz KM, Froelich SC, Dagnew E, Jean W, Breneman JC, Zuccarello M, et al. Large sphenoid wing meningiomas involving the cavernous sinus: conservative surgical strategies for better functional outcomes. *Neurosurgery.* 2004 Jun; 54(6): 1375-83; discussion 83-4.

[107] Tobias S, Kim CH, Kosmorsky G, Lee JH. Management of surgical clinoidal meningiomas. *Neurosurg. Focus.* 2003 Jun 15; 14(6):e5.

[108] Francel PC, Bhattacharjee S, Tompkins P. Skull base approaches and gamma knife radiosurgery for multimodality treatment of skull base tumors. *J. Neurosurg.* 2002 Dec; 97(5 Suppl): 674-6.

[109] Iwai Y, Yamanaka K, Ishiguro T. Gamma knife radiosurgery for the treatment of cavernous sinus meningiomas. *Neurosurgery.* 2003 Mar; 52(3): 517-24; discussion 23-4.

[110] Lee JY, Niranjan A, McInerney J, Kondziolka D, Flickinger JC, Lunsford LD. Stereotactic radiosurgery providing long-term tumor control of cavernous sinus meningiomas. *J. Neurosurg.* 2002 Jul; 97(1): 65-72.

[111] Pendl G, Schrottner O, Friehs GM, Feichtinger H. Stereotactic radiosurgery of skull base meningiomas. *Stereotact. Funct. Neurosurg.* 1995; 64 Suppl 1: 11-8.

[112] Greenberg MS. Pituitary adenomas. In: Thieme, editor. *Handbook of neurosurgery* 2006. p. 438-56.

[113] Chen Y, Wang CD, Su ZP, Chen YX, Cai L, Zhuge QC, et al. Natural History of Postoperative Nonfunctioning Pituitary Adenomas: A Systematic Review and Meta-Analysis. *Neuroendocrinology.* 2012 Aug 28.

[114] Fernandez-Balsells MM, Murad MH, Barwise A, Gallegos-Orozco JF, Paul A, Lane MA, et al. Natural history of nonfunctioning pituitary adenomas and incidentalomas: a systematic review and metaanalysis. *J. Clin. Endocrinol. Metab.* 2011 Apr; 96(4): 905-12.

[115] Karavitaki N, Collison K, Halliday J, Byrne JV, Price P, Cudlip S, et al. What is the natural history of nonoperated nonfunctioning pituitary adenomas? *Clin. Endocrinol.* (Oxf). 2007 Dec; 67(6): 938-43.

[116] Racadot J. [Adenomas of the anterior pituitary : natural history, classification and histopathology]. *Rev. Prat.* 1980 Oct 11; 30(45): 2981-4, 7-92, 97-8 passim.

[117] Babey M, Sahli R, Vajtai I, Andres RH, Seiler RW. Pituitary surgery for small prolactinomas as an alternative to treatment with dopamine agonists. *Pituitary*. 2011 Sep; 14(3): 222-30.

[118] Colao A, Savastano S. Medical treatment of prolactinomas. *Nat. Rev. Endocrinol*. 2011 May; 7(5): 267-78.

[119] Delemer B. [Prolactinomas: diagnosis and treatment]. *Presse Med.* 2009 Jan; 38(1): 117-24.

[120] Iglesias P, Bernal C, Villabona C, Castro JC, Arrieta F, Diez JJ. Prolactinomas in men: a multicentre and retrospective analysis of treatment outcome. *Clin. Endocrinol.* (Oxf). 2012 Aug; 77(2): 281-7.

[121] Mann WA. Treatment for prolactinomas and hyperprolactinaemia: a lifetime approach. *Eur. J. Clin. Invest*. 2011 Mar; 41(3): 334-42.

[122] Regis J, Castinetti F. Radiosurgery: a useful first-line treatment of prolactinomas? *World Neurosurg*. 2010 Jul; 74(1): 103-4.

[123] Wu ZB, Yu CJ. [Treatment strategy of pituitary invasive prolactinomas]. *Zhonghua Wai Ke Za Zhi*. 2009 Jan 15; 47(2): 123-7.

[124] Capatina C, Christodoulides C, Fernandez A, Cudlip S, Grossman AB, Wass JA, et al. Current treatment protocols can offer a normal or near normal quality of life in the majority of patients with non-functioning pituitary adenomas. *Clin. Endocrinol.* (Oxf). 2012 May 29.

[125] Cappabianca P, Solari D. The endoscopic endonasal approach for the treatment of recurrent or residual pituitary adenomas: widening what to see expands what to do? *World Neurosurg*. 2012 Mar-Apr; 77(3-4): 455-6.

[126] Dai CX, Yao Y, Cai F, Liu XH, Ma SH, Wang RZ. Advances in medical treatment of clinically nonfunctioning pituitary adenomas. *Zhongguo Yi Xue Ke Xue Yuan Xue Bao*. 2012 Jun; 34(3): 298-302.

[127] Huo G, Feng QL, Tang MY, Li D. Diagnosis and treatment of hemorrhagic pituitary adenomas. *Acta Neurochir. Suppl*. 2011;111:361-5.

[128] Pereira AM, Biermasz NR. Treatment of nonfunctioning pituitary adenomas: what were the contributions of the last 10 years? A critical view. *Ann. Endocrinol.* (Paris). 2012 Apr; 73(2): 111-6.

[129] Zhou T, Wei SB, Meng XH, Xu BN. [Pure endoscopic endonasal transsphenoidal approach for 375 pituitary adenomas]. *Zhonghua Wai Ke Za Zhi*. 2010 Oct 1; 48(19): 1443-6.

[130] Santos AR, Fonseca Neto RM, Veiga JC, Viana Jr J, Scaliassi NM, Lancellotti CL, et al. Endoscopic endonasal transsphenoidal approach for pituitary adenomas: technical aspects and report of casuistic. *Arq. Neuropsiquiatr*. 2010 Aug; 68(4): 608-12.

[131] Ceylan S, Koc K, Anik I. Endoscopic endonasal transsphenoidal approach for pituitary adenomas invading the cavernous sinus. *J. Neurosurg*. 2010 Jan; 112(1): 99-107.

[132] Zhang X, Fei Z, Zhang JN, Liu WP, Fu LA, Song SJ, et al. [Endoscopic endonasal transsphenoidal approach for pituitary adenomas]. *Zhonghua Wai Ke Za Zhi*. 2006 Nov 15; 44(22): 1551-4.

[133] Cappabianca P, Cavallo LM, Colao A, de Divitiis E. Surgical complications associated with the endoscopic endonasal transsphenoidal approach for pituitary adenomas. *J. Neurosurg.* 2002 Aug; 97(2): 293-8.

[134] Guiot G, Thibaut B, Bourreau M. [Extirpation of hypophyseal adenomas by trans-septal and trans-sphenoidal approaches]. *Ann. Otolaryngol.* 1959 Dec; 76: 1017-31.

[135] Hardy J. Transphenoidal microsurgery of the normal and pathological pituitary. *Clin. Neurosurg.* 1969; 16: 185-217.

[136] Mehta RP, Cueva RA, Brown JD, Fliss DM, Gil Z, Kassam AB, et al. What's new in skull base medicine and surgery? Skull Base Committee Report. *Otolaryngol. Head Neck Surg.* 2006 Oct; 135(4): 620-30.

[137] Messerer M, De Battista JC, Raverot G, Kassis S, Dubourg J, Lapras V, et al. Evidence of improved surgical outcome following endoscopy for nonfunctioning pituitary adenoma removal. *Neurosurg. Focus.* 2011 Apr; 30(4): E11.

[138] de Divitiis E. Endoscopic transsphenoidal surgery: stone-in-the-pond effect. *Neurosurgery.* 2006 Sep; 59(3): 512-20; discussion -20.

[139] D'Haens J, Van Rompaey K, Stadnik T, Haentjens P, Poppe K, Velkeniers B. Fully endoscopic transsphenoidal surgery for functioning pituitary adenomas: a retrospective comparison with traditional transsphenoidal microsurgery in the same institution. *Surg. Neurol.* 2009 Oct; 72(4): 336-40.

[140] Ferrante E, Ferraroni M, Castrignano T, Menicatti L, Anagni M, Reimondo G, et al. Non-functioning pituitary adenoma database: a useful resource to improve the clinical management of pituitary tumors. *European Journal of endocrinology* / European Federation of Endocrine Societies. 2006 Dec; 155(6): 823-9.

[141] Saito K, Kuwayama A, Yamamoto N, Sugita K. The transsphenoidal removal of nonfunctioning pituitary adenomas with suprasellar extensions: the open sella method and intentionally staged operation. *Neurosurgery.* 1995 Apr; 36(4): 668-75; discussion 75-6.

[142] Kabil MS, Eby JB, Shahinian HK. Fully endoscopic endonasal vs. transseptal transsphenoidal pituitary surgery. *Minimally invasive neurosurgery*: MIN. 2005 Dec; 48(6): 348-54.

[143] Fahlbusch R, Buchfelder M. Transsphenoidal surgery of parasellar pituitary adenomas. *Acta neurochirurgica.* 1988; 92(1-4): 93-9.

[144] Dehdashti AR, Ganna A, Karabatsou K, Gentili F. Pure endoscopic endonasal approach for pituitary adenomas: early surgical results in 200 patients and comparison with previous microsurgical series. *Neurosurgery.* 2008 May; 62(5): 1006-15; discussion 15-7.

[145] Frank G, Pasquini E, Farneti G, Mazzatenta D, Sciarretta V, Grasso V, et al. The endoscopic versus the traditional approach in pituitary surgery. *Neuroendocrinology.* 2006; 83(3-4): 240-8.

[146] Tabaee A, Anand VK, Barron Y, Hiltzik DH, Brown SM, Kacker A, et al. Predictors of short-term outcomes following endoscopic pituitary surgery. *Clinical neurology and neurosurgery.* 2009 Feb; 111(2): 119-22.

[147] Jho HD. Endoscopic transsphenoidal surgery. *Journal of neuro-oncology.* 2001 Sep; 54(2): 187-95.

[148] Dolenc VV, Lipovsek M, Slokan S. Traumatic aneurysm and carotid-cavernous fistula following transsphenoidal approach to a pituitary adenoma: treatment by transcranial operation. *British journal of neurosurgery.* 1999 Apr; 13(2): 185-8.

[149] Kono Y, Prevedello DM, Snyderman CH, Gardner PA, Kassam AB, Carrau RL, et al. One thousand endoscopic skull base surgical procedures demystifying the infection potential: incidence and description of postoperative meningitis and brain abscesses. Infection control and hospital epidemiology : *the official Journal of the Society of Hospital Epidemiologists of America.* 2011 Jan; 32(1): 77-83.

[150] Semple PL, Laws ER, Jr. Complications in a contemporary series of patients who underwent transsphenoidal surgery for Cushing's disease. *Journal of neurosurgery.* 1999 Aug; 91(2):175-9.

[151] Ciric I, Ragin A, Baumgartner C, Pierce D. Complications of transsphenoidal surgery: results of a national survey, review of the literature, and personal experience. *Neurosurgery.* 1997 Feb; 40(2): 225-36; discussion 36-7.

[152] Cottier JP, Destrieux C, Brunereau L, Bertrand P, Moreau L, Jan M, et al. Cavernous sinus invasion by pituitary adenoma: *MR imaging. Radiology.* 2000 May; 215(2): 463-9.

[153] Raymond J, Hardy J, Czepko R, Roy D. Arterial injuries in transsphenoidal surgery for pituitary adenoma; the role of angiography and endovascular treatment. *AJNR American Journal of neuroradiology.* 1997 Apr; 18(4): 655-65.

[154] Benoit BG, Wortzman G. Traumatic cerebral aneurysms. Clinical features and natural history. *Journal of neurology, neurosurgery, and psychiatry.* 1973 Feb; 36(1): 127-38.

[155] Dolenc VV. Surgery of vascular lesions of the cavernous sinus. *Clin. Neurosurg.* 1990; 36: 240-55.

[156] Shaffrey ME, Dolenc VV, Lanzino G, Wolcott WP, Shaffrey CI. Invasion of the internal carotid artery by cavernous sinus meningiomas. *Surg. Neurol.* 1999 Aug; 52(2): 167-71.

[157] Iwai Y, Yamanaka K, Honda Y, Matsusaka Y. Radiosurgery for pituitary metastases. *Neurol. Med. Chir.* (Tokyo). 2004 Mar; 44(3): 112-6; discussion 7.

[158] Jagannathan J, Yen CP, Pouratian N, Laws ER, Sheehan JP. Stereotactic radiosurgery for pituitary adenomas: a comprehensive review of indications, techniques and long-term results using the Gamma Knife. *J. Neurooncol.* 2009 May; 92(3): 345-56.

[159] Losa M, Valle M, Mortini P, Franzin A, da Passano CF, Cenzato M, et al. Gamma knife surgery for treatment of residual nonfunctioning pituitary adenomas after surgical debulking. *J. Neurosurg.* 2004 Mar; 100(3): 438-44.

[160] Mingione V, Yen CP, Vance ML, Steiner M, Sheehan J, Laws ER, et al. Gamma surgery in the treatment of nonsecretory pituitary macroadenoma. *J. Neurosurg.* 2006 Jun; 104(6): 876-83.

[161] Pollock BE, Cochran J, Natt N, Brown PD, Erickson D, Link MJ, et al. Gamma knife radiosurgery for patients with nonfunctioning pituitary adenomas: results from a 15-year experience. *Int. J. Radiat. Oncol. Biol. Phys.* 2008 Apr 1; 70(5): 1325-9.

[162] Sheehan JP, Kondziolka D, Flickinger J, Lunsford LD. Radiosurgery for residual or recurrent nonfunctioning pituitary adenoma. *J. Neurosurg.* 2002 Dec; 97 (5 Suppl): 408-14.

[163] Landolt AM, Haller D, Lomax N, Scheib S, Schubiger O, Siegfried J, et al. Stereotactic radiosurgery for recurrent surgically treated acromegaly: comparison with fractionated radiotherapy. *J. Neurosurg.* 1998 Jun;88(6):1002-8.

[164] Mokry M, Ramschak-Schwarzer S, Simbrunner J, Ganz JC, Pendl G. A six year experience with the postoperative radiosurgical management of pituitary adenomas. *Stereotact. Funct. Neurosurg.* 1999;72 Suppl 1:88-100.

[165] Pan L, Zhang N, Wang E, Wang B, Xu W. Pituitary adenomas: the effect of gamma knife radiosurgery on tumor growth and endocrinopathies. *Stereotact. Funct. Neurosurg.* 1998 Oct; 70 Suppl 1: 119-26.

[166] Pollock BE, Nippoldt TB, Stafford SL, Foote RL, Abboud CF. Results of stereotactic radiosurgery in patients with hormone-producing pituitary adenomas: factors associated with endocrine normalization. *J. Neurosurg.* 2002 Sep; 97(3): 525-30.

[167] Pollock BE, Kondziolka D, Lunsford LD, Flickinger JC. Stereotactic radiosurgery for pituitary adenomas: imaging, visual and endocrine results. *Acta Neurochir. Suppl.* 1994; 62:33-8.

[168] Sheehan JP, Jagannathan J, Pouratian N, Steiner L. Stereotactic radiosurgery for pituitary adenomas: a review of the literature and our experience. *Front Horm. Res.* 2006; 34:185-205.

[169] Sheehan JP, Pouratian N, Steiner L, Laws ER, Vance ML. Gamma Knife surgery for pituitary adenomas: factors related to radiological and endocrine outcomes. *J. Neurosurg.* 2011 Feb; 114(2): 303-9.

[170] Feigl GC, Bonelli CM, Berghold A, Mokry M. Effects of gamma knife radiosurgery of pituitary adenomas on pituitary function. J. Neurosurg. 2002 Dec; 97(5 Suppl): 415-21.

[171] Lim YL, Leem W, Kim TS, Rhee BA, Kim GK. Four years' experiences in the treatment of pituitary adenomas with gamma knife radiosurgery. *Stereotact Funct. Neurosurg.* 1998 Oct; 70 Suppl 1: 95-109.

[172] Feigl GC, Pistracher K, Berghold A, Mokry M. Pituitary insufficiency as a side effect after radiosurgery for pituitary adenomas: the role of the hypothalamus. *J. Neurosurg.* 2010 Dec; 113 Suppl: 153-59.

[173] Marek J, Jezkova J, Hana V, Krsek M, Bandurova L, Pecen L, et al. Is it possible to avoid hypopituitarism after irradiation of pituitary adenomas by the Leksell gamma knife? *Eur. J. Endocrinol.* 2011 Feb; 164(2): 169-78.

[174] Castinetti F, Brue T. Gamma Knife radiosurgery in pituitary adenomas: Why, who, and how to treat? *Discov. Med.* 2010 Aug; 10(51): 107-11.

[175] Castinetti F, Regis J, Dufour H, Brue T. Role of stereotactic radiosurgery in the management of pituitary adenomas. *Nat. Rev. Endocrinol.* 2010 Apr; 6(4): 214-23.

[176] Langsenlehner T, Stiegler C, Quehenberger F, Feigl GC, Jakse G, Mokry M, et al. Long-term follow-up of patients with pituitary macroadenomas after postoperative radiation therapy: analysis of tumor control and functional outcome. *Strahlenther. Onkol.* 2007 May; 183(5): 241-7.

[177] Sheehan JP, Niranjan A, Sheehan JM, Jane JA, Jr., Laws ER, Kondziolka D, et al. Stereotactic radiosurgery for pituitary adenomas: an intermediate review of its safety, efficacy, and role in the neurosurgical treatment armamentarium. *J. Neurosurg.* 2005 Apr; 102(4):678-91.

In: Minimally Invasive Skull Base Surgery
Editor: Moncef Berhouma
ISBN: 978-1-62808-567-9
© 2013 Nova Science Publishers, Inc.

Chapter XXIII

Anterior Craniovertebral Junction Tumors: Successful Resection Through Simple Approaches

Mario Ammirati[1] and Varun R. Kshettry[1,2]
[1]Dardinger Microneurosurgical Skull Base Laboratory, Department of Neurological Surgery, Ohio State University Medical Center, Columbus, OH
[2]Department of Neurological Surgery, Cleveland Clinic, Cleveland, OH

Abstract

Tumors at the craniovertebral junction often present a challenge due to proximity to vital neural and vascular structures. In the last few decades, many authors have proposed complex surgical approaches to correct anterior or anterolateral craniovertebral junction pathologies while attempting to reduce neurological and surgical complications. These include the far lateral, transcondylar, supracondylar, paracondylar, anterolateral and extreme lateral approaches. Based on our experience, we believe that the simple lateral suboccipital approach can be used to resect these lesions safely and effectively. This chapter reviews the development of surgical approaches to these lesions, discusses the authors' surgical technique, and presents illustrative cases. Finally, this chapter reviews pertinent anatomical studies in the literature and discusses pathology and anatomic specific factors that help guide approach.

Introduction

Tumors positioned anterior and anterolateral to the brainstem and spinal cord at the craniovertebral junction (CVJ) have posed a significant surgical challenge for the neurosurgeon. In this location, lesions are frequently intimately associated with the vasculature of the posterior fossa, the lower cranial nerves, and often at times adherent to the brainstem or upper spinal cord. Earlier reports of surgical treatment of these lesions detailed high morbidity and mortality [1-3]. As a result, neurosurgeons have devised various surgical approaches to attempt safer removal of these lesions.

In the last several decades, there have been an abundance of reports illustrating various approaches to anterior and anterolateral CVJ lesions. The shear number of reports has led to confusion in the terminology of approaches. The first type of approach is a midline posterior suboccipital approach [1-6]. This type of approach is generally done in prone or sitting position with the head neutral and is used for posterior or posterolateral CVJ lesions. Initial attempts to resect anterolateral or anterior lesions led to significant morbidity and mortality due to the need for brainstem retraction [1, 3, 6]. The second type of approach is the transoral approach. This approach has been attempted for anteriorly based CVJ lesions [7-9]. This approach has not gained wide acceptance due to multiple shortcomings: contaminated field with increased risk for meningitis and wound infection, cerebrospinal fluid (CSF) leak, visualization of tumor-brainstem interface late in tumor dissection, poor resectability of lateral tumor extension, craniocervical instability, and velopalatine insufficiency [8-11].

The third category of approaches is the posterolateral or lateral suboccipital approach, also roughly equivalent to the "basic" far lateral or retrocondylar far lateral approach. These types of approaches have been developed to address anterior and anterolateral lesions from a more lateral position to visualize brain-tumor interface and decrease the need for brainstem retraction. These approaches can be done in lateral decubitus, park bench position, or sitting position. Bone removal entails lateral suboccipital bone down to the foramen magnum and extending laterally up to the occipital condyle. Depending on caudal extension of tumor, this approach generally requires removal of the posterior arch of the atlas toward the exposed vertebral artery [12-14].

This approach has been advocated by numerous neurosurgeons. Heros first published on this technique for vertebrobasilar vascular lesions [14]. Samii *et al.* presented a series of 38 patients with CVJ meningiomas operated in Germany between 1977-1995 [11]. He used partial condylar resection in 4 of 10 anterior based meningiomas and 2 of 12 laterally based meningiomas. While achieving gross total resection (GTR) in 63%, he stated that the limiting factor for complete resection was vertebral artery encasement or en plaque pattern of meningioma rather than operative exposure. Nanda *et al.* presented a series of 10 patients with anterior foramen magnum lesions demonstrating successful resection in all 10 without the need for any condylar resection [13]. Wu *et al.* presented a series of 114 foramen magnum meningiomas (FMMs) operated in China between 1993-2008 [15]. They used a retrocondylar far lateral approach in 65 or 80 anteriorly based tumors and 22 of 24 anterolateral tumors. They specifically note that the first 10 anterior/anterolateral cases were all done using a transcondylar approach, but with increasing experience, they found that they could achieve GTR safely in the vast majority of anterior and anterolateral meningiomas using a retrocondylar approach. Bassiouni *et al.* presented a series of 25 anterior or anterolateral FMMs using a retrocondylar far lateral approach and achieved a Simpson Grade 2 in 96% with 8% permanent morbidity and 4% mortality [16]. Others have used a similar approach and provided good results [17-20].

In 1978, Seeger first published on the technique of resection of the occipital condyle and jugular tubercle for a more lateral approach [21]. Rhoton categorized approaches into basic far lateral and far lateral transcondylar [12]. Basic far lateral approaches did not involve condylar drilling. The far lateral transcondylar approach was further divided into a transcondylar versus an atlanto-occipital transarticular approach, the latter in which part of the superior articulating facet of C1 was removed in addition to partial condyle resection. Finally, he described two extensions of this approach. The first is a supracondylar transtubercular extension, which entails removal of the jugular tubercle to better visualize the area anterior to cranial nerves (CN) 9, 10, and 11 and allow visualization more distally on the vertebral artery. He also described a paracondylar extension lateral to the condyle along the jugular process to reach lesions involving the jugular foramen. Numerous authors have published clinical series using these types of lateral approaches [22-27]. Bertalanffy *et al.* presented a series of 19 anterior or anterolateral FMMs using a partial transcondylar approach with transtubercular extension. This was done with vertebral artery identification, but without mobilization and transposition. They achieved GTR in all 19 cases with no mortality and minimal morbidity [22, 23]. Other authors have used this approach utilizing ¼ to ½ condyle resection with success [24-27].

The final category is the extreme lateral or anterolateral approach. Some overlap exists between the extreme lateral and far lateral transcondylar approach. The far lateral transcondylar approach, when combined with unroofing of the vertebral artery in the C1 transverse foramen and medial transposition to allow working angle in front of the vertebral artery, would be considered by many to be part of the extreme lateral category. This approach was first published in 1990 by Sen and Sekhar, followed by multiple subsequent reports detailing slight modifications [28-31]. This group demonstrated 66% GTR in 24 patients with anterior or anterolateral FMMs [30]. Arnautovic and Al-Mefty reported on 18 patients with anterior or anterolateral FMMs utilizing an extreme lateral transcondylar approach with vertebral artery transposition and 1/3 to ½ condylar resection [32]. They achieve GTR in 12/16 (75%) virgin cases, but suffered 1 vertebral artery rupture that was repaired and 3 cases of major bleeding from the venous plexus surrounding the vertebral artery. They demonstrated transient worsening or new deficits of CN 9 and 10 in 10/18 patients, all of which resolved over weeks to months. Other authors have also published clinical series using the extreme lateral approach [33].

The anterolateral approach was described by George *et al.* and is similar to the extreme lateral approach, but is done in the supine position with contralateral head rotation and approaches anterior to the sternocleidomastoid muscle [34, 35]. This group has published extensively on a large series of patients undergoing both posterolateral and anterolateral approaches with good results [35-38]. Despite extensive literature on this topic, there is no clear consensus on the best approach to anterior and anterolateral lesions at the craniovertebral junction and there is no data to demonstrate that one approach results in greater resection or less morbidity. We advocate that the majority of these lesions can be resected successfully using simple approaches. The purpose of this chapter is to demonstrate surgical technique and present illustrative cases. We will then review literature regarding anatomical analyses comparing the different techniques and anatomical and clinical factors that can be used to help select the optimal approach for these types of lesions.

Surgical Technique

Preoperative MRI is evaluated for tumor relation to neural and vascular structures, degree of brainstem shift from mass effect, and for any signs of violation of brainstem or spinal cord pial plane on T2 weighted sequence. Preoperative embolization is not advocated, but balloon test occlusion may be performed in cases with vertebral artery encasement.

Semisitting position was used for the posterolateral approach. Stereotaxis and electrophysiological monitoring were frequent surgical adjuncts. A midline incision from the inion down to the upper cervical spine was used when C1 or C2 laminectomy was required. In cases solely requiring lateral suboccipital exposure, a curvilinear incision along the mastoid was used. Musculature was dissected off bone in subperiosteal fashion. The vertebral artery was identified in the C1 sulcal groove, but not transposed. Hemilaminectomy of C1 was performed up to the lateral mass. A curvilinear dural flap was made medial to the dural entrance of the vertebral artery and retracted laterally. A lateral suboccipital craniotomy was performed up to the occipital condyle. In cases where craniectomy was performed, closure was supplemented with porex-titanium mesh cranioplasty. No condyle was resected in any case.

CSF was drained from the cisterna magna and lateral cerebellar medullary cistern for brain relaxation. Standard microsurgical instruments and techniques were used. In certain cases, endoscope-assisted technique was used for better visualization, to identify surrounding arteries or nerves, or to ensure complete resection. We use a Moeller-Wedel microscope (Moeller-Wedel GmbH, Wedel, Germany) with integrated binocular data injection system (BiOpix, CLA Medical, Milford, OH). This allowed us to use various rigid 0°, 30°, and 45° endoscopes with picture-in-picture visualization in the microscope oculars.

Illustrative Cases

Case #1

A 65-year-old lady presented with a long history of mild suboccipital pain, chronic progressive numbness and weakness of the left arm and gait imbalance. Exam was significant for unsteady gait, left arm weakness (4/5), and hyperreflexia. She additionally had loss of gag reflex and weakness of the left trapezius and sternocleidomastoid muscles. MRI revealed a 2.7 x 2.0 x. 1.8cm homogenously enhancing anteriorly based extra-axial lesion with ventral attachment and severe medulla and upper cervical spinal cord compression (figure 1).

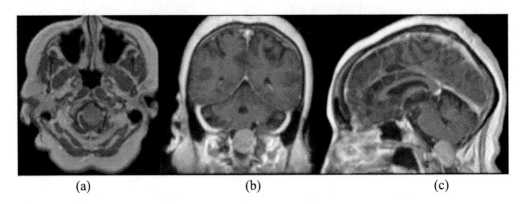

Figure 1. Preoperative MRI with contrast with axial (a), coronal (b), and sagittal (c) views demonstrating large anteriorly based extra-axial enhancing lesion with severe neural compression.

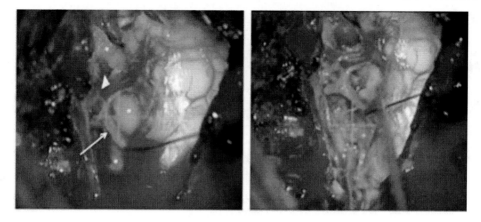

Figure 2. Intraoperative photographs. Both photographs demonstrate tumor anteriorly displacing brainstem posteriorly and to the right. The vagus nerve (arrow head) and accessory nerve (arrow) created three working windows (asterisks) for tumor resection.

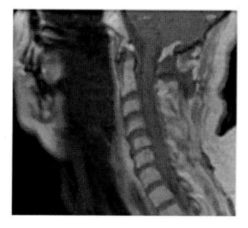

Figure 3. Postoperative MRI demonstrating gross total resection and re-expansion of neural structures.

The lesion was near, but below the vertebral artery. The patient underwent a left posterolateral suboccipital craniectomy, left C1 and partial C2 hemilaminectomy, nuchal ligament duraplasty and porex-titanium mesh cranioplasty (figure 2). Postoperative MRI revealed gross total resection (figure 3) and pathology was WHO I meningioma. Postoperative course was unremarkable and patient was discharged after 8 days. In follow-up, patient had nearly complete resolution of preoperative gait imbalance and weakness after three months of physical therapy.

Case #2

A 54-year-old lady with poorly controlled diabetes presented with several months of progressive dysphagia, right-sided hearing loss and gait imbalance. Preoperative exam was notable for right tongue deviation and gait imbalance. MRI revealed a 2.7 x 2.1cm enhancing lesion anterolaterally based from the lower clivus to C1 with some extension into the right hypoglossal canal (figure 4).

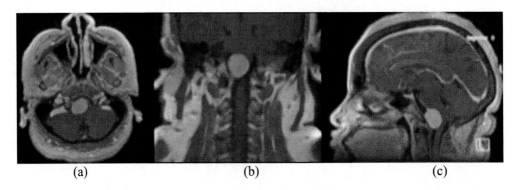

Figure 4. Preoperative MRI with contrast with axial (a), coronal (b), and sagittal (c) views demonstrating large anterolateral extra-axial enhancing lesion with extension into right hypoglossal canal, causing severe neural compression.

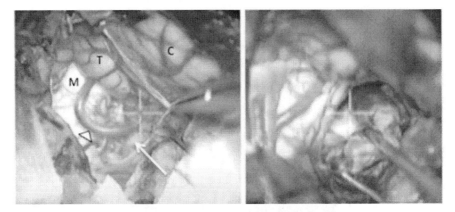

Figure 5. Intraoperative photographs. Left view demonstrates initial view after exposure with tonsils (T), cerebellar hemisphere (CH), medulla (M), right PICA (arrowhead) and tumor (arrow). Right view demonstrating microsurgical dissection.

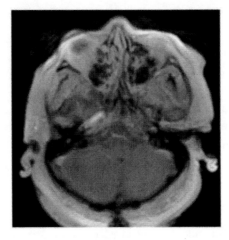

Figure 6. Postoperative MRI demonstrating gross total resection and re-expansion of the neural structures.

Patient underwent a right posterolateral suboccipital craniectomy, right C1 hemilaminectomy, porex-titanium mesh cranioplasty, and resection of lesion (figure 5, 6). Pathology revealed WHO I meningioma. Patient had some transient worsening of swallowing difficulties, but in follow-up had complete resolution. Her gait imbalance improved postoperatively as well.

Case #3

A 75-year-old man with a previous history of occipito-cervical synovial cyst drainage presented with neck pain, dysphagia, gait imbalance, bilateral arm numbness and weakness, greater on the left side. His exam was significant for myelopathic gait and loss of gag reflex. MRI revealed a 1.7 x 1.4cm anterolateral synovial cyst arising from the occiput – C1 joint on the left side with compression of the caudal medulla and upper cervical spinal cord (figure 7). Patient underwent a left C1 hemilaminectomy and resection of cyst (figure 8, 9). Postoperative course was unremarkable and patient was discharged after three days. At follow-up, patient had improvement in neck pain and motor strength. His gait imbalance was nearly resolved and was back to usual activities.

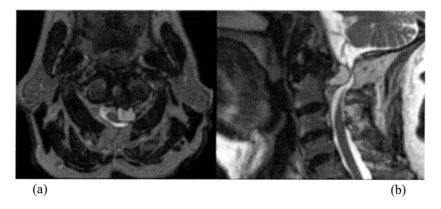

(a)   (b)

Figure 7. Preoperative MRI with contrast with axial (a), and sagittal (b) views demonstrating left sided anterolateral synovial cyst with moderate neural compression.

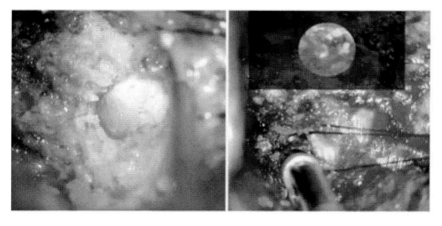

Figure 8. Intraoperative views. Left view demonstrates white synovial cyst. Significant adhesions made resection difficult and endoscopic-assisted technique (right view) along with microdoppler was used to ensure vertebral artery was not in immediate vicinity when resecting cyst walls.

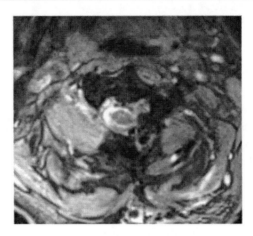

Figure 9. Postoperative view after complete resection of left sided synovial cyst.

Discussion

Numerous cadaveric studies have been performed to compare the retrocondylar and transcondylar far lateral approaches [13, 39-41]. Nanda et al. found that an additional 15.9° and 19.9° of lateral exposure were gained via resection of 1/3 and ½ of the occipital condyle respectively using an extreme lateral approach [13]. Wanebo and Chicoine found the angle of exposure increased laterally by 10.7° and 15.9° after resection of ¼ and ½ condyle respectively using a far lateral approach [39]. They noted that the transcondylar approach provided contralateral exposure of the foramen magnum. Spektor et al. found that the area of petroclival exposure increased from 21 to 28% after resection of the posteromedial condyle up to the hypoglossal canal [41]. The main deterrant to further rostral visualization was the jugular tubercle, which when removed increased rostral clival exposure from 28 to 71%. They noted that the condylar resection doubled surgical freedom in working about a point on the anterior rim of the foramen magnum (FM). Dowd et al. removed an average of 17 mm (56%) of condyle to expose the contralateral jugular tubercle without any brainstem retraction [40]. This amount of condyle resection gained an average of 41° of additional lateral exposure using a slightly different measurement technique. Of note, none of these studies calculate working area on the brainstem, nor do the account for the effect tumor displacement has on working area and surgical freedom.

Numerous studies demonstrate there is a large variation in the normal anatomy of the FM and occipital condyles [39, 42-45]. Therefore, the need to study the preoperative CT and individualize each approach is of paramount importance. Wanebo and Chicoine found that in some specimens, the condyle length consisted of 25% of the anterior-posterior (AP) foramen magnum diameter whereas in other specimens it was >50% [39]. Muthukumar et al. and Barut et al. found that in 20% and 21% of skulls respectively, one or both occipital condyles protruded into the foramen magnum [42, 44]. Hypoglossal canals were septated in 25-30% of normal skulls [42, 44].

In deciding on the approach, one must also factor in the risks of each approach. More extensive approaches that include drilling of the condyle, jugular tubercle, and transposition of the vertebral artery increase operative time, blood loss, postoperative pain, and risk of injury to lower cranial nerves and vertebral arteries. Arnautovic *et al.* noted 3 of 18 cases required staging due to troublesome bleeding from the venous plexus surrounding the vertebral artery, which they title the suboccipital cavernous sinus, during transposition [32, 46]. Margalit *et al.* suffered 4 vertebral artery injuries during transposition [27]. Biomechanic testing has demonstrated mild instability particularly with flexion-extension even with 25% condylar resection and the authors conclude that resection greater than 50% necessitates occipitocervical fusion [47].

Given that tumors naturally open the surgical corridor, we agree with the experience of Samii, George, Nanda, and Goel that the majority of intradural anterior and anterolateral tumors can be resected using a simpler posterolateral technique [11, 13, 17, 20]. The need for more extensive lateral approaches can be reserved for certain indications. Some pertinent pathological considerations that may necessitate greater exposure include vertebral artery encasement [13, 15, 27], contralateral extension [15, 27, 39], small ventral tumors without any brainstem shift [15, 27], en plaque pattern or reoperations cases where there is loss of arachnoid sheath between tumor and surrounding structures [36, 39, 48]. Some pertinent anatomical factors that may necessitate greater exposure include short foramen magnum [39] or protruding occipital condyles [42, 44]. Although not discussed, extradural tumors frequently require more extensive approaches not for exposure, but for tumor resection [20, 27, 49].

Finally, we have added to our surgical armamentarium the endoscopic-assisted technique, which we have described elsewhere [50]. We have utilized rigid 0°, 30°, and 45° and flexible endoscopes for greater visualization of lower cranial nerves and arterial branches. In tumors above or encasing the vertebral artery, position of the lower cranial nerves and arterial branches is more variable and these structures are at greater risk of injury [15, 20]. In these tumors, small cuts in the tumor can create a corridor for endoscope exploration and earlier identification of lower cranial nerves and arterial branches.

The endoscope has also been of benefit to assess completeness of resection. Others have had some early positive experience with the addition of endoscopes to assist in microsurgical resection of craniovertebral junction lesions [51, 52]. As endoscopic or endoscopic-assisted instruments evolve, the amount that can be achieved from the posterolateral approach will increase.

Conclusion

Various surgical approaches have been developed over the last few decades to approach anterior and anterolateral craniovertebral junction tumors from increasingly lateral angles. Numerous anatomical studies have been performed to calculate the increase in lateral exposure and surgical freedom afforded by these approaches. However, these fail to account for natural brainstem retraction afforded by tumor displacement. Therefore, the large majority of these lesions can be safely resected from a simpler lateral suboccipital approach. There still remains a small subset of cases that require a further lateral approach due to pathology-related and patient anatomical factors. Lastly, further development of endoscopic-assisted microsurgical technique will enhance the neurosurgeon's ability to treat these lesions safely and effectively.

References

[1] Love JG, Thelen EP, Dodge HW, Jr. Tumors of the foramen magnum. *J. Int. Coll. Surg.* 1954 Jul; 22(1:1): 1-17.

[2] Stein BM, Leeds NE, Taveras JM, Pool JL. Meningiomas of the Foramen Magnum. *J. Neurosurg.* 1963 Sep; 20: 740-751.

[3] Zoltan L. Tumours at the foramen magnum (author's transl). *Acta Neurochir.* (Wien) 1974; 30(3-4): 217-225.

[4] Smolik EA, Sachs E. Tumors of the foramen magnum of spinal origin. *J. Neurosurg.* 1954 Mar; 11(2): 161-172.

[5] Yasuoka S, Okazaki H, Daube JR, MacCarty CS. Foramen magnum tumors. Analysis of 57 cases of benign extramedullary tumors. *J. Neurosurg.* 1978 Dec; 49(6): 828-838.

[6] Guidetti B, Spallone A. Benign extramedullary tumors of the foramen magnum. *Surg. Neurol.* 1980 Jan; 13(1): 9-17.

[7] Mullan S, Naunton R, Hekmat-Panah J, Vailati G. The use of an anterior approach to ventrally placed tumors in the foramen magnum and vertebral column. *J. Neurosurg.* 1966 Feb; 24(2): 536-543.

[8] Miller E, Crockard HA. Transoral transclival removal of anteriorly placed meningiomas at the foramen magnum. *Neurosurgery* 1987 Jun; 20(6): 966-968.

[9] Crockard HA, Sen CN. The transoral approach for the management of intradural lesions at the craniovertebral junction: review of 7 cases. *Neurosurgery* 1991 Jan; 28(1): 88-97; discussion 97-8.

[10] Menezes AH, Traynelis VC, Gantz BJ. Surgical approaches to the craniovertebral junction. *Clin. Neurosurg.* 1994; 41: 187-203.

[11] Samii M, Klekamp J, Carvalho G. Surgical results for meningiomas of the craniocervical junction. *Neurosurgery* 1996 Dec; 39(6): 1086-94; discussion 1094-5.

[12] Rhoton AL, Jr. The far-lateral approach and its transcondylar, supracondylar, and paracondylar extensions. *Neurosurgery* 2000 Sep; 47(3 Suppl): S195-209.

[13] Nanda A, Vincent DA, Vannemreddy PS, Baskaya MK, Chanda A. Far-lateral approach to intradural lesions of the foramen magnum without resection of the occipital condyle. *J. Neurosurg.* 2002 Feb; 96(2): 302-309.

[14] Heros RC. Lateral suboccipital approach for vertebral and vertebrobasilar artery lesions. *J. Neurosurg.* 1986 Apr; 64(4): 559-562.

[15] Wu Z, Hao S, Zhang J, Zhang L, Jia G, Tang J, et al. Foramen magnum meningiomas: experiences in 114 patients at a single institute over 15 years. *Surg. Neurol.* 2009 Oct; 72(4): 376-82; discussion 382.

[16] Bassiouni H, Ntoukas V, Asgari S, Sandalcioglu EI, Stolke D, Seifert V. Foramen magnum meningiomas: clinical outcome after microsurgical resection via a posterolateral suboccipital retrocondylar approach. *Neurosurgery* 2006 Dec; 59(6): 1177-85; discussion 1185-7.

[17] Goel A, Desai K, Muzumdar D. Surgery on anterior foramen magnum meningiomas using a conventional posterior suboccipital approach: a report on an experience with 17 cases. *Neurosurgery* 2001 Jul; 49(1): 102-6; discussion 106-7.

[18] Gupta SK, Khosla VK, Chhabra R, Mukherjee KK. Posterior midline approach for large anterior/anterolateral foramen magnum tumours. *Br. J. Neurosurg.* 2004 Apr; 18(2): 164-167.

[19] Pritz MB. Evaluation and treatment of intradural tumours located anterior to the cervicomedullary junction by a lateral suboccipital approach. *Acta Neurochir.* (Wien) 1991; 113(1-2): 74-81.

[20] Bruneau M, George B. Foramen magnum meningiomas: detailed surgical approaches and technical aspects at Lariboisiere Hospital and review of the literature. *Neurosurg. Rev.* 2008 Jan; 31(1): 19-32; discussion 32-3.

[21] Seeger W. *Atlas of Topographic Anatomy of the Brain and Surrounding Structures.* Wien: Springer-Verlag; 1978.

[22] Bertalanffy H, Gilsbach JM, Mayfrank L, Klein HM, Kawase T, Seeger W. Microsurgical management of ventral and ventrolateral foramen magnum meningiomas. *Acta Neurochir.* Suppl. 1996; 65:82-85.

[23] Bertalanffy H, Seeger W. The dorsolateral, suboccipital, transcondylar approach to the lower clivus and anterior portion of the craniocervical junction. *Neurosurgery* 1991 Dec; 29(6): 815-821.

[24] David CA, Spetzler RF. Foramen magnum meningiomas. *Clin. Neurosurg.* 1997; 44: 467-489.

[25] Kratimenos GP, Crockard HA. The far lateral approach for ventrally placed foramen magnum and upper cervical spine tumours. *Br. J. Neurosurg.* 1993; 7(2): 129-140.

[26] Parlato C, Tessitore E, Schonauer C, Moraci A. Management of benign craniovertebral junction tumors. *Acta Neurochir.* (Wien) 2003 Jan; 145(1): 31-36.

[27] Margalit NS, Lesser JB, Singer M, Sen C. Lateral approach to anterolateral tumors at the foramen magnum: factors determining surgical procedure. *Neurosurgery* 2005 Apr; 56(2 Suppl): 324-36; discussion 324-36.

[28] Sen CN, Sekhar LN. An extreme lateral approach to intradural lesions of the cervical spine and foramen magnum. *Neurosurgery* 1990 Aug; 27(2): 197-204.

[29] Sen CN, Sekhar LN. Surgical management of anteriorly placed lesions at the craniocervical junction--an alternative approach. *Acta Neurochir.* (Wien) 1991; 108(1-2): 70-77.

[30] Salas E, Sekhar LN, Ziyal IM, Caputy AJ, Wright DC. Variations of the extreme-lateral craniocervical approach: anatomical study and clinical analysis of 69 patients. *J. Neurosurg.* 1999 Apr; 90(2 Suppl): 206-219.

[31] Babu RP, Sekhar LN, Wright DC. Extreme lateral transcondylar approach: technical improvements and lessons learned. *J. Neurosurg.* 1994 Jul; 81(1): 49-59.

[32] Arnautovic KI, Al-Mefty O, Husain M. Ventral foramen magnum meninigiomas. *J. Neurosurg.* 2000 Jan; 92(1 Suppl): 71-80.

[33] Banerji D, Behari S, Jain VK, Pandey T, Chhabra DK. Extreme lateral transcondylar approach to the skull base. *Neurol. India* 1999 Mar; 47(1): 22-30.

[34] Bruneau M, Cornelius JF, George B. Antero-lateral approach to the V3 segment of the vertebral artery. *Neurosurgery* 2006 Feb; 58(1 Suppl): ONS29-35; discussion ONS29-35.

[35] George B, Lot G. Anterolateral and posterolateral approaches to the foramen magnum: technical description and experience from 97 cases. *Skull. Base Surg.* 1995; 5(1): 9-19.

[36] George B, Dematons C, Cophignon J. Lateral approach to the anterior portion of the foramen magnum. Application to surgical removal of 14 benign tumors: technical note. *Surg. Neurol.* 1988 Jun; 29(6): 484-490.

[37] George B, Lot G, Boissonnet H. Meningioma of the foramen magnum: a series of 40 cases. *Surg. Neurol.* 1997 Apr; 47(4): 371-379.

[38] Lot G, George B. The extent of drilling in lateral approaches to the cranio-cervical junction area from a series of 125 cases. *Acta Neurochir.* (Wien) 1999; 141(2): 111-118.

[39] Wanebo JE, Chicoine MR. Quantitative analysis of the transcondylar approach to the foramen magnum. *Neurosurgery* 2001 Oct; 49(4): 934-41; discussion 941-3.

[40] Dowd GC, Zeiller S, Awasthi D. Far lateral transcondylar approach: dimensional anatomy. *Neurosurgery* 1999 Jul; 45(1):95-9; discussion 99-100.

[41] Spektor S, Anderson GJ, McMenomey SO, Horgan MA, Kellogg JX, Delashaw JB,Jr. Quantitative description of the far-lateral transcondylar transtubercular approach to the foramen magnum and clivus. *J. Neurosurg.* 2000 May; 92(5): 824-831.

[42] Muthukumar N, Swaminathan R, Venkatesh G, Bhanumathy SP. A morphometric analysis of the foramen magnum region as it relates to the transcondylar approach. *Acta Neurochir.* (Wien) 2005 Aug; 147(8): 889-895.

[43] Tubbs RS, Griessenauer CJ, Loukas M, Shoja MM, Cohen-Gadol AA. Morphometric analysis of the foramen magnum: an anatomic study. *Neurosurgery* 2010 Feb; 66(2): 385-8; discussion 388.

[44] Barut N, Kale A, Turan Suslu H, Ozturk A, Bozbuga M, Sahinoglu K. Evaluation of the bony landmarks in transcondylar approach. *Br. J. Neurosurg.* 2009 Jun; 23(3): 276-281.

[45] Acikbas SC, Tuncer R, Demirez I, Rahat O, Kazan S, Sindel M, et al. The effect of condylectomy on extreme lateral transcondylar approach to the anterior foramen magnum. *Acta Neurochir.* (Wien) 1997; 139(6): 546-550.

[46] Arnautovic KI, al-Mefty O, Pait TG, Krisht AF, Husain MM. The suboccipital cavernous sinus. *J. Neurosurg.* 1997 Feb; 86(2): 252-262.

[47] Vishteh AG, Crawford NR, Melton MS, Spetzler RF, Sonntag VK, Dickman CA. Stability of the craniovertebral junction after unilateral occipital condyle resection: a biomechanical study. *J. Neurosurg.* 1999 Jan; 90(1 Suppl): 91-98.

[48] Yasargil M. *Microneurosurgery of CNS Tumors.* New York: Thieme Verlag; 1996.

[49] al-Mefty O, Borba LA, Aoki N, Angtuaco E, Pait TG. The transcondylar approach to extradural nonneoplastic lesions of the craniovertebral junction. *J. Neurosurg.* 1996 Jan; 84(1): 1-6.

[50] Salma A, Ammirati M. Real Time Parallel Intraoperative Integration of Endoscopic, Microscopic, and Navigation Images: A Proof of Concept Based on Laboratory Dissections. *Journal of Neurological Surgery Part B: Skull. Base* 2012(01): 36-41.

[51] Velat GJ, Spetzler RF. The far-lateral approach and its variations. *World Neurosurg.* 2012 May; 77(5-6): 619-620.

[52] Sekhar LN, Ramanathan D. Evolution of far lateral and extreme lateral approaches to the skull base. *World Neurosurg.* 2012 May; 77(5-6): 617-618.

In: Minimally Invasive Skull Base Surgery
Editor: Moncef Berhouma

ISBN: 978-1-62808-567-9
© 2013 Nova Science Publishers, Inc.

Chapter XXIV

Endoscopic-Assisted Transoral Approach to the Clivus and the Craniovertebral Junction. Transnasal or Transoral? A Clinical and Experimental Issue

Massimiliano Visocchi

Institute of Neurosurgery, Catholic University, Largo Gemelli, Rome, Italy

Abstract

The introduction of the endoscopy in spine surgery has strongly updated the classical microsurgical transoral decompression strategy, as supported by the most recent literature dealing this topic. In this chapter all the reported experiences on the surgical approaches to anterior cranioveretbral junction (CVJ) compressive pathologies, with or without endoscopy, open the discussion on the so called open accesses, the microsurgical techniques, the neuronavigation and the video-assisted procedures.

Endoscopy represents a useful complement to the standard microsurgical approach to the anterior CVJ. Endoscopy can be used by transnasal, transoral and transcervical routes; it provides information for a better decompression *with no* need for soft palate splitting, hard palate resection or extended maxillotomy. Although neuronavigation allows a better orientation on the surgical field, intraoperative fluoroscopy helps to recognize residual compression.

Virtually, in normal anatomic conditions, no surgical limitations exist for endoscopically assisted transoral approach, compared with the pure endonasal and transcervical endoscopic approaches.

According to the personal experience in the cadaver lab, the endoscope deserves an interesting role as "support" to the standard transoral microsurgical approach, since 30° angulated endoscopy strongly increase the surgical area exposed over the posterior pharyngeal wall and the extent of the clivus. Moreover, compared to the pure transnasal endoscopic procedure, it deserves the main role due to the wider linear and angled surgical route exposure.

Keywords: Cranio-vertebral junction, transoral approach, transnasal approach, transcervical approach, endoscopy

Introduction

The transoral approach to the posterior pharyngeal wall has been used for years to drain retropharyngeal abscesses, but it was only in the 1940s that it was first used in the treatment of spinal abnormalities [18]. In 1962, Fang and Ong (5) published the first series of patients that undergo transoral decompression for irreducible atlantoaxial abnormalities. The high rate of morbidity and mortality caused poor acceptance of the transoral approach as a means for decompression of cervicomedullary junction abnormality.

Popularized by Crockard, the microsurgical ventral approach to the CVJ has been widely described for decompression of irreducible extradural pathology [3]. The shortest and most physiological route to the ventral aspect of the CVJ is represented by an anterior approach through the pharynx. The use of the operating microscopes, high-speed drills, self-retaining mouth retractors, flexible oral endotracheal tubes, intraoperative fluoroscopy and electrophysiological monitoring has made this procedure much safer [17]. A number of anterior approaches have been described to allow exposure to the midline and lateral aspects of both the cranial base and upper cervical spine [16]. The transoral-transpharyngeal approach, a technique that is well known to many spine surgeons, provides surgical access to the anterior clivus, C1 and C2. Transoral approaches provide the fundamental anatomy and technique upon which the more complex jaw-splitting approaches are based (i.e. "transoral extended approaches" with transmaxillary and transmandibular extensions). The transoral-transpharyngeal approach historically remains the "gold standard" for anterior approaches to the cervical spine. However, there are still technical difficulties with the operating microscope, such as the need to see and work through a narrow opening in a deep cavity; to improve visualization, soft palate splitting and even hard-palate resection along with extended maxillotomy are occasionally required.

To overcome such complications, endoscopic-assisted procedures for CVJ decompression have been developed, starting from the experience with the use of the endoscope for transsphenoidal pituitary surgery and cervical spine. An update to the concept of classical transoral microsurgical decompression is now strongly provided by the most recent literature dealing with the introduction of the endoscopy in spine surgery.

Microsurgery (vs) Endoscopy

The Classic Transoral Microsurgical Approach

Historically Menezes outlined several factors influencing the specific treatment of anterior CVJ compressive abnormalities. These included:

1) the reducibility of the lesion, i.e., whether anatomic alignment be restored thus alleviating the compression,
2) the direction and the mechanics of the compression,
3) the aetiology of the compression, and
4) the presence of ossification centres.

The approach to the lesion is dictated by the location and nature of the compression [11]. When preoperative dynamic neuroradiological examinations demonstrate that the CVJ compression is reducible, neural decompression may be obtained by simply reducing the dislocation, as well as by stabilizing the CVJ with a posterior instrumentation, either with wires, claws or screws ("functional decompression"); otherwise anterior decompression is required [11 19; 20; 21; 22].

The huge Menezes' experience on transoral approach started in 1977 and up to the 2008 the number of the microsurgical procedures has been calculated to be 732 (280 children) [12]. This author in his papers concluded that the ventral transoral–transpalatopharyngeal approach has evolved into a safe, rapid, effective and direct approach to the ventral irreducible pathology of CVJ with minimal morbidity and mortality. Although there have been recent attempts at obtaining better visualization and reducing the surgical morbidity with endoscopically assisted procedures, Menezes has not felt the need for any of those. In his opinion, in addition, intra- operative fluoroscopy or the use of "Stealth Technology" have been of little value because of the marked improvement in the three-dimensional imaging. Menezes concludes that the advantages of the transoral-transpalatine approach to the CVJ, compared with other operative approaches in irreducible pathology, are:

1) the impinging bony pathology and granulation tissue that accompanies chronic instability is easily accessible;
2) the patient is placed in the extended position as opposed to the flexed position, thus, decreasing the angulation on the brain stem during surgery;
3) surgery is performed through the avascular median raphe and through the clivus [11; 12; 23].

The Endoscopy (dealing with "minimally invasive surgery") means *looking inside* and typically refers to looking inside the body for medical reasons using an endoscope, an instrument used to examine the interior of a hollow organ or cavity of the body. It was used as early as the ancient Greek and Roman periods. An instrument considered a prototype of endoscopes was evidenced and discovered in the ruins of Pompei. It was Philip Bozzini who in 1805 made the first attempt to observe the living human body directly through a tube he created known as a Lichtleiter (light guiding instrument) to examine the urinary tract, rectum and pharynx. Unlike most other medical imaging devices, endoscopes are inserted directly into the organ. In the early 1950s it was first designed a "fibroscope" (a coherent bundle of flexible glass fibres able to transmit an image), which led to further improvements in image quality. Further innovations included using additional fibres to chanel light to the objective end from a powerful external source along with and $0° - 30° - 45°$ lenses, thereby achieving the high level of full spectrum illumination and oriented vision that was needed for detailed viewing and colour photography. It was the beginning of key-hole surgery as we know it today [1; 6; 7; 8; 10; 13; 14; 28].

Rationale

Contrary to Menezes' experience, some papers claim significant oropharyngeal morbidity from splitting the soft palate associated with the transoral approach. Jones reported a striking difference in oropharyngeal complications when analyzed with regard to splitting of the soft palate (no splitting vs splitting complication rate: 1/5); oropharyngeal complications dropped to 15.4% in those patients who did not undergo splitting of the soft palate, as compared with 75% in the split soft palate group. The author concludes that this procedure should be discontinued where it is not absolutely necessary [9]. The surgical risks dealing with the lateral exposure (toughly 15 to 20 mm bilaterally off the midline from the inferior clivus to the C3 body) consists of trauma to 1) the Eustachian tube orifice, 2) the hypoglossal nerve, 3) the vidian nerve 4) the vertebral artery at the C1 – C2 interface; those dealing with the longitudinal exposure (due to soft palatal splitting with velopalatine incompetence) consist of 1) nasal speech 2) dysphagia, 3) regurgitation of liquids [15].

Endoscopic-assisted Procedures

Endoscopic endonasal, transoral and transcervical approaches developed recently as promising alternatives to the classic microsurgical transoral approach to the CVJ that may become more mainstream as experience with these approaches increases (cons: learning curve, loss of 3-dimensional visualization).

Endonasal Approach

The increased diffusion in the use of the endoscope for transsphenoidal pituitary surgery led some studies to explore the possibility of applying the endoscopic endonasal approach in the surgical treatment of skull base lesions other than pituitary tumors. In recent years some papers have reported anatomical studies and surgical experience in the endoscopic endonasal approach to different areas of the midline skull base, from the olfactory groove to the CVJ [13]. In 2002 Alfieri was the first to perform a cadaveric study on totally transnasal endocopic odontoidectomy through one or two nostril routes, by following the Jho's endonasal paraseptal technique [8]. Rodlens endoscopes, which were 2.7 or 4 mm in diameter, 18 cm in length with 0-, 30-, and 70-degree lenses, were used. The surgical landmarks leading to the CVJ were the inferior margin of the middle turbinate, nasopharynx and the Eustachian tubes. The nasopharynx was readily identified following the inferior margin of the middle turbinate. The line drawn between the Eustachian tubes indicated the juncture between the clivus and the atlas. The Author concluded that

> "..contrary to a conventional transoral approach, this endoscopic endonasal approach provides unlimited access to the midline clivus and a potential of carrying out surgical decompression at the ventral craniocervical junction without adding C1-2 instability" [28].

Three years later Cavallo confirmed that observation on a cadaveric study [2].

After the intuition of Alfieri, in 2005 Kassam operated the first case through a fully transnasal endoscopic resection of the odontoid in a 73- year old woman affected by rheumatoid arthritis [1; 14]. In his historical report, Kassam's recommended equipment consisted of

1) navigation system;
2) a zero degree endoscope;
3) long angled endonasal drill,
4) ultrasonic aspirator;
5) bayoneted handheld microinstrumentation and concluded

"The transoral approach remains the "gold standard" but in contrast with this "... the defect created by transnasal approach is above the level of soft palate and should not be exposed to the same degree of bacterial contamination".

THE NASOPALATINE LINE

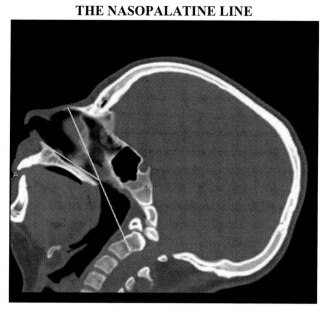

Figure 1. The nasopalatine line is measured by connecting the most inferior point on the nasal bone to the most posterior point on the hard palate in the midsagittal plane (see text).

Further anatomic studies performed by Messina one year later concluded that as the transoral approach, the endoscopic endonasal provides a direct route to the surgical target, but it seems related to less morbidity. Nevertheless, as matter of fact, things are less simple. The group of Kassam published in 2009 the concept of the "Nasopalatine line" (NPL) which is the line created by connecting the most inferior point on the nasal bone to the most posterior point on the hard palate in the midsagittal plane. Intersection of this line with the vertebral column is measured relative to the inferior aspect of the body of C2 along its posterior surface (Figure 1) [4]. The NPL is a reliable predictor of the maximal extent of inferior dissection, and odontoid surgery can reliably be performed according to the preoperative radiological study of the possible anatomical limitations of the endonasal approach. This approach is recommended by the authors in selected cases as valid alternative to the transoral microscopic approach for the resection of the odontoid process of C2 and should be performed only by surgeons very skilled in endoscopic endonasal surgery and in endoscopic cadaver- dissections [21; 23 – 27].

Indications, Advantages/Disadvantages Side-Effects, Putative Complications

According to Kassam, the approach, originally described, was applicable to the selected group of rheumatoid patients presenting with brainstem compression who had clinical progression of disease despite posterior spinal fixation, significant bony compression from pannus formation or a significant anterior vector of pannus. Furthermore, in the author's indications, the associated pathologies for endoscopic transnasal resection of the odontoid included also tumors in the region of the foramen magnum, vertebrobasilar aneurysms not ablated by endovascular treatment, dens displacement secondary to C1/C2 traumatic fracture and other occipitocervical anomalies associated with anterior cervicomedullary compression, such as os odontoideum, atlantal assimilation and basilar invagination. To be highlighted that the endonasal approach to the odontoid can even be performed in the presence of the retro pharyngeal location of internal carotid arteries. Relative contraindications to a transnasal endoscopic odontoid resection include tumors lateral to, or encasing, the extracranial vertebral arteries, or pathology existing inferior to C2. In general, the expanded endonasal approach offers a number of advantages to the traditional open transoral approach, including improved visualization, decreased airway and swallowing morbidity, preservation of palatal function, decreased postoperative pain and reduced duration of hospitalization. With the incision performed above the soft palate, should limit postoperative swallowing dysfunction and minimize exposure to oral bacterial flora; moreover it is possible to remove the odontoid process without disturbing the C1 ring due to the more caudal surgical route. Of course, there are putative risks with this surgery, which include possible cerebrospinal fluid leak from aggressive pannus resection or dural tear, cervical instability and vascular injury. These risks are shared by other approaches and can be effectively managed with the endonasal.

Summary

Pros: partial isolation of the oral cavity, no needs of tracheostomy and reduced need of feeding tube.

Cons: oblique approach, only piecemeal removal of CVJ pathology is allowed, not recommended for large tumors and low sited CVJ pathologies.

Transoral Approach

The 30-degree endoscope was proposed for transoral approach to avoid full soft-palate splitting, hard-palate splitting, or extended maxillo/mandibulotoy (7). Using the endoscope, the operator is able to look in all directions, by rotating the instrument. Because the light source is at the level of the abnormality, superior illumination can be obtained. With the aid of an endoscope, abnormalities, such as high as the midclivus, can be visualized without extensive soft- or hard-palate manipulation.

The last high profile cadaveric study recently available in the literature is the one of the Ammirati's group which quantifies the surgical volume gained by this approach: the surgical area exposed over the posterior pharyngeal wall is significantly improved using the endoscope (606.5 -127.4 mm^3) compared with the operating microscope (425.7 - 100.8 mm^3), without any compromise of surgical freedom (p 0.05). The extent of the clivus exposed with the endoscope (9.5 - 0.7 mm), without splitting the soft palate, is significantly improved, compared with that associated with microscopic approach (2.0 - 0.4 mm) (P 0.05). [17]. With this paper it is well demonstrated that with the aid of the endoscope and image guidance, is it possible to approach the ventral CVJ transorally with minimal tissue dissection, no palatal splitting and no compromise of surgical freedom. In addition, the use of an angled-lens endoscope can significantly improve the exposure of the clivus without splitting the soft palate (Figure 2).

Moreover, personal unpublished data on the comparison between the transnasal and transoral pure endoscopic exposition in the cadaver, confirm that a better exposure of the CVJ (C0-C1-C2) can be obtained by using the transoral rather than the transnasal approach, both in anterior and lateral projection. More in detail, the transoral endoscopic surgical exposition is wider compared to the transnasal 1) in the sagittal plane, from a minimum of 23% to a maximum of 94,11% (average 64,1%) as vertical surgical route length and from a minimum of 78% to a maximum of 22,58% as vertical surgical route angle (average 33.33 %) 2) in the coronal plane, from a minimum of 16,12% to a maximum of 49,23% (average 29.65%) as coronal surgical route length and from a minimum of 13,29% to a maximum of 34,42 % as coronal surgical route angle (average 19.14 %) (Figure 3).

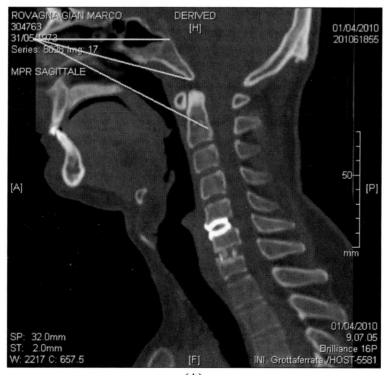

(A)

Figure 2. (Continued).

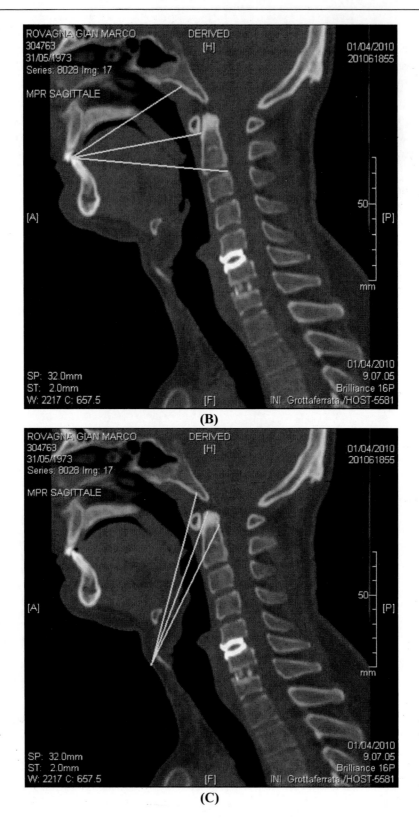

Figure 2. (Continued).

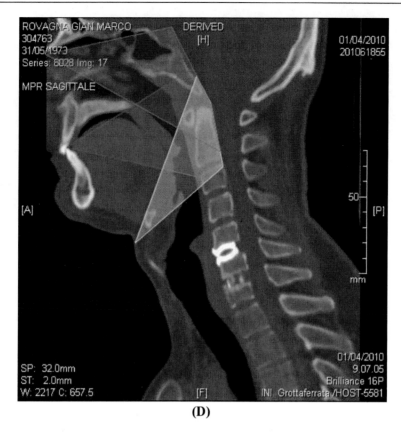

(D)

Figure 2. A) Computed tomographic scans demonstrating the surgical trajectory and angles for the endonasal approach (personal observation); B) Computed tomographic scans demonstrating the surgical trajectory and angles for the the transoral approach (personal observation); C) Computed tomographic scans demonstrating the surgical trajectory and angles for the transcervical approach (personal observation); D) Common surgical area of the 3 approaches is represented by the overlapping illumination.

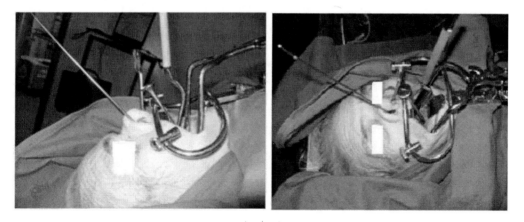

Figure 3. Cadaver lateral view of transnasal and transoral probes with the oral distractor: left lateral view; right AP view.

Indications, Advantages/Disadvantages Side-Effects, Putative Complications

Virtually no surgical limitations exist for the endoscopically assisted transoral approach, compared with the pure endonasal and transcervical approaches (Figure 4; 5).

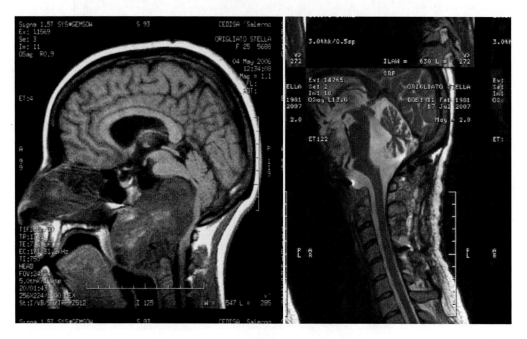

Figure 4. Huge chordoma in 26 yrs lady before (left) and after (right) endoscopic-assisted transoral microsurgical approach, not suitable for endoscopic endonasal and transcervical approach (personal observation).

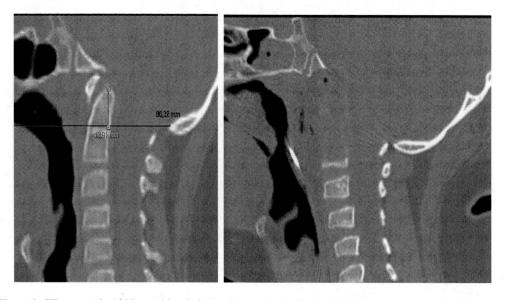

Figure 5. CT scan study of 18 yrs old male harboring severe platibasia. CVJ before transoral endocopic assisted approach (left); after surgery (right). To note the almost complete demolition of the CVJ including the inferior and middle third of the Clivus along with C1 and C2.

Of course, there are putative risks with this surgery, which include possible cerebrospinal fluid leak from aggressive pannus resection or dural tear, cervical instability and vascular injury. These risks are shared by other approaches and can be effectively managed with the endonasal. To be highlighted that alternative procedures must be required (i.e. endonasal or transcervical endoscopic approach) in the presence of the retro pharyngeal location of internal carotid arteries.

Summary

Pros: direct approach, radical removal of huge tumors, good visualization and comfortable mobilization of surgical tools.

Cons: possible need of tracheostomy, need of feeding tube, difficult management of very high invagination conditions with platibasia.

Transcervical Approach

Wolinsky first described in 2007 an alternative endoscopic route to the anterior CVJ with the endoscopic transcervical approach [23 – 27; 29]. The need of this option deals with the limitation of transpharyngeal approaches above mentioned. When the pharynx is traversed, the operative field is virtually contaminated with oral flora. The risk of infection, poor pharyngeal healing and meningitis (if the dura is transgressed) can all be increased. Moreover the transcervical exposure is familiar to neurosurgeons and the trajectory proposed by the author, allows deep-seated basilar invaginations to be decompressed [10]. The endoscopic odontoidectomy via a standard anterior cervical approach has been described as the evolution of the procedure used for a transodontoid screw.

Indications, Advantages/Disadvantages Side-Effects, Putative Complications

According to Wolinski, the endoscopic transcervical odontoidectomy has many advantages over the conventional approaches to odontoid resection: the exposure is familiar to neurosurgeons. It does not require traversing the oral mucosa and therefore theoretically decreases the chance of postoperative meningitis in the setting of an inadvertent or intentional breach of the dura mater. In addition, the trajectory of the approach should allow even the deepest of basilar invaginations to be decompressed. The postoperative recovery time is shorter compared to other techniques. Patients are able to ingest food orally shortly after removing the endotracheal tube. In patients without preoperative dysphagia, there is no need for a tracheostomy or gastric or duodenal feeding tube, as a result of the procedure. The risk of postoperative phonation difficulty, that is present in a transoral approach, is avoided with a transodontoid approach. The risk of injury to the recurrent laryngeal nerve is present but is the same as in an anterior cervical approach. Using a transodontoid approach, more caudal vertebral body resection (below the odontoid) is possible through the same incision, because the technique exposes C-1 through C-4 ventrally and the exposure can be easily extended to provide access caudal to C-4. Not all patients are candidates for this approach. As in the case of transodontoid screw placement, the trajectory may not be achieved in patients who are obese, barrel-chested, or severely kyphotic. Nevertheless the odontoid decompression is too oblique and partial although without disturbing the C1 ring. To gain access to the lower clivus, C1 ring has to be removed but the angle of attack makes this portion of dissection most difficult or impossible. Finally, in our opinion, in cases of impression basilaris or other high pathologies, such an approach could be uncomfortable and challenging. Of course, possible cerebrospinal fluid leak from aggressive pannus resection or dural tear, cervical instability and vascular injury must be put into consideration.

Summary

Pros: complete isolation of the oral cavity, no needs of tracheostomy and feeding tube.

Cons: oblique approach, only piecemeal removal of CVJ pathology is allowed, not recommended for large tumors, obese, barrel chested and severely kyphotic patients. (Figure 6).

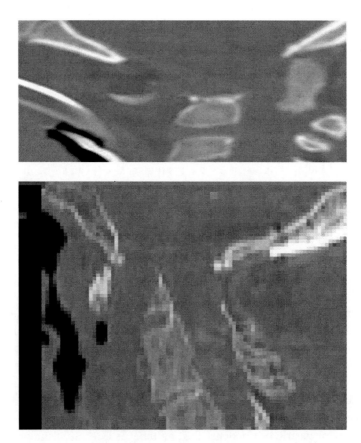

Figure 6. Comparison between transnasal (upper) and transcervical (lower) postoperative CT scan in odontoidectomy performed in two different patients. To note the horizontal decompression allowed by the transnasal compared to the oblique sharp decompression by the transcervical approach.

Considerations

Endoscopy represents the future of neurosurgery, both for the ventricular and skull base surgery and the spinal surgery.

The progressive worldwide blooming of transoral and transnasal procedures, thanks to the intensive care and the intraoperative neurophysiological monitoring techniques improvements (once considered pioneering and very selective), are spreading the expertise in this surgery to a new population of Surgeons.

New trends in technology drive from the "old fashioned referenced" micro surgeons to the young "endo" spine surgeons, more committed in video- assisted and minimally invasive procedures.

As far as possible to summarize from the literature and conclude according to personal experience, although blooming in the worldwide literature, pure endonasal and cervical endoscopic approach deserve consideration but still has some disadvantages:

1) the learning curve,
2) the lack of 3-dimensional perception of the surgical field which could be an operationally limiting factor. Image clarity will be diminished when endoscopes smaller than 2.7 mm are used. Standard 4-mm endoscopes give a good image quality, but 2.7-mm scopes provide better maneuverability;.
3) a limited working channel, according to the variability of the nasopalatine line, which can make difficult to remove huge tumors like the one shown in Figure 4.

The endoscopically assisted transoral surgery with 30 degrees endoscopes represents an emerging alternative to standard microsurgical techniques for transoral approaches to the anterior CVJ. Used in conjunction with traditional microsurgery and intraoperative fluoroscopy, it provides a safe and improved method for anterior decompression without or with a reduced need for extensive soft palate splitting, hard palate resection, or extended maxillotomy. Although we use in a limited number of patients transnasal approach, virtually no surgical limitations do exist for endoscopically assisted transoral approach, compared with the pure endonasal and transcervical approaches (Table 1).

So far, the endoscope deserves an interesting role as a "support" to the standard transoral microsurgical approach since 30° angulated endoscopy strongly improve the *visual but not the working channel and volume*.

Consequently, although we take advantage by endoscopy, we continue to perform the soft palate splitting, since at the maximum follow up, no one patient complained nasal speech, dysphagia or regurgitation of liquids.

Transoral (videoassisted) approach still remains the gold standard compared to the "pure" transnasal and transcervical approaches due to the wider working channel provided by the former technique. Experience is required with greater numbers of patients and long-term follow-up to further validate this promising technique.

Furthermore, the use of image guidance systems before surgery allows a correct planning and during endoscopic procedures gives the surgeon a constant orientation in the surgical field, thus increasing the accuracy and the safety of the approach, although the use of contrast medium fluoroscopy "per se" represents an "ever green" old fashion image guidance system still effective.

Table 1.

Patient Initials Case N°	Age (Sex)	Primary disease	1. Radiology	1. Pre-op 2. C1-C2 shift 3. (X Rays)	1. Treatment	Post-op shift (X Rays)	Frankel Scale & Di Lorenzo Grade changes	External orthosis	Follow-up (months)
SO 1	26 (F)	CVJ Chordoma	C0-C2 anterior compression CVJ instability	Virtual	1) Transoral decompression 2) C0-C3 reduction, lateral masses screws instrumentation and heterologous bone fusion	No	E/E I/I	Philadelphia (1 month)	76
FF 2	33 (M)	CVJ Chordoma	C0-C2 anterior compression CVJ instability	Virtual	1) Transoral decompression 2) C0-C3 reduction, C2 pedicles and C3 lateral masses screws instrument-ation and heterologous bone fusion	No	E/E I/I	Philadelphia (1 month)	70
CO 3	68 (F)	Rheumatoid Arthritis	Anterior C1-C2 compression C1-C2 instability	>5 mm	1)Transoral decompression 2) C0, C2 pedicles and lateral masses screws instrumentation	No	D/E II/I	Soft collar (1 month)	53
CL 4	15 (M)	Developmental anomaly C0-C1	C0-C1 anterior compression C1-C2 instability	>5 mm	2) Transoral decompression 2) C1 laminectomy, C0 double vertical screws, C2 pedicles and C3 lateral masses screws instrumentation	No	D/E II/I	Soft collar (1 month)	48

Patient Initials Case N°	Age (Sex)	Primary disease	1. Radiology	1. Pre-op 2. C1-C2 shift 3. (X Rays)	1. Treatment	Post-op shift (X Rays)	Frankel Scale & Di Lorenzo Grade changes	External orthosis	Follow-up (months)
CA 5	78 (M)	Chordoma (Chondroid)	CVJ instability C0-C2 anterior compression	Virtual	Transoral C1 - odontoidecotmy and clivectomy C0 double vertical screws, C2 C3 C4 C5 lateral masses screws instrumentation	No	D/E II/I	Soft Collar (1 month)	35
EA 6	11 (M)	Impressio Basilaris Os odontoideum (Down s.)	C1- C2 anterior compression	Virtual, previously documented	Transoral C1 - odontoidecotmy and clivectomy in C0 – C2 – C3 screwing instrumentation and heterologous bone fusion (previously implanted)	No	D/E II/I	Soft Collar (1 month)	34
RR 7	14 (M)	C2 fracture and dislocation	C2 fracture and C1 – C2 dislocation with cerivcomedullary contusion	> 7 mm	1) Transoral C1 – C2 decompression 2) C0 – C3 C5 screwing instrumentation and heterologous bone fusion	No	D/E II/I	Soft Collar (1 month)	31
PF 8	18 (M)	Impressio Basilaris	C0-C2 anterior compression CVJ instability	> 7 mm	1) Transoral decompression 2) C0-C3 reduction, lateral masses screws instrumentation and heterologous bone fusion	No	D/E II/I	Philadelphia (1 month)	16

Table 1. (Continued)

Patient Initials Case N°	Age (Sex)	Primary disease	1. Radiology	1. Pre-op 2. C1-C2 shift 3. (X Rays)	1. Treatment	Post-op shift (X Rays)	Frankel Scale & Di Lorenzo Grade changes	External orthosis	Follow-up (months)
PS 9	7 (M)	Platibasia Impressio Basilaris	C0-C2 anterior compression CVJ instability	Virtual	1) Transoral decompression 2) C0-C3 reduction, C2 pedicles and C3 lateral masses screws instrumentation and heterologous bone fusion	No	D/E II/I	Philadelphia (1 month)	12
PM 10	8 (M)	Platibasia Impressio Basilaris	C0-C2 anterior compression CVJ instability	Virtual	1) Transoral decompression 2) C0-C3 reduction, C2 pedicles and C3 lateral masses screws instrumentation and heterologous bone fusion	No	D/E II/I	Soft collar (1 month)	11
OS 11	30 (F)	Chordoma C0-C1 – C2	C0-C1 anterior compression C1-C2 instability	>5 mm	1) Transnasal decompression 2) Previously fused and fixed	No	D/D II/II	Soft collar (1 month)	10

Patient Initials Case N°	Age (Sex)	Primary disease	1. Radiology	1. Pre-op 2. C1-C2 shift 3. (X Rays)	1. Treatment	Post-op shift (X Rays)	Frankel Scale & Di Lorenzo Grade changes	External orthosis	Follow-up (months)
SD 12	5 (M)	Os odontoideum	CVJ instability C0-C2 anterior compression	>5 mm	1) Transoral C1 - odontoidecotmy 2) C0 C2 C3 lateral masses screws instrumentation	No	E/E I/I	Soft Collar (1 month)	6
RS 13	49 (F)	Platibasia Impressio Basilaris	C0 - C2 anterior compression	No instability	1) Transnasal clivectomy, odontoidectomy 2) Previously fixed and fused with C0 – C2 – C3 screwing instrumentation and heterologous bone fusion	No	D/E II/I	Soft Collar (1 month)	5

References

[1] Alfieri A, Jho HD, Tschabitscher M: Endoscopic endonasal approach to the ventral cranio-cervical junction: anatomical study. *Acta Neurochir* (Wien) 144:219-225 (2002).

[2] Cavallo LM, Messina A, Cappabianca P, Esposito F, de Divitiis E, Gardner P, Tschabitscher M: Endoscopic endonasal surgery of the midline skull base: Anatomical study and clinical considerations. *Neurosurg Focus* 19:E2, (2005).

[3] Crockard HA. Ventral approaches to the upper cervical spine. *Orthopadie*;20(2):140-146, (1991).

[4] de Almeida JR, Zanation AM, Snyderman CH, Carrau RL, Prevedello DM, Gardner PA, Kassam AB. Defining the nasopalatine line: the limit for endonasal surgery of the spine. *Laryngoscope*, 119:239–244, (2009).

[5] Fang HSY, Ong GB: Direct anterior approach to the upper cervical spine. *J Bone Joint Surg Am* 44A:1588–1604, (1962).

[6] 6) Frempong-Boadu AK, Faunce WA, Fessler RG. Endoscopically assisted transoral-transpharyngeal approach to the craniovertebral junction. *Neurosurgery.*;51(5 Suppl):S60-S66 (2002).

[7] Husain M, Rastogi M, Ojha BK, Chandra A, Jha DK. Endoscopic transoral sur- gery for craniovertebral junction anomalies. *J Neurosurg Spine.*;5(4):367-373, (2006).

[8] Jho HD, Ha HG: Endoscopic endonasal skull base surgery: Part 3--The clivus and posterior fossa. *Minim Invasive Neurosurg.* Feb;47(1):16-23, (2004).

[9] Jones DC, Hayter JP, Vaughan ED, Findlay GF: Oropharyngeal morbidity following transoral approaches to the upper cervical spine. *Int J Oral Maxillofac Surg* 27:295-298, (1998).

[10] Mc Girt MJ, Attenello RJ, Sciubba DM, Gokaslan ZL, Wolinsky JP: Endoscopic transcervical odontoidectomy for pediatric basilar invagination and cranial settling *J Neurosurg Pediatrics* 1:337–342, (2008).

[11] Menezes AH: Occipito-cervical fusion: indications, technique and avoidance of complications. In Hitchon PW (ed): *Techniques of spinal fusion and stabilisation.* New York, Thieme, pp 82 – 91(1994).

[12] Menezes A. H.: Surgical approaches: postoperative care and complications "transoral–transpalatopharyngeal approach to the raniocervical. junction" Child' *Nerv Syst* 24: 1187 – 1193 (2008).

[13] Messina A, Bruno MC, Decq P, Coste A, Cavallo LM, de Divittis E, Cappabianca P, Tschabitscher M: Pure endoscopic endonasal odontoidectomy: Anatomical study. *Neurosurg Rev* 30:189–194, (2007).

[14] Kassam AB, Snyderman C, Gardner P, Carrau R, Spiro R: The expanded endonasal approach: a fully endoscopic transnasal approach and resection of the odontoid process: technical case report. *Neurosurgery* 57:E213; discussion E213, (2005).

[15] Mummaneni PV, Haid RW: Transoral odontoidectomy. *Neurosurgery* 56:1045–1050, (2005).

[16] Oya S, Tsutsumi K, Shigeno T Takahashi H: Posterolateral odontoidectomy for irreducible atlantoaxial dislocation: a technical case report. *Spine J* 4: 591 – 594 (2004)

[17] Pillai P, Baig MN, Karas CS, Ammirati M. Endoscopic image-guided transoral approach to the craniovertebral junction: an anatomic study comparing surgical

exposure and surgical freedom obtained with the endoscope and the operating microscope. *Neurosurgery.* May;64(5 Suppl 2):437-42, (2009)

[18] Symonds CP, Meadows SP: Compression of the spinal cord in the neigh borhood of the foramen magnum with a note on the surgical approach by Julian Taylor. *Brain* 60:52–84, (1937)

[19] Visocchi M, Di Rocco F, Meglio M: Craniocervical junction instability: instrumentation and fusion with titanium rods and sublaminar wires. Effectiveness and failures in personal experience. *Acta Neurochir* 145: 265 – 272, (2003).

[20] Visocchi M, Pietrini D, Tufo T, Fernandez E, Di Rocco C Pre-operative irreducible C1-C2 dislocations: intra-operative reduction and posterior fixation. The "always posterior strategy". *Acta Neurochir* (Wien) 151(5):551-60, (2009)

[21] Visocchi M, Fernandez EM, Ciampini A, Di Rocco C, Reducible and irreducible os odontoideum treated with posterior wiring, instrumentation and fusion. Past or present? *Acta Neuroch* (Wien) Oct;151(10):1265-74, (2009).

[22] Visocchi M: Response to Wang et al., re Letter re Visocchi M, Pietrini D, Tufo T, Fernandez E, Di Rocco C Pre-operative irreducible C1-C2 dislocations: intra-operative reduction and posterior fixation. The "always posterior strategy". *Acta Neurochir* (Wien) 151(5):551-9; discussion 560, (2009).

[23] Visocchi M: Transoral approach to the skull base and the upper cervical spine. Essential in Neurosurgery, Y Kato Editor, Printed in Japan *WFNS* 781 – 789 (2009)

[24] Visocchi M: The craniovertebral junction: posterior and anterior approaches. State of art. Crit Rev Neurosurg *WFNS* Vol 1, 1 – 11; (2010).

[25] Visocchi M, Della Pepa GM, Doglietto F, Esposito G, La Rocca G, Massimi L.: Video-assisted microsurgical transoral approach to the craniovertebral junction: personal experience in childhood. *Childs Nerv Syst.* 2011 May;27(5):825-31(2011)

[26] Visocchi M, Doglietto F, Della Pepa GM, Esposito G, La Rocca G, Di Rocco C, Maira G, Fernandez E.: Endoscope-assisted microsurgical transoral approach to the anterior craniovertebral junction compressive pathologies. *Eur Spine J.* 1518 - 1525– 7 (2011).

[27] Visocchi M: Advances in videoassisted anterior surgical approach to the craniovertebral junction. *Adv Tech Stand* 37: 97 – 110(2011).

[28] Vougioukas VI, Hubbe U, Schipper J, Spetzger U: Navigated transoral approach to the cranial base and the craniocervical junction: technical note. *Neurosurgery* 52: 247 – 251, (2003)

[29] Wolinsky JP, Sciubba DM, Suk I, Gokaslan ZL. Endoscopic image-guided odontoidectomy for decompression of basilar invagination via a standard anterior cervical approach. Technical note. *J Neurosurg Spine* 6(2):184-191 (2007).

In: Minimally Invasive Skull Base Surgery
Editor: Moncef Berhouma

ISBN: 978-1-62808-567-9
© 2013 Nova Science Publishers, Inc.

Chapter XXV

Endoscopic Endonasal Approaches to the Paramedian Posterior Skull Base

Jun Muto[1], Danielle de Lara[1], Leo F. S. Ditzel Filho[1], Pornthep Kasemsiri[2], Bradley A. Otto[2], Ricardo L. Carrau[2] and Daniel M. Prevedello[1,*]

[1]Departments of Neurosurgical Surgery, and [2]Otolaryngology- Head and Neck Surgery, The Ohio State University, Columbus, Ohio, USA

Abstract

Endonasal Endoscopic Approaches (EEA) are minimal invasive surgical techniques to the ventral skull base lesions that develop as part of continuous evolution of skull base surgery. It is useful to categorize paramedian approaches into those that provide access to the anterior, middle, and posterior cranial fossa. Each approach is limited by tumor pathology, location, extension and relationship with surrounding nerves and arteries.

The purpose of this study is to describe anatomy-based surgical modules to the paramedian posterior cranial fossa, especially those to the lower clivus, occipital condyle and jugular foramen.

EEAs offer a potential option to access lesions in these areas, such as schwannomas, meningiomas, paragangliomas, chordomas and chondrosarcomas. Based on tumor location and crucial anatomic relationships, such as that with the ICA, the endonasal corridor to the jugular foramen may be used as the only access or it may be considered in combination with conventional approaches.

A precise understanding of anatomical landmarks and of pertinent technical nuances is indispensable to achieve an optimal resection and to minimize morbidity. Our operative technique is hereby described in stepwise fashion using cadaveric photographs.

---

[*] Corresponding Author: Daniel M. Prevedello, M.D.,The Ohio State University, Department of Neurosurgery, N-1049 Doan Hall, 410 W. 10th Avenue, Columbus, OH, 43210. Phone: 614-293-7190. E-mail: dprevedello@gmail.com.

Keywords: Endonasal endoscopic approach, skull base, jugular tubercle, condyle, transcondylar approach, transjugular foramen approach

Abbreviations

CT	computed tomography;
ICA	internal carotid artery;
CS	Cavernous sinus;
IPS	inferior petrosal sinus;
EEA	Endoscopic endonasal approach;
ET	Eustachian tube

Introduction

In the past decade endoscopic endonasal approaches (EEAs) have progressed greatly, becoming a feasible, and often optimal, alternative approach to the ventral skull base. As in any surgical approach to the skull base, safe implementation of EEAs requires a thorough understanding of anatomical landmarks and of pertinent technical nuances. Pathologies originating or involving the lateral segment of the posterior fossa include chondrosarcomas, chordomas, meningiomas, schwannomas, and paragangliomas. EEAs should be weighed against other approaches such as the lateral suboccipital approach, far lateral approach, transpetrosal approach and combined approach [1, 13, 14] during surgical planning. [2, 8] As in open approaches to the skull base, an EEA is ideal when provides the most direct corridor into the target considering tumor pathology, location, attachment to the meninges as well as the relationship to surrounding nerves and arteries.

Despite advances in technique over the past 40 years, choosing the most appropriate surgical approach to access jugular foramen tumors remains a complex decision-making process. [15] In general, these procedures are exhausting and long procedures that often require surgery staging. An initial decompression usually involves a retrosigmoid craniotomy to address the intracranial disease. A second staged procedure addresses the extracranial disease. EEAs to address jugular foramen tumors is a relatively new surgical alternative. [10] EEAs have the advantage of addressing the pathology from a midline approach avoiding cranial nerve manipulations. However, any intracranial component of the tumor in the posterior fossa can only be addressed for brainstem decompression after the extracranial component is resected. Thus, although EEAs can be effectively staged if needed without the downside of reopen skin that is already healing, as it occurs in standard approaches, attention should be given to this simple aspect in terms of surgical goals.

Kassam AB et al. reported a large case series of surgery involving EEAs in single institution demonstrating progressively lower rates of postoperative cerebrospinal fluid leakage following the introduction of vascularized pedicled flaps. [9] Along with progress related to skull base reconstruction, a better understanding of the anatomy from the endoscopic perspective opened the possibility to expand the transnasal endoscopic approach beyond the sella.

Approaches to lesions of the lateral posterior fossa, jugular foramen and condyle are challenging due to the complex anatomy and difficulty access. Here we describe the stepwise approaches to the jugular tuberculum, occipital condyle and jugular foramen with remarks to the various anatomical landmarks.

Surgical Considerations

The anatomical region of the occipital condyle /hypoglossal canal is also known as Zone 6. [4] The approaches to Zone 6 can be organized into the supracondylar approach and the transcondylar approach. During the supracondylar approach, the supracondylar groove of the occipital bone is drilled laterally to the petroclival synchondrosis and the condyle is preserved exposing the dura dorsal to the removed jugular tubercle. The jugular tubercle is part of the occipital bone and it is located immediately dorsal to the supracondylar groove.

During a transcondylar approach, drilling of the medial anterior half of condyle exposes the hypoglossal canal .

Suitable indications for these approaches include schwannomas, meningiomas, paragangliomas, chordomas and chondrosarcomas. Certain tumors, especially chordomas and chondrosarcomas can invade the hypoglossal canal and further extend laterally to jugular foramen.

Surgical Anatomy

Surgical Corridor

The occipital condyle is located behind the fossa of Rosenmüller. The medial aspect of the occipital condyle is lateral to the foramen magnum. Therefore, full access to the condyle from an endonasal approach requires exposure of the foramen magnum, and often, removal of the medial Eustachian tube. When accessing the inferolateral clivus ipsilateral to the lesion of interest, a nasoseptal flap should ideally be elevated on the contralateral side; otherwise, the requisite transpterygoid approach on the side ipsilateral to the disease will likely compromise the pedicle of the flap.

A wide sphenoidotomy and maxillary antrostomy are performed, facilitating the remainder of the transpterygoid approach. Removal of the medial posterior wall of the maxillary sinus exposes the periosteum of the medial pterygopalatine fossa. The palatovaginal canal and its neurovascular bundle are identified and transfixed. The pterygopalatine fossa is then retracted laterally exposing the vidian neurovascular bundle and canal, which are located at the level of floor of the sphenoid sinus. The vidian canal is drilled out in an anteroposterior direction toward the ICA. [3, 6] The vidian nerve is an important landmark to define the level of the petrous ICA. [11] If needed, the periosteum around the paraclival and foramen lacerum ICA can be exposed to allow its mobilization and better tumor exposure during resection.

Transcondylar Approach

The lower part of foramen lacerum is composed of fibrous tissue related to the pharyngobasilar fascia adjacent to the Eustachian tube and petroclival synchondrosis. The pharyngobasilar mucosa is trimmed exposing the *rectus capitis anterior* and *longus capitis muscles*. These muscles are attached to the inferior clivus, and there are well-defined lines of attachment on the ventral surface of the inferior clival bone.

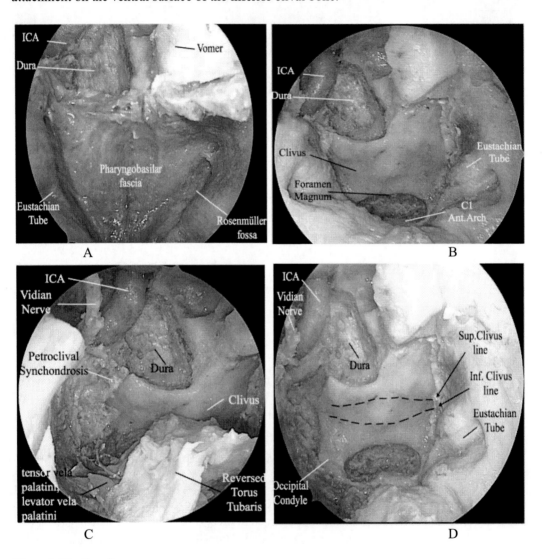

Figure 1. (Continued).

Endoscopic Endonasal Approaches to the Paramedian Posterior Skull Base 393

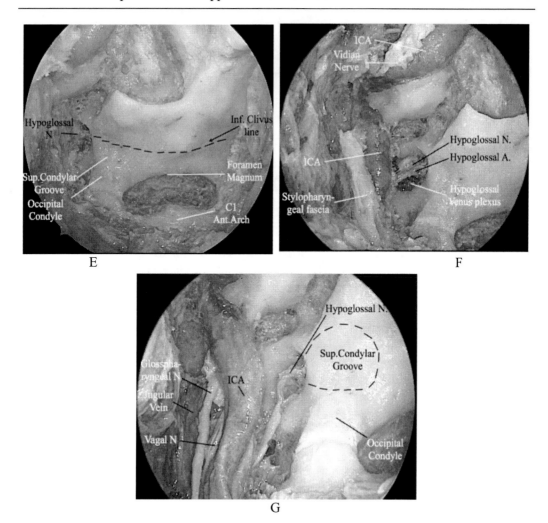

Figure 1. Stepwise exposure of the endoscopic endonasal transclival and transcondylar approaches, and approach to the jugular foramen: (A) The vidian canal is drilled out in an anteroposterior direction toward the paraclival ICA and anterior genu of the ICA was exposed. The right half of the ventral clivus bone was drilled and exposed the dura. Then the right half of the vomer was drilled and the pharyngobasilar fascia was exposed. (B) The pharyngobasilar fascia has been removed with the underlying muscles, rectus capitis anterior, and longus capitis. The superior clival line is attachment of the longus capitis and inferior line is rectus capitis anterior. Note that the inferior clival line is an important landmark to estimate the hypoglossal canal. Then the inferior clivus bone and anterior part of the foramen magnum was shown. The bilateral Eustachian tubes are located at the floor of the nasopharyngeal space. (C) The right side of torus tubarius of the Eustachian tube is inverted. The petroclival synchondrosis is exposed. The inferior petrosal sinus runs from cavernous sinus to jugular bulb along with petroclival synchondrosis and forms the superolateral limit of the jugular tuberculum. The foramen lacerum is located lateral to the floor of the sphenoid sinus. The lower portion of the foramen lacerum is composed of fibrocartilaginous tissue connected to the pharyngobasilar fascia petroclival synchondrosis, Eustachian tube, and underlying muscles. Resection of the Eustachian tube is tough because it is connective cartilage tissue. (D) The torus tubarius of the Eustachian tube with muscles, tensor vela palatini, and levator vela palatini has been transected. The tensor and lavatory vela palatini are attached to the Eustachian tube in the area below the horizontal segment of the petrous portion of ICA. The condyle joint is exposed and located at the 2 and 10 o'clock positions. The anterior arch of C1 is positioned at the level of the soft palate. The occipital condyle is located behind Rosenmüller fossa. (E) The parapharyngeal ICA, jugular vein and vagus nerve are surrounded by carotid sheath. The stylopharyngeal fascia is an important landmark, which is located in front of parapharyngeal ICA. [12]. (F) The vagus nerve (X) and parapharyngeal ICA and jugular vein are surrounded by carotid sheath, which is the

dense fibrous tissue. The carotid sheath is resected and the neurovascular structures are exposed. The glossopharyngeal nerve (IX), vagus nerve (X), accessory nerve (XI) are situated in front of jugular vein. The jugular vein is identified lateral to the parapharyngeal ICA. The glossopharyngeal nerve (IX) lies between the stylopharyngeal fascia and the ICA. (G) The carotid sheath is removed and the lower nerves and the parapharyngeal ICA and the jugular vein are exposed. The stylopharyngeal fascia has been resected. The glossopharyngeal nerve runs lateral side of ICA. The supracondylar groove is shown. (dot line) The hypoglossal (XII) nerve runs medial side of the ICA.

The superior clival line marks the *longus capitis* insertion, whereas the inferior line is the equivalent for the *rectus capitis anterior*. The inferior clival line is an important landmark to estimate the position of the hypoglossal canal. Removal of these muscles exposes the foramen magnum and inferior clival bone. The fibrous pharyngobasilar fascia extends laterally on the ventral surface of the clivus, petroclival synchondrosis and petrous bone parallel and dorsal to the Eustachian tube. Removal of this fibrous fascia exposes the occipital condyle and hypoglossal canal. It should be noted that there is a venous plexus around the hypoglossal canal that directly drains into the jugular bulb and that can produce significant bleeding. The Eustachian tube is an important landmark for identifying the parapharyngeal ICA.

The supracondylar groove provides a reliable landmark for the superior limit of the hypoglossal canal. The distance between the supracondylar groove and the articular surface of the occipital condyle averages 10mm. [7] Therefore, a ventromedial condyle resection can be accomplished with preservation of the synovial joint; thus, defining a supra-articular medial condylectomy. [8] A medial condylectomy is limited laterally by the hypoglossal canal. The cortical layer of the bony canal is easily recognized during lateral drilling and can be preserved, leaving the nerve untouched. However, this cortical bone can also be removed exposing the hypoglossal canal periosteum, which protects the nerve.

Supracondylar Approach (Jugular Tubercle)

The jugular tubercle is a bony prominence, in the posterior fossa, arising from the inferolateral clivus. The hypoglossal canal forms its lower limit and the jugular foramen is immediately lateral to it. The medial margin of the jugular tubercle is located posteroinferior to the lower aspect of the foramen lacerum, which is located lateral to a well pneumatized floor of the sphenoid sinus.

In order to drill the jugular tubercle, at least a hemi-clivectomy is required on the ipsilateral side. Furthermore, skeletonization and lateralization of the ipsilateral paraclival ICA facilitates further lateral exposure as well as exposure of the area behind the ICA.

As the clivus is drilled, the basopharyngeal fascia is exposed and followed laterally at the level of the foramen lacerum. Once the ICA is identified just above the foramen lacerum, further drilling is performed in the occipital bone medial to the petroclival synchondrosis, inferior to the petrous apex. The use of angled drills facilitates this maneuver.

For bony lesions as such as cholesterol granulomas and some select chordomas, this access may be sufficient; thus, removal of the Eustachian tube may not be necessary. In these cases, usually the tumor has eroded the petrous bone allowing the surgeon to progress laterally following and resecting the tumor. However, in patients presenting further lateral involvement, which is frequent in aggressive chordomas or lateral chondrosarcomas, removal of the Eustachian tube should be considered early in the surgery to facilitate the exposure. In patients with intradural disease dorsal to the jugular tubercle, the lateral extension of the disease will determine if removal of the Eustachian tube will be helpful. Extirpation of jugular tubercle meningiomas with lateral extensions would likely benefit from ipsilateral Eustachian tube removal as well.

Jugular Foramen Approach

The Eustachian tube is an important landmark in endoscopic skull base surgery as it guards the parapharyngeal ICA and jugular vein from a medial to lateral approach. Although it is possible to navigate in the direction of the jugular foramen during an EEA preserving the Eustachian tube, it requires the use of angled-lens endoscopes and angled instruments. This may be appropriate when navigating through the jugular tubercle of the occipital bone; however, it will not guarantee adequate exposure for disease on the jugular foramen. Thus, we recommend the removal of the Eustachian tube to allow a direct visualization and maneuverability on the jugular foramen area.

The jugular foramen is divided into three components: two venous and one neural, or intrajugular compartment. [12] The two venous compartments consist of sigmoid sinus and inferior petrosal sinus. The intrajugular or neural portion, through which the glossopharyngeal, vagus, and accessory nerves run, is located ventral to the intrajugular processes of the temporal and occipital bones. Hence, access to the jugular foramen via fully endonasal endoscopic approach requires complete exposure of the structures located below the level of the foramen lacerum and the petrosal ICA. Interstingly, it is ourcexperience that most jugular foramen tumors expand the area dislocating the jugular pars nervosa inferiorly; thus, creating a large corridor direct into the jugular foramen above the cranial nerves, which are kept protected. Is very common to encounter the XII cranial nerve still protected by cortical bone in situation where the entire occipital condyle was destroyed by erosive tumor. The approach to jugular foramen is also referred to as Zone7. [4] This approach is lateral to zone 6 and requires the proximal control of parapharyngeal ICA.

Wide sphenoidotomy and medial maxillectomy are required to access lesions in the lateral posterior fossa. A transpterygoid approach is performed and the pterygoid process is completely removed with exposure of the tensor veli palatini, levator veli palatini, lateral pterygoid and medial pterygoid muscles. The foramen ovale and the mandibular branch of the trigeminal nerve should be identified. This wide exposure provides the space for instrumentation and full visualization of the structures; thus, improving safety.

The vidian nerve leads to the petrous ICA and foramen lacerum. The pharyngobasilar fascia is removed from the surface of the inferior clivus and the paraclival ICA is skeletonized at the foramen lacerum and paraclival areas. Removal of medial wall of paraclival ICA exposes the inferior petrosal sinus, which drains from cavernous sinus to jugular bulb and runs parallel to the petroclival synchondrosis. This represents the superolateral margin of the jugular tuberculum. The Eustachian tube is an important landmark for the parapharyngeal ICA. Its removal allows the identification and control of the parapharyngeal ICA proximally. The torus tubarius of the Eustachian tube with muscle attachments and fibrous connections to foramen lacerum and petrous bone are resected; thus, providing a direct view of the parapharyngeal ICA, which can be confirmed with the use of intraoperative acoustic Doppler sonography. The jugular vein can be identified lateral to the parapharyngeal ICA, and the cranial nerve IXX and XI are located between the jugular vein and the ICA. The the jugular foramen can be directly accessed above the cranial nerves at the point they angulate inferiorly coming out of the skull base. Suitable indications for this approach include malignant tumors, chondrosarcomas, chordomas, schwannomas, meningiomas, and paragangliomas.

Discussion

Endoscopic approaches to posterior fossa are not only feasible for median lesions through a transclival approach, but also for paramedian pathologies related to the occipital condyle and jugular foramen. A supra-condylar approach is the lateral extension of a transclival approach preserving the occipital condyle and pars articularis. In this specific approach the occipital bone is drilled above the hypoglossal canal accessing the lateral aspect of the posterior fossa ventral to the cerebello-medullary cistern. The portion of the occipital bone that is drilled in the supra-condylar approach is the jugular tubercle. The jugular tuberculum is a bony prominence located in the lateral margin of the clivus. The inferior limit of jugular tuberculum is the hypoglossal nerve and canal, and upper limit is the foramen lacerum and petroclival synchondrosis. The glossopharyngeal nerve, vagus nerve, accessory nerve arising from the medulla cross the posterior portion of jugular tuberculum on their way to the jugular foramen, which is slightly more lateral. The glossopharyngeal nerve (IX) courses superior to the vagus and accessory nerves and is separated by the dural septum. The vagus nerve bundle (X) is located behind the glossopharyngeal nerve and courses straight toward the medulla. Neuromonitoring to lower cranial nerves and the hypoglossal nerve are indispensable when accessing this areas. The accessory nerve (XI) is positioned beside the jugular bulb (JB) and courses on the medial side of the jugular tubercle. [14] During an EEA, the glossopharyngeal (IX), vagus (X) and accessory (XI) nerves are located in front of the jugular vein and are dislocated inferiorly when there is tumor expansion in the jugular foramen. The glossopharyngeal nerve descends along the lateral side of the ICA. The accessory nerve passes at the lateral surface of the jugular vein. The hypoglossal nerve passes through the hypoglossal canal and descends with IX, X, and XI. The stylopharyngeal muscle is an important landmark, which is situated in front of parapharyngeal ICA. [12]

The main advantage of the EEA to approach posterior fossa lesions located in the coronal plane is the ability to access directly the lower clivus, the anterior part of foramen magnum and jugular tuberculum with the minimal manipulation to the neurovascular structures. Foramen magnum and lower clivus are located behind nasopharynx, which can be suitable for endonasal access.

The angle of approach and the caudal limit of access for EEA to C1 is the nasopalatine line, from bony nasal ridge to the bony hard palate extended into the depth. [5] The endonasal route for the lateral part of posterior fossa requires exposure of paraclival and petrosal part of the ICA and may require transection of the Eustachian tube in selected cases. These maneuvers need technical skills and experience and surgeons should be well familiar not only with this anatomy, but also aware of the nuances of the endoscopic ventral perspective. Image guided surgical navigation and intraoperative Doppler are indispensable to avoid the injury of ICA, and neuromonitoring is recommended in all cases . The supracondylar, transcondylar and jugular approaches allow a direct access to the ventral lateral lesions in the inferior clival regions without retracting the cerebral or cerebellar tissues also minimizing the need for cranial nerve manipulation.

Conclusion

Fully endoscopic endonasal approach to the lateral part of posterior fossa related to the occipital condyle and jugular foramen is feasible. However, endoscopic endonasal approaches are not a substitute to conventional approaches; rather, they should be seen as complementary.

Evaluation of tumor pathology, location, relationship with surrounding nerves and arteries plays the most important role in the decision-making process for the preferred surgical approach to a lesion in the posterior fossa. Lesions that are lateral to the cranial nerves are better treated by lateral conventional skull base approaches.

References

[1] Bruneau M, George B: The juxtacondylar approach to the jugular foramen. *Neurosurgery* 62: 75-78; discussion 80-71, 2008.

[2] Fernandez-Miranda JC, Morera VA, Snyderman CH, Gardner P: Endoscopic EndonasalTransclival Approach to the Jugular Tubercle. *Neurosurgery,* 2012.

[3] Fortes FS, Sennes LU, Carrau RL, Brito R, Ribas GC, Yasuda A, Rodrigues AJ, Jr., Snyderman CH, Kassam AB: Endoscopic anatomy of the pterygopalatine fossa and the transpterygoid approach: development of a surgical instruction model. *Laryngoscope* 118: 44-49, 2008.

[4] Kassam AB, Prevedello DM, Carrau RL, Snyderman CH, Thomas A, Gardner P, Zanation A, Duz B, Stefko ST, Byers K, Horowitz MB: Endoscopic endonasal skull base surgery: analysis of complications in the authors' initial 800 patients. *J. Neurosurg.* 114: 1544-1568, 2010.

[5] Kassam AB, Snyderman C, Gardner P, Carrau R, Spiro R: The expanded endonasal approach: a fully endoscopic transnasal approach and resection of the odontoid process: technical case report. *Neurosurgery* 57: E213; discussion E213, 2005.

[6] Kassam AB, Vescan AD, Carrau RL, Prevedello DM, Gardner P, Mintz AH, Snyderman CH, Rhoton AL: Expanded endonasal approach: vidian canal as a landmark to the petrous internal carotid artery. *J. Neurosurg.* 108: 177-183, 2008.

[7] Morera VA, Fernandez-Miranda JC, Prevedello DM, Madhok R, Barges-Coll J, Gardner P, Carrau R, Snyderman CH, Rhoton AL, Jr., Kassam AB: "Far-medial" expanded endonasal approach to the inferior third of the clivus: the transcondylar and transjugular tubercle approaches. *Neurosurgery* 66: 211-219; discussion 219-220.

[8] Morera VA, Fernandez-Miranda JC, Prevedello DM, Madhok R, Barges-Coll J, Gardner P, Carrau R, Snyderman CH, Rhoton AL, Jr., Kassam AB: "Far-medial" expanded endonasal approach to the inferior third of the clivus: the transcondylar and transjugular tubercle approaches. *Neurosurgery* 66: 211-219; discussion 219-220, 2010.

[9] Prevedello DM, Barges-Coll J, Fernandez-Miranda JC, Morera V, Jacobson D, Madhok R, dos Santos MC, Zanation A, Snyderman CH, Gardner P, Kassam AB, Carrau R: Middle turbinate flap for skull base reconstruction: cadaveric feasibility study. *Laryngoscope* 119: 2094-2098, 2009.

[10] Prevedello DM, Doglietto F, Jane JA, Jr., Jagannathan J, Han J, Laws ER, Jr.: History of endoscopic skull base surgery: its evolution and current reality. *J. Neurosurg.* 107: 206-213, 2007.

[11] Prevedello DM, Pinheiro-Neto CD, Fernandez-Miranda JC, Carrau RL, Snyderman CH, Gardner PA, Kassam AB: Vidian nerve transposition for endoscopic endonasal middle fossa approaches. *Neurosurgery* 67: 478-484, 2010.

[12] Rhoton AL, Jr.: Jugular foramen. *Neurosurgery* 47:S267-285, 2000.

[13] Sanna M, Bacciu A, Falcioni M, Taibah A, Piazza P: Surgical management of jugular foramen meningiomas: a series of 13 cases and review of the literature. *Laryngoscope* 117: 1710-1719, 2007.

[14] Sutiono AB, Kawase T, Tabuse M, Kitamura Y, Arifin MZ, Horiguchi T, Yoshida K: Importance of preserved periosteum around jugular foramen neurinomas for functional outcome of lower cranial nerves: anatomic and clinical studies. *Neurosurgery* 69: ons230-240; discussion ons 240, 2012.

[15] Tran Ba Huy P, Chao PZ, Benmansour F, George B: Long-term oncological results in 47 cases of jugular paraganglioma surgery with special emphasis on the facial nerve issue. *J. Laryngol. Otol.* 115: 981-987, 2001.

Editor Editor Contact Information

Dr. Moncef Berhouma
Consultant Neurosurgeon
Associate Professor
Department of Neurosurgery B
Neurological and Neurosurgical Hospital - Lyon
59 Boulevard Pinel – 69500 Bron France
Tel: 0033 6 77 96 03 44
berhouma.moncef@yahoo.fr

Index

#

20th century, xviii, xxi, xxiii, 2, 3, 14

A

accessibility, 166
acid, 304
acoustic neuroma, xviii, xix, 16, 178, 270, 346, 347, 348
acoustics, 329
acromegaly, 15, 341, 342, 344, 355
ACTH, 341, 343
AD, 5, 16, 32, 50, 108
adaptability, 106, 270
adenocarcinoma, 72, 79, 232
adenoidectomy, 208
adenoma, 152, 219, 220, 342, 344, 353, 354
adhesion(s), 184, 191, 222, 264, 265, 272, 301, 312, 314, 320, 363
adhesives, 54, 130
adrenaline, 40, 54
adulthood, 20, 205
adults, xii, 68, 119, 194, 196, 197, 199, 200, 202, 208
advancements, xxiii, 123, 124
adverse effects, 53
affirming, 283
age, xii, 131, 193, 195, 196, 197, 200, 202, 205, 206, 207, 253, 260, 272, 277, 330, 333, 335, 338, 342
airline industry, xxii
Algeria, 311, 319
algorithm, 39, 50, 67, 69, 130, 138, 149, 321
allergy, 70
American Heart Association, 10
amplitude, 294, 295, 296
anaesthesiologist, 302
anastomosis, 146

anatomic site, 51
anatomical separation, 22
anesthesiologist, 144, 199, 341
aneurysm, 11, 17, 152, 228, 250, 344, 351, 354
aneurysm clipping, 11
angiofibroma, 163, 202
angiography, xxv, 1, 13, 14, 87, 143, 152, 199, 228, 297, 298, 299, 303, 305, 337, 354
angulation, 301
annual rate, 332
antibiotic, 54, 203
antrum, 166, 171
apex, 33, 37, 38, 41, 42, 44, 45, 47, 48, 72, 73, 77, 80, 84, 91, 124, 147, 155, 156, 182, 183, 185, 186, 187, 188, 189, 190, 191, 217, 218, 223, 226, 227, 230, 250, 264, 279, 336, 338, 339
apoplexy, 341
apoptosis, 328
arachnoiditis, 265
arteries, 19, 23, 29, 31, 32, 39, 44, 45, 46, 47, 88, 96, 98, 99, 102, 113, 156, 158, 160, 162, 197, 258, 260, 274, 297, 298, 300, 315, 316, 317, 318, 328, 336, 338, 360
arteriovenous malformation, xxii, 228, 254, 345, 346
Asia, 133
aspirate, 12, 228
aspiration, xxv, 236, 241, 336
assessment, 12, 39, 73, 114, 138, 143, 210
Astrocytomas, 320
asymmetry, 104, 329
atrium, 314, 316
atrophy, 52, 168, 246, 343
attachment, 31, 38, 155, 315, 317, 328, 333, 336, 360
audiograms, 325
auditory evoked potentials, 308
auditory nerve, 297, 306
audits, 10
Austria, 97, 98, 102
authorities, xvi

autonomic nervous system, 124
avoidance, 36, 61, 268, 310, 343
axilla, 39

B

bacteria, 54
barriers, 101, 199
basilar artery, 85, 187, 191, 199, 254, 302, 314, 333
bending, 24, 217
beneficial effect, 263
benefits, 63, 116, 124, 237
benign, 44, 84, 112, 131, 137, 138, 149, 154, 169, 193, 198, 202, 225, 230, 234, 235, 318, 319, 325, 326, 327, 328, 345
benign tumors, 230, 234, 326, 328
bias, 118, 131, 279, 285
Bilateral, 148
biopsy, xii, 45, 46, 137, 138, 145, 147, 149, 150, 152, 198, 202, 208, 225, 226, 227, 228, 229, 231, 232, 233, 235, 236, 311, 312, 318, 319, 320, 321, 322, 323
biopsy needle, 226, 229
bleeding, 13, 32, 89, 96, 97, 98, 99, 103, 113, 114, 131, 144, 186, 187, 188, 194, 199, 202, 219, 220, 319, 338, 345, 359
blindness, 338
blood, xxi, xxii, 32, 52, 63, 76, 102, 112, 130, 144, 162, 170, 199, 200, 253, 254, 328, 337
blood pressure, 144
blood supply, 102, 199, 337
blood vessels, 112, 328
bone form, 37, 155
bone marrow, 195
bones, 124, 142, 154, 199, 333
bradycardia, 268
brain abscess, 354
brain contusion, 241
brain herniation, 165
brain stem, 185, 190, 321
brain structure, x, 312
brain tumor, 345
brainstem, 84, 139, 182, 253, 258, 261, 273, 274, 276, 278, 279, 280, 281, 282, 283, 284, 285, 288, 294, 301, 303, 304, 308, 310, 312, 313, 318, 319, 320, 326, 327, 329, 330, 333, 334, 342, 357, 358, 359, 361
brainstem auditory evoked potentials, 308
brainstem glioma, 319
branching, 162

C

cadaver, 11, 24, 153
calcifications, 119, 228
calcium, 54, 241
calculus, 119
caliber, 23
calvarium, 74, 166, 172
canals, 40, 98, 104, 155, 156, 162, 199
cancer, 72, 75, 230, 268
candidates, 76, 114, 197, 249, 285, 307
capillary, 130
capsule, 94, 99, 112, 113, 169, 328, 329, 330, 331, 336
carcinoma, 79
caries, 344
carotid arteries, 72, 84, 85, 93, 98, 101, 108, 113, 117, 195, 197
cartilage, 54, 86, 138
cartilaginous, 138
case examples, 71, 83, 203
catheter, 63, 317
cauda equina, xxi
causal relationship, 254
cauterization, 113
cell death, 328
cellulitis, 231
cellulose, 187, 188, 265
central nervous system (CNS), 112, 323
cephalosporin, 40
cerebellum, 294, 295, 297, 300, 306, 307, 330
cerebral aneurysm, 152, 354
cerebral arteries, 199, 315, 336
cerebral edema, 61, 63
cerebrospinal fluid, x, 20, 33, 68, 69, 70, 72, 76, 100, 107, 130, 165, 177, 178, 179, 185, 200, 250, 284, 297, 341, 358
cervical radiculopathy, 68
CFR, 72, 74, 75, 76, 77, 81
challenges, 72, 336
chemical, 13, 17, 174, 274
chemotherapy, 232
Chicago, xxi
childhood, 195
children, xii, 119, 131, 134, 198, 200, 202, 206, 209, 210, 211
China, 358
chloroform, xxi
cholesterol, 84, 250
chondrosarcoma, 147
choroid, xxi, 303, 304, 314, 315, 316, 317, 318
circulation, 189, 190, 258, 265, 271
classes, 281

classification, 51, 72, 94, 102, 108, 157, 197, 223, 250, 267, 275, 287, 329, 335, 343, 352
cleavage, 304, 336
clinical application, 11, 33, 90, 322, 328
clinical assessment, 275
clinical examination, 333
clinical neurophysiology, 308
clinical presentation, 166, 325
clinical symptoms, 329
clinical trials, 10
closure, 54, 55, 58, 69, 81, 88, 102, 103, 108, 111, 113, 114, 144, 147, 174, 175, 176, 185, 188, 189, 200, 239, 242, 247, 259, 264, 265, 302, 329, 330, 337, 338, 341, 360
clothing, xxi
cobalt, 327
cochlea, 168, 182, 188, 190, 294, 331
collaboration, xxiii, 1, 20, 97, 111, 119, 345
collagen, 54, 74, 165, 174, 177, 259
collateral, 160, 271, 314
commercial, xxii
commissure, 226, 227, 229, 315, 316
common presenting symptoms, 176
common symptoms, 329
communication, 2, 50, 202, 337
community, xvi, xxi
comparative analysis, 235
compatibility, 271
complement, xii
complexity, 20, 318
comprehension, 11
compression, 42, 44, 52, 55, 97, 112, 123, 125, 131, 185, 254, 258, 266, 267, 271, 274, 275, 281, 284, 285, 286, 287, 293, 296, 297, 298, 302, 305, 307, 309, 321, 329, 335, 341, 360, 362, 363
computed tomography, 20, 24, 53, 155, 158, 166, 193, 199, 337
computer, xxii, 276
conduction, 302
conductive hearing loss, 171, 173, 174
configuration, 104
conflict, 214, 221, 242, 251, 252, 253, 254, 255, 258, 259, 261, 262, 265, 266, 269, 274, 277, 298, 299, 300, 303, 304, 309
conflict resolution, 252, 261, 262, 266
conformity, 327, 331, 337
congenital malformations, 50
Congress, 135, 290, 291
consensus, 131, 165, 265, 359
consent, 262
conservation, 145, 297
construction, 3
contamination, 170

contracture, 329
controlled trials, 10, 131
controversial, 42, 62, 63, 226
controversies, xii, 131, 168
convergence, 93, 328
cooperation, xii, 232, 321
cornea, 243, 248
correlation(s), 91, 131, 149, 231, 267, 269, 283, 286, 294, 309
cortex, 240
cortical bone, 188
corticosteroid therapy, 130
corticosteroids, 42, 130
cosmetic, x, 20, 103, 154, 238, 241, 248, 264, 265
cost, 194, 307
coughing, 42, 52
covering, 24, 25, 88, 97, 127, 146, 168, 222, 282, 297, 301, 304, 330
cranial nerve, xvi, xviii, 2, 3, 4, 37, 38, 40, 72, 86, 87, 89, 112, 182, 188, 189, 190, 194, 202, 218, 247, 253, 259, 262, 269, 293, 294, 306, 308, 309, 326, 329, 333, 335, 340, 344, 357, 359
craniectomies, 13
craniopharyngioma, 98, 100, 114, 115, 118, 119, 131, 132, 206, 207
craniotomy, xxv, 9, 12, 13, 33, 36, 45, 50, 73, 75, 76, 78, 80, 90, 112, 116, 118, 149, 171, 172, 173, 186, 194, 226, 233, 238, 239, 240, 242, 243, 244, 245, 248, 249, 264, 284, 294, 297, 300, 301, 302, 304, 330, 336, 339, 360
cranium, 169, 195
CT, xxii, xxv, 14, 20, 21, 23, 24, 25, 27, 33, 39, 49, 53, 62, 64, 65, 66, 90, 96, 119, 125, 132, 143, 144, 148, 156, 165, 166, 168, 169, 170, 176, 178, 184, 189, 193, 199, 200, 203, 204, 208, 210, 218, 228, 242, 276, 277, 279, 321, 337
CT scan, xxii, xxv, 14, 23, 25, 27, 49, 64, 65, 66, 96, 132, 165, 168, 169, 170, 184, 189, 200, 203, 210, 242, 337
CTA, 87, 199
cues, 199
culture, 14
cure(s), 72, 274, 275, 284, 302
cyst, 103, 131, 186, 189, 209, 312, 363
cytology, 236

D

damages, 262
danger, 318
data analysis, 82
database, 10, 353
debridement, 44

decision-making process, 137, 150, 226
defects, 32, 34, 49, 50, 51, 52, 53, 54, 56, 58, 60, 62, 69, 107, 151, 165, 166, 167, 169, 170, 171, 172, 173, 174, 175, 176, 177, 179, 200, 202, 241, 337
deficit, 262, 326, 342, 343, 344
dehiscence, 52, 62, 194
demyelination, 299, 335
depressed fracture, 50
depression, 183, 259
depth, 128, 166, 181, 188, 197, 220, 240, 248
dermis, 54, 81
dermoid cyst, 138
designers, xxii
destruction, 235
detachment, 186
detection, 53, 178, 309
deviation, 39, 138, 228, 361
diabetes, 326, 342, 344, 361
diabetes insipidus, 326, 342, 344
diagnostic criteria, 253
diaphragm, 335
differential diagnosis, 226
diffusion, 138
dilation, 130
diplopia, 41, 42, 48, 339
disaster, 219
discomfort, 203
discordance, 53
discrimination, 39, 329
diseases, xviii, 20, 41, 163, 181, 186, 190, 191, 293, 307
disequilibrium, 329
disinfection, 143
dislocation, 44
disorder, 346
displacement, 36, 41, 44, 45, 221
distribution, 155, 333, 340
diversity, 77
dopamine, 352
dopamine agonist, 352
doppler, 214
dosage, 283
drainage, 22, 31, 42, 102, 114, 144, 148, 149, 172, 178, 185, 204, 238, 239, 240, 241, 246, 255, 258, 262, 263, 264, 363
drawing, xvii, 21, 23, 25, 29, 155, 156, 161
dressings, 200
drugs, 275, 277
dry eyes, 282
dry-eye syndrome, 283
Duma, 234, 288, 349
dura mater, xxii, 38, 146, 148, 239, 240, 241, 243, 245, 257, 258, 259, 265, 338, 341
dysphagia, 361, 363
dysplasia, 167, 168, 202

E

edema, 154, 238, 241, 246, 284
editors, 104, 105, 106, 107, 290, 292, 309, 310
education, 1, 10, 11, 12, 14, 17
educators, 11
EEA, 83, 84, 88, 89, 90, 94, 193, 194, 196, 197, 198, 199, 200, 202, 208, 209
effusion, 177
electrocautery, 46, 256
electrodes, 99, 147
electromyography, 87
electron, 343, 345
electron microscopy, 343
emboli, 188, 199, 360
embolization, 188, 199, 360
embryology, 133
EMG, 40, 329, 333
emphysema, 42
empyema, 211
encephalitis, 166
endocrine, 123, 131, 148, 344, 355
endocrinologist, 199
endocrinology, 353
endoscopy, x, xii, xxv, 3, 50, 55, 69, 105, 137, 149, 179, 237, 238, 246, 248, 260, 312, 319, 353
enlargement, 88, 98, 116, 318
environment, 182, 280
environmental conditions, 194
ependymal, 17
epidemiology, 354
epidermoid cyst, 186, 230
epidural abscess, 203, 204, 205, 206, 211
epidural hematoma, 176, 189
epilepsy, 288, 292
epinephrine, 40
epistemology, 1
epithelium, 200
equipment, 103
erosion, 50, 51, 168
essential tremor, 346
etiology, 51, 53, 62, 166, 168, 170, 255
Europe, xi
Eustachian tube, 49, 52, 62, 90, 91, 151, 372, 390, 391, 392, 393, 394, 395, 396
evacuation, 94, 99, 176
evidence, 1, 2, 6, 10, 14, 16, 69, 81, 82, 123, 129, 130, 176, 252, 262, 271, 272, 274, 280, 284, 285, 292, 294, 336
evoked potential, 40, 87, 307, 308, 310, 333

evolution, 2, 11, 13, 14, 15, 16, 20, 47, 94, 95, 105, 114, 214, 220, 250, 273, 343
examinations, 2, 228
excision, 75, 106, 120, 250, 270, 334, 347, 348
exophthalmos, 133
expertise, 10, 119, 124, 181
exposure, x, xx, xxiii, 19, 25, 41, 45, 58, 60, 62, 74, 75, 78, 84, 88, 89, 97, 102, 104, 113, 124, 154, 159, 160, 161, 172, 173, 175, 182, 188, 190, 191, 194, 199, 207, 213, 214, 219, 222, 226, 256, 258, 260, 264, 337, 358, 360, 362
extraocular muscles, 38, 41
extravasation, 64
extrusion, 81
eye movement, 38

F

facial asymmetry, 194
facial nerve, 156, 167, 168, 177, 188, 189, 190, 243, 244, 254, 260, 261, 294, 302, 303, 304, 307, 309, 326, 328, 330, 331, 332, 333, 347, 348
facial pain, 253, 268, 275
facial palsy, 284
fascia, 31, 43, 45, 50, 54, 61, 74, 81, 113, 144, 160, 174, 175, 176, 189, 207, 265, 302, 304, 330, 336
fat, 31, 38, 41, 42, 43, 44, 46, 54, 55, 56, 61, 64, 65, 74, 87, 101, 102, 123, 127, 128, 144, 148, 157, 160, 163, 174, 175, 185, 189, 202, 216, 265, 304, 330, 336, 338, 341, 344
fetus, 21
fiber(s), 125, 156, 235, 255, 262, 301
fibrin, 47, 54, 64, 65, 102, 108, 148, 165, 174, 189, 239, 241, 252, 258, 261, 264, 265, 330, 337, 338
fibrinogen, 54, 220
fibroblasts, 174
fibrosis, 54, 265
fibrous dysplasia, 123, 202
fibrous tissue, 54
films, 278
filters, 53
fistulas, 69, 149, 328, 344
fixation, 40, 96, 213, 214, 255
flex, 238
flexibility, xxi, 81
flight, 297, 298, 299, 303, 305
fluid, 49, 50, 53, 65, 66, 68, 166, 169, 177, 178, 211, 330
foramen ovale, 137, 138, 149, 150, 152, 187, 225, 226, 227, 228, 229, 230, 231, 233, 235, 236, 274
force, 169
formation, 54, 58, 174
fovea, 20, 195

fractures, 69, 130, 167, 169, 178
fragments, 42, 56, 232, 259, 321
France, vii, xxv, 1, 137, 181, 225, 228, 229, 237, 251, 273, 275, 293, 322
freedom, 277, 279, 281, 284, 285
frontal lobe, 49, 61, 76, 99, 240, 241, 245
fusion, 156, 209, 298, 299, 303, 305, 337

G

gadolinium, 178, 231, 232, 271, 297, 298, 299, 321, 342
gait, 330, 331, 360, 361, 363
gamma rays, 328
ganglion, 85, 86, 89, 151, 155, 156, 160, 161, 162, 173, 181, 182, 184, 188, 235, 273, 274, 278
general anaesthesia, 255
general anesthesia, xxi, 40, 112, 127, 284
general surgery, 11, 12
Germany, 95, 358, 360
gestation, 167
gigantism, 15
gill, 123
gland, 43, 154, 341
glial cells, 125
glossopharyngeal nerve, 293, 294, 305, 310
glucose, 53, 170, 178
glue, 47, 54, 64, 65, 102, 108, 148, 165, 174, 189, 239, 241, 252, 258, 261, 264, 265, 330, 337, 338
glycerol, 273, 274, 275, 278, 284, 287
glycol, 174, 176, 177
GPS, xxiii
grades, 277, 335
grading, 213, 214, 346, 350
graft technique, 56
granulomas, 84, 265
gray matter, 314
growth, 194, 195, 199, 200, 210, 221, 222, 329, 332, 341, 347
growth hormone, 341
growth rate, 347
guidance, 40, 87, 103, 127, 199, 205, 210, 308, 326, 328
guidelines, 10, 33, 129, 291

H

haemostasis, 99, 102, 259
haemostatic agent, 97, 98
hair, 238, 240, 248, 329
hair loss, 238, 240
harbors, 181

harvesting, 144
HE, 291
head injury, 166, 167, 169
head trauma, 50
headache, 49, 169, 264, 271, 329
healing, 54, 63, 174
health, 33, 273, 274
hearing impairment, 293
hearing loss, 165, 166, 168, 169, 172, 176, 272, 284, 294, 296, 303, 307, 308, 309, 329, 330, 331, 333, 361
height, 95, 154, 243
hemangioma, 36, 123
hematoma, 29, 32, 42, 174, 186, 231, 241, 284
hemiparesis, 333
hemiplegia, 340
hemisphere, 300, 303, 304, 307, 362
hemodynamic instability, 254
hemorrhage, xxii, 2, 39, 41, 45, 46, 61, 76, 149, 156, 162, 238, 344
hemostasis, 13, 46, 54, 91, 113, 148, 188, 219, 241, 265, 302
heterogeneity, 275, 279
histological examination, 320
histology, 229, 333
history, xvii, xxi, xxiii, 2, 15, 16, 17, 47, 52, 73, 105, 152, 166, 169, 176, 184, 199, 203, 206, 207, 250, 253, 254, 287, 335, 346, 349, 351, 352, 354, 360, 363
hormone(s), 341, 342
hospitalization, 154
House, 10, 234, 263, 326, 329, 333, 334, 346
human, xvi, xviii, xxi, xxii, 2, 11, 13, 14, 166, 210, 284, 319
human body, xviii, xxii, 166
human brain, xxi, 14
human resources, 284
Hunter, xvi
hybrid, 42, 213, 223
hydrocephalus, xxi, 51, 52, 53, 118, 119, 284, 311, 312, 314, 317, 319, 321, 323, 329
hydroxyapatite, 174, 176, 177, 178, 179
hyperactivity, 309
hyperplasia, 138
hypertension, 123, 134, 168, 178
hypertrophy, 341
hypotension, 52, 254
hypothalamus, 112, 116, 355
hypothesis, 294

I

IAM, 330

iatrogenic, 41, 51, 165, 177
ID, 104
ideal, 102, 118, 174, 216, 263, 264, 319
identification, 22, 32, 36, 42, 52, 53, 78, 88, 93, 102, 144, 160, 163, 256, 261, 303, 329, 336, 337, 359
idiopathic, 51, 123, 151, 234, 236, 252, 253, 269, 274, 275, 284, 287, 288, 289, 290, 292, 309, 346
illumination, xviii, xxv, 13, 79, 112
IMA, 156, 157, 161, 162
image(s), xvi, xxii, 73, 78, 79, 87, 96, 103, 144, 199, 203, 205, 210, 216, 228, 231, 232, 326, 328, 261, 276, 297, 321, 328, 339
imaging modalities, 165, 166
immunohistochemistry, 343
impairments, 343
implants, 265, 302
improvements, 103, 130
in vivo, 280
incidence, 41, 51, 63, 72, 75, 76, 89, 111, 169, 170, 171, 200, 230, 253, 265, 291, 308, 335, 354
incus, 171
indentation, 93, 95
individuals, xxi, xxiii, 51, 123, 131, 139
induction, 255, 294
industry, xxii
infancy, 202
infants, xxi
infarction, 76, 130, 173, 284, 301
infection, xxiii, 39, 42, 44, 63, 72, 81, 114, 169, 170, 174, 248, 258, 354
inflammation, 39, 42, 50, 177, 237
inflammatory disease, 138
inflation, 31, 318
infundibulum, 21, 94, 108, 316, 335
initiation, 3
injure, 25, 301
injury(s), 29, 39, 41, 42, 48, 63, 72, 75, 89, 99, 103, 118, 124, 128, 130, 131, 143, 149, 152, 154, 166, 170, 190, 194, 197, 202, 210, 214, 218, 219, 231, 240, 263, 274, 283, 295, 296, 319, 336, 343, 344, 354
insertion, 22, 45, 205, 219, 229, 256, 266
inspections, xii
institutions, 73
integration, 10
integrity, 102, 294
intensive care unit, 148, 189
interface, 174, 327, 334, 358
interference, 258
interlamellar cells, 22
interrogations, 2
intervention, 130, 131, 202, 278, 285, 294, 331, 333
intracranial pressure, 51, 53, 55, 62, 112, 168, 341

intraocular, 47, 125
intraocular pressure, 47
investments, 11
ionizing radiation, 327
ipsilateral, 112, 123, 127, 128, 144, 145, 240, 255, 256, 263, 284, 337
Ireland, 259
irradiation, 281, 284, 327, 328, 337, 355
irrigation, 44, 99, 102, 123, 128, 187, 304, 317
ischemia, 294, 304, 335, 336
issues, 280, 333, 335
Italy, xi, xiii, xvi, 93, 251, 259

J

Japan, 213, 214, 216
Jordan, 19, 153
justification, 116
juvenile angiofibroma, 33, 84

L

lacerate, 169
lack of control, 2
lamella, 21, 22, 23, 25, 26, 27, 37, 38, 59, 60
lamina papyracea, 22, 23, 25, 26, 27, 29, 37, 40, 41, 42, 44, 46, 119, 123, 127, 128
laminectomy, 360
latency, 277, 295, 296
layered closure, 165, 174, 175, 177
lead, 2, 11, 29, 41, 49, 50, 52, 53, 62, 63, 75, 145, 167, 168, 173, 188, 219, 233, 239, 247, 254, 303
leakage, 32, 50, 54, 94, 100, 147, 149, 165, 166, 169, 176, 178, 265, 294, 297, 341
leaks, x, xii, 20, 33, 49, 50, 51, 52, 54, 56, 58, 62, 63, 68, 69, 89, 107, 114, 131, 144, 149, 166, 167, 168, 169, 170, 174, 176, 177, 178, 179, 189, 191, 200, 202, 211, 247, 248, 250, 338, 344
learning, 16, 49, 114, 119, 307, 319, 322
legs, 95
lens, 41, 88, 214, 317
life quality, 230
lifetime, 352
ligament, 98, 113, 189, 336, 361
light, xxi, xxv, 10, 12, 14, 22, 48, 53, 55, 112, 116, 130, 229, 248, 279, 340
local anesthesia, xx, 229, 276, 285, 328
local conditions, 345
localization, xxiii, 12, 33, 37, 38, 47, 282
locus, 258
low risk, 284, 332, 337
lumbar puncture, xxi, 148, 170
lumen, 188
lying, 23
lymphoma, 45, 138, 151

M

magnetic resonance, 53, 64, 87, 166, 193, 210, 276, 289, 290, 297, 309, 337
magnetic resonance imaging, 53, 64, 87, 166, 193, 210, 276, 289, 290, 297, 309, 337
majority, 1, 50, 52, 95, 114, 116, 126, 202, 255, 282, 283, 284, 285, 294, 341, 343, 352, 358, 359
malignancy, 3, 44, 71, 75, 76, 79, 80, 82, 119, 154, 230, 344
malignant melanoma, 72
malignant tumors, 72, 75, 81, 82, 235, 248, 318, 319
malocclusion, 194
man, xvi, 63, 363
mandible, 163
manipulation, 19, 20, 35, 36, 40, 84, 89, 94, 95, 102, 113, 116, 154, 163, 190, 295, 304, 306, 319, 333
manufacturing, xxi, xxii
mapping, 113
marrow, 195
mass, 103, 112, 188, 190, 230, 235, 271, 321, 327, 329, 333, 341, 359, 360
mastoid, 50, 54, 156, 166, 167, 168, 171, 175, 177, 256, 257, 258, 264, 269, 297, 300, 301, 302, 304, 306, 329, 330, 360
materials, xii, xxi, 54, 103, 166, 171, 173, 174, 176, 177
matrix, 54, 74, 177
matter, 95, 103, 263, 283, 302, 317, 335
maxilla, 37, 85
maxillary sinus, 25, 26, 27, 41, 53, 60, 65, 72, 88, 97, 140, 141, 144, 145, 153, 154, 158, 159, 160, 161, 196, 229, 337, 338
MB, 234, 250, 308, 346
measurements, 193, 195, 257
median, 35, 36, 84, 98, 102, 276, 277, 278, 282, 283, 284, 285, 332
medical, xxiii, 10, 11, 12, 41, 42, 73, 133, 149, 207, 226, 268, 274, 285, 341, 343, 352
medical history, 207, 285
medication, 277, 281, 284
medicine, 10, 16, 353
medulla, 304, 306, 360, 362, 363
medulla oblongata, 304
medulloblastoma, 321
melanoma, 73
membranes, 156, 162, 264
memory, 190, 311, 314, 319
memory loss, 311

meninges, 125, 233
meningioma, 2, 6, 12, 36, 42, 44, 99, 101, 123, 131, 132, 181, 186, 187, 230, 231, 235, 244, 326, 333, 334, 335, 338, 339, 348, 349, 350, 351, 358, 361, 362
meningitis, 2, 42, 44, 49, 50, 52, 62, 63, 76, 81, 118, 149, 165, 166, 167, 168, 169, 170, 176, 194, 202, 263, 284, 354, 358
mentorship, 14
meridian, 36
mesentery, 23
meta-analysis, 10, 62, 106, 130, 179, 285
metals, xxi
metastasis, 225, 230, 321
methodology, 281
microscope, xii, xxi, xxv, 2, 3, 13, 14, 20, 112, 116, 237, 238, 246, 248, 252, 258, 259, 260, 262, 264, 294, 296, 300, 360
midbrain, 316
migration, 176
Minneapolis, xxiii
misconceptions, 2
models, 10, 276, 328
modifications, 271, 359
modules, 17, 35, 36, 195
morbidity, xvii, xx, 13, 47, 63, 68, 71, 73, 95, 111, 112, 116, 119, 124, 137, 147, 149, 162, 173, 194, 198, 200, 202, 214, 225, 226, 230, 239, 240, 246, 247, 308, 311, 312, 319, 320, 322, 326, 335, 340, 344, 358, 359
morphology, 348
mortality, xvii, xxi, xxiii, 2, 13, 63, 68, 73, 225, 226, 233, 262, 312, 319, 320, 322, 334, 340, 344, 358, 359
mortality rate, xxiii, 63, 344
movement disorders, 328
MR, xxii, 16, 49, 62, 64, 65, 66, 78, 79, 134, 138, 144, 165, 178, 190, 199, 232, 252, 254, 259, 261, 267, 279, 291, 321, 339, 347, 349, 354
mucosa, xx, 25, 26, 28, 54, 56, 58, 61, 97, 156, 157, 166, 341
multiple factors, 166
multiple sclerosis, 254, 275, 276, 282, 284, 289, 290
muscles, 36, 38, 45, 46, 147, 160, 254, 259, 330, 360
myelin, 130, 258
myoclonus, 254
myringotomy, 169

N

nasopharynx, 88, 200, 268, 337
natural evolution, xxv
nausea, 203
necrosis, 130, 344
neglect, 62
neoplasm, 42
nerve fibers, 42
neural function, 333
neuralgia, 226, 235, 251, 253, 255, 266, 267, 268, 269, 271, 273, 274, 286, 287, 289, 290, 293, 304, 305, 307, 309, 310, 332
neuritis, 151, 234
neuroblastoma, 72, 78
neuroimaging, 149, 150, 170, 226, 335
neuroma, 346
neurons, 156
neuropathic pain, 267
neuropathy, 41, 42, 48, 123, 130, 131, 133, 134, 135, 151, 234
neuroprotection, 13
neuroscience, 14, 15
neurosurgery, xx, xxi, xxii, xxiii, xxv, 1, 2, 10, 11, 12, 13, 14, 16, 17, 68, 69, 70, 90, 145, 151, 152, 225, 234, 235, 252, 269, 274, 287, 292, 307, 308, 309, 310, 319, 322, 323, 346, 351, 353, 354
neurosurgical approaches, 11, 13
neutral, 24, 40, 159, 248, 317, 358
nitrous oxide, xxi
North America, 90, 134, 151, 210, 234
Norway, 267
NPL, 199
nuclei, 314, 319
nucleus, 294, 315, 316

O

obesity, 62
obstruction, 203
occlusion, 143, 146, 360
oculomotor, 220, 232, 314, 319
oculomotor nerve, 220, 232, 314
OH, 35, 71, 83, 357, 360, 389
old age, 263
olfaction, 124
olfactory nerve, 19, 30, 31, 72, 80, 240, 241
oligodendrocytes, 280
opacification, 65, 66
operations, xi, 15, 160, 269, 270, 271
ophthalmologist, 199
opportunities, 10
opthalmoplegia, 133
optic chiasm, 99, 102, 103, 113, 115, 339
oral cavity, 156, 162, 253
orbit, 20, 22, 23, 25, 29, 30, 32, 36, 37, 38, 40, 41, 43, 44, 46, 47, 72, 77, 85, 98, 124, 126, 127, 133, 195, 200, 243, 338

organ, xxv
ossicles, 174
ossification, 174
osteogenic sarcoma, 72
osteotomy, 336
ostium, 25, 26, 41, 45, 88, 138, 139, 159, 341
otitis media, 169, 176, 177
otolaryngologist, 2, 194, 198
otorrhea, 49, 52, 62, 165, 166, 167, 169, 175, 176, 177, 178, 179
overlap, 343, 359
overlay, 49, 55, 56, 61, 177, 189, 200, 338
overweight, 255, 263

P

pain, 154, 194, 207, 235, 252, 253, 254, 263, 264, 265, 266, 267, 268, 269, 274, 275, 277, 278, 279, 280, 281, 282, 283, 284, 285, 286, 287, 289, 290, 291, 302, 309, 345, 360, 363
palate, 26, 145, 155, 162, 199
papilledema, 42
paradigm shift, xxv
parallel, 2, 21, 41, 43, 167, 171, 172, 255, 301, 338, 341
paralysis, 40
parenchyma, 72, 73, 77, 80
paresis, 333
parotid, 227
parotid duct, 227
pathogenesis, 52, 286
pathologist, xii, xviii, 321
pathology, x, xviii, 32, 36, 47, 89, 95, 112, 113, 116, 118, 132, 133, 159, 169, 194, 202, 205, 207, 235, 260, 286, 345, 357, 361
pathophysiology, 22, 254, 293
pathways, 102, 294, 337, 339, 342
patient care, 17
PCM, 332, 333, 334
pediatrician, 199
penicillin, 40
periosteum, 42, 85, 86, 88, 98, 123, 127, 171, 239, 243, 265, 329
permission, 28, 29, 30, 31, 196, 198, 200, 262, 277
permit, xxv, 44, 321
personal communication, 202
PET, 143
pH, 304
pharynx, 155
Philadelphia, 104, 105, 210, 266, 267
phosphate, 174
photographs, 205, 217, 220, 262, 361, 362
physical therapy, 361

physiology, 14
pilot study, 267
pineal gland, 316
pituitary gland, xx, 28, 99, 104, 114, 148, 194, 335, 340, 341, 343, 344
pituitary tumors, xxi, 15, 94, 105, 207, 223, 337, 341, 353
plaque, 333, 358
plexus, xxi, 113, 163, 257, 274, 279, 280, 283, 303, 304, 314, 315, 316, 317, 318, 359
point of origin, 335
policy, 10
polymer, 16, 54
polymethylmethacrylate, 54
pons, 183, 184, 186, 187, 189, 191, 266, 274, 280, 282, 286, 307, 328
population, 167, 194, 198, 199, 202, 207, 253
posterior ethmoidal arteries, 45, 98
posterior ethmoidectomy, 25, 44, 46, 107
postoperative outcome, 326
potential benefits, 162
prejudice, 272
preparation, 25, 26, 45, 54, 127, 256
preservation, 36, 42, 46, 47, 48, 58, 106, 112, 116, 120, 124, 185, 199, 250, 308, 325, 326, 330, 332, 335, 345, 346, 347, 348, 349
prevalence rate, 253, 273, 274
prevention, x, 17, 174, 271
principles, ix, 10, 36, 99, 105, 113, 250, 305, 308, 331
probability, 277, 278, 284, 285
probe, 41, 96, 214, 316
prognosis, 130, 147, 308
project, 112
prolactin, 341
prolapse, 41
proliferation, 328
promote innovation, xii
prophylactic, 40, 42, 62, 63, 170
prophylaxis, 179
proptosis, 41
prosthesis, 295, 300, 301, 305, 307
protection, 328, 337
prototype, 275
psychiatry, 68, 152, 354
puberty, 206
pulmonary edema, 53
pus, xxi

Q

quality of life, xxv, 33, 162, 277, 332, 335, 348, 349, 352

question mark, 172, 185

R

radiation, 62, 81, 89, 114, 207, 235, 274, 278, 279, 285, 288, 291, 325, 327, 328, 337, 344, 346, 355
radiation therapy, 81, 89, 114, 207, 235, 344, 355
radio, 319, 321, 322
radiotherapy, 73, 75, 112, 119, 226, 237, 330, 350, 355
real time, xxii
reality, xxii, 11, 14, 16, 17, 47, 48, 126
recognition, 157, 257
recommendations, 176
reconstruction, x, xii, 31, 32, 34, 62, 69, 71, 74, 78, 80, 81, 89, 91, 94, 96, 100, 101, 104, 107, 108, 114, 144, 147, 148, 151, 179, 193, 194, 199, 200, 202, 210, 211, 228, 337
recovery, 76, 89, 124, 130, 194, 199, 335
rectus abdominis, 74, 200
recurrence, 55, 61, 62, 63, 70, 75, 118, 171, 173, 176, 202, 248, 265, 277, 278, 281, 282, 283, 284, 285, 286, 340, 343
regulations, 16
relatives, 184
relaxation, 240, 241, 262, 360
relevance, 87
relief, 252, 254, 263, 277, 278, 280, 281, 282, 283, 284, 285, 289, 291, 292
remission, 219, 222, 321, 341, 343
repair, x, 33, 49, 50, 51, 53, 54, 55, 56, 57, 58, 60, 61, 62, 63, 65, 68, 69, 70, 102, 107, 130, 165, 166, 171, 173, 174, 175, 176, 177, 178, 179, 193, 200, 202, 205, 207, 250, 326
requirements, 10, 11
residual disease, 198
resistance, xii
resolution, 13, 41, 53, 143, 170, 178, 204, 205, 252, 261, 263, 266, 267, 290, 297, 298, 299, 303, 305, 309, 337, 361, 363
response, 131, 253, 275, 281, 283, 287, 348
restoration, 42
restrictions, 10
RH, 14, 68, 308, 352
rhino, 68, 149
rhinoplasty, 196
rhinorrhea, 49, 50, 51, 52, 53, 54, 55, 62, 64, 65, 67, 68, 70, 176, 178, 179
rhythm, xxv
risk factors, 73
robotics, 11

root(s), xxi, 125, 172, 186, 187, 188, 235, 251, 253, 254, 260, 265, 267, 271, 280, 282, 286, 288, 293, 294, 298, 300, 301, 302, 307
rotations, 11
routes, xii, 84, 102, 124, 163, 191, 238, 243, 246, 248, 294, 312, 318, 319
rules, 183, 241

S

safety, xxii, 11, 75, 93, 174, 182, 190, 200, 217, 274, 275, 278, 285, 290, 296, 340, 344, 345, 355
salts, 174
SAS, 228
scarcity, 132
schema, 3, 221
scientific publications, xxv
sclerosis, 281
scope, x, xix, xxi, 38, 53, 74, 114, 202, 214, 219, 221, 240, 262
security, 304
seeding, 54, 154
seizure, 76, 114, 118, 190
selectivity, 327, 331, 337
self-control, 277
sensitivity, 53, 55, 149, 170, 230, 298
sensorineural hearing loss, 168
septum, 19, 24, 26, 27, 40, 46, 54, 81, 94, 96, 97, 98, 102, 113, 138, 139, 157, 160, 200, 244, 316, 318
services, 307
shape, 22, 93, 98, 101, 103, 124, 160, 161, 183, 186, 194, 258, 264, 330, 341
shear, 358
shock, 274, 275
showing, 21, 23, 24, 26, 29, 37, 39, 64, 65, 66, 100, 101, 125, 132, 168, 169, 184, 186, 187, 206, 231, 232, 256, 259, 261, 297, 315, 318, 321, 339
side effects, 162, 194
signals, 55, 296, 297
signs, 53, 112, 253, 359
silk, 46, 189
silver, 13
simulation, 11, 16, 17
sinuses, xi, xxi, 26, 32, 33, 82, 84, 85, 125, 127, 144, 145, 149, 188, 193, 196, 197, 206, 246, 248, 258, 264, 301, 306, 337
sinusitis, 42, 149, 154, 199, 203
skeleton, 20, 36, 154, 194, 196
skin, xx, 17, 20, 87, 185, 190, 194, 239, 241, 244, 248, 256, 257, 259, 260, 264, 294, 300, 329, 330, 337
skull fracture, 14
solid tumors, 328

solution, 40, 53, 54, 112, 144, 259, 304, 317
SPA, 156, 157, 158, 161, 162
specialists, 1
specific gravity, 53
spectroscopy, 138
speculation, 283
speech, 329
speech discrimination, 329
spinal cord, xxi, 357, 359, 360, 363
spinal tap, 31
spine, xxi, 193, 199, 209, 211, 360
splenius capitis, 256
sponge, 12, 265
spontaneous recovery, 130
squamous cell, 72
squamous cell carcinoma, 72
SS, 14, 25, 30, 47, 141, 252, 256, 257, 258, 264, 347, 350
stability, 22, 209
stabilization, 40
standardization, 72
stapes, 171
state(s), 105, 124, 131, 168, 195, 199, 250
steel, xxi
stent, 204
sterile, 256
sternocleidomastoid, 359, 360
steroids, 130, 131, 340
stimulation, 220, 235, 253, 329, 330, 333
stretching, 46, 293, 294, 295, 296, 297, 301, 303, 306, 317
stroke, 118
structure, xii, 19, 20, 25, 31, 84, 85, 93, 99, 138, 140, 196, 294, 314, 328, 331, 344
style, 214
subcutaneous tissue, 171
substitutes, 81, 174, 176, 265
substitution, 342
success rate, 61, 62, 130, 170
succession, 328
suppression, 216
supraorbital cells, 22
surface area, 54
surgical intervention, xxii, 63, 131, 165, 169, 170, 175, 202, 314, 319, 320, 330, 331
surgical removal, 94, 149, 222, 226, 348
surgical resection, 78, 81, 112, 115, 162, 166, 200, 230, 319, 332, 348
surgical technique, xviii, xx, 32, 36, 78, 81, 82, 90, 95, 119, 181, 182, 198, 214, 251, 255, 262, 271, 285, 286, 312, 327, 351, 357, 359
survival, 75, 76, 112, 119, 269, 287
survival rate, 75

suture, 39, 41, 46, 55, 172, 241
Sweden, 276, 328
swelling, 130, 301, 327, 340, 343
Switzerland, 137, 225, 273, 311, 322, 325
symptoms, 42, 63, 112, 131, 169, 170, 203, 230, 253, 254, 277, 329, 330, 335, 340, 341, 342
syndrome, 149, 151, 253, 275, 346
systolic pressure, 144

T

target, xi, xii, xxv, 11, 36, 84, 181, 182, 190, 215, 216, 220, 221, 273, 274, 275, 276, 278, 279, 280, 281, 282, 283, 284, 285, 288, 327, 328, 331, 333, 334, 339, 342, 344, 346
teams, xxi, 119, 193, 209, 247, 283
technological advances, xxv, 2
technological progress, xxv, 2, 95
technological revolution, 13
technology(s), ix, xii, xxi, xxii, xxv, 14, 16, 36, 50, 124, 133, 337
teeth, 253
telephone, 277
temporal lobe, 84, 166, 171, 173, 188, 190, 336, 346
temporal lobe epilepsy, 346
tendon, 38
tension, 52, 54, 60
testing, 206
testosterone, 206
texture, 184, 191
thalamus, 311, 315, 316, 317, 328
theatre, xxii, 296
therapy, 41, 131, 170, 191, 203, 207, 267, 277, 291, 319, 321, 326, 335, 341, 344, 346
thinning, 52
thorax, 96
thrombin, 54
thrombosis, 319
thyroid, 47
tic douloureux, 267, 268, 271, 273, 274, 291
time frame, xxi
tinnitus, 329, 331
tissue, 38, 39, 53, 54, 55, 57, 60, 61, 65, 77, 81, 84, 88, 112, 119, 130, 131, 138, 157, 163, 165, 166, 170, 174, 177, 199, 200, 216, 220, 227, 228, 230, 265, 271, 302, 304, 327, 330, 344
titanium, 173, 189, 241, 242, 259, 265, 360, 361, 362
tonsils, 362
torus, 22
toxicity, 274, 280, 282, 283, 337, 343
toxin, 255
training, xxi, 1, 2, 10, 11, 14, 16, 17, 89, 127, 183, 222, 248

trajectory, xii, 96, 113, 221, 227, 228, 229, 231, 235, 236, 255, 263, 306
transferrin, 53, 68, 169, 177, 178
translocation, 2, 36, 191, 194
transverse section, 188
trapezius, 74, 360
trauma, 11, 42, 50, 51, 52, 56, 63, 69, 135, 169, 170, 176, 178, 295, 306, 314, 320
tremor, xxii
trial, 274, 278, 279, 281, 287, 346
trigeminal nerve, 37, 84, 86, 172, 182, 186, 187, 189, 230, 253, 258, 259, 262, 263, 265, 266, 267, 269, 270, 271, 274, 276, 278, 279, 280, 281, 282, 283, 284, 285, 286, 287, 288, 293, 294, 298, 299, 300, 301, 306, 307, 328, 330
trigeminal neuralgia, x, 138, 226, 236, 251, 253, 266, 267, 268, 269, 270, 271, 272, 275, 277, 280, 282, 283, 284, 286, 287, 288, 289, 290, 291, 292, 293, 294, 298, 302, 307, 308, 309, 310, 328, 332, 340, 346
trochlear nerve, 189, 300, 301
TSH, 206
tumor development, 336
tumor growth, 146, 147, 328, 332, 334, 355
tumor progression, 326
tumor removal, 11, 45, 78, 79, 89, 94, 103, 113, 119, 143, 146, 149, 217, 220, 333, 338, 340
tumours, 107, 234, 323, 347
turbinates, 19, 24, 40, 97, 127, 138, 144, 158, 159, 160, 337, 341
Turkey, 71
tympanic membrane, 52, 156, 169

U

UK, 259, 289, 350
United States (USA), xi, xix, 16, 258, 259
urologist, xxi

V

vagus, 255, 293, 294, 305, 361
vagus nerve, 255, 294, 361
valence, xxv
vancomycin, 40
variables, 130, 274, 289
variations, 33, 39, 75, 126, 139, 153, 185, 193, 259, 263, 269
vascular diseases, 190
vascular occlusion, 143
vascularization, 102, 228
vasculature, 30, 357

vasospasm, 294, 304, 336
vein, 46, 55, 173, 185, 188, 227, 257, 262, 269, 274, 299, 300, 301, 304, 315, 316, 318, 330
ventilation, 25
ventricle, 94, 100, 106, 112, 113, 114, 115, 116, 304, 311, 312, 313, 314, 315, 316, 317, 318, 319, 321, 322, 329
ventriculoperitoneal shunt, 323
vertebral artery, 250, 254, 261, 304, 358, 359, 360, 361, 363
vertigo, 307, 329
vessels, x, 32, 38, 103, 104, 116, 140, 160, 168, 182, 219, 269, 274, 297, 301, 302, 304, 305, 306, 315, 328, 336
vestibular schwannoma, 190, 270, 327, 328, 331, 332, 346, 347, 348
vestibulocochlear nerve, 260, 269
vision, 45, 46, 94, 130, 131, 199, 237, 238, 240, 243, 247, 248, 255, 311, 312, 319, 320, 339, 342, 343
visual acuity, 130, 131, 339, 340, 342
visual field, 131, 207, 342
visualization, x, xxi, 11, 20, 24, 31, 36, 44, 49, 53, 55, 61, 62, 76, 95, 100, 103, 112, 113, 116, 124, 128, 173, 194, 197, 214, 216, 251, 252, 255, 256, 260, 261, 341, 343, 358, 359, 360
vomiting, 203

W

water, 50, 54
weakness, 263, 329, 333, 360, 361, 363
white matter, 125
WHO, 333, 338, 361, 362
windows, 112, 146, 196, 333, 361
wires, 241, 246
wood, xxi
worldwide, xxv
wound dehiscence, 176
wound healing, 54
wound infection, 176, 194, 263, 358

Y

yield, 101, 260, 280, 319

Z

Zone 1, 127
Zone 2, 128
Zone 3, 128
zygoma, 186, 226, 229
zygomatic arch, 117, 185, 234, 243, 256